PLANNING & SCHE

USING

MICROSOFT® PROJECT 2003

BY

PAUL EASTWOOD HARRIS

DISCLAIMER

The information contained in this book is to the best of the author's knowledge true and correct. The author has made every effort to ensure accuracy of this publication, but cannot be held responsible for any loss or damage arising from any information in this book.

AUTHOR AND PUBLISHER

Paul E Harris
Eastwood Harris Pty Ltd
PO Box 4032
Doncaster Heights 3109
Victoria
Australia

harrispe@eh.com.au
http://www.eh.com.au
Tel: +61 (0)4 1118 7701
Fax: +61 (0)3 9846 7700

Please send any comments on this publication to the author.

ISBN 0-9751503-3-2 B5 Paperback
ISBN 0-9751503-4-0 A4 Spiral Bound
27 August 2005

SUMMARY

The book was written so it may be used as:

- A training manual for a two-day training course, or
- A self teach book, or
- A reference manual.

The screen shots for the book are mainly taken from Microsoft Project Standard 2003 but the book may be used to learn Microsoft Project Professional 2003.

The book has been written to be used as the basis for a two-day training course and includes exercises for the students to complete at the end of each chapter. Unlike many training course publications the course book may then be used by the students as a reference book.

This publication is ideal for people who would like to quickly gain an understanding of how the software operates and explains how the software differs from Primavera P3 and SureTrak, thus making it ideal for people who wish to convert from these products.

CUSTOMISATION FOR TRAINING COURSES

Training organisations or companies that wish to conduct their own training may have the book tailored to suit their requirements. This may be achieved removing, reordering or adding content to the book and by writing their own exercises. This book is available in both A4 Spiral Bound, which lies flat on the desk for training and/or self teaching and in B5 paperback as a reference manual. Please contact the author to discuss this service.

AUTHOR'S COMMENT

As a professional project planner and scheduler I have used a number of planning and scheduling software packages for the management of a range of project types and sizes.

The first books I published were user guides/training manuals for Primavera SureTrak and P3 users. These were well received by professional project managers and schedulers, so I decided to turn my attention to Microsoft Project 2000, 2002 and 2003. This book follows the same proven layout of my Microsoft Project 2002 and 2003 book. I trust this book will assist you in understanding how to use Microsoft Project 2003 on your projects. Please contact me if you have any comments on this book.

I would like thank my wife Susan Harris, my daughter Samantha Harris and Susan Aaron for their assistance in the production of this publication.

CURRENT BOOKS PUBLISHED BY EASTWOOD HARRIS

Planning Using Primavera Project Planner P3® Version 3.1
ISBN 0-9577783-7-6 A4 Paperback
ISBN 0-9577783-8-4 A4 Spiral Bound
First Published March 2000

Planning Using Primavera SureTrak Project Manager® Version 3.0
ISBN 0-9577783-9-2 A4 Paperback
ISBN 0-9751503-0-8 A4 Spiral Bound
First Published June 2000

Planning and Scheduling Using Microsoft® Project 2002
ISBN 0-9751503-1-6 B5 Paperback
ISBN 0-9751503-2-4 A4 Spiral Bound
First Published January 2002

Planning and Scheduling Using Microsoft® Project 2003
ISBN 0-9751503-3-2 B5 Paperback
ISBN 0-9751503-4-0 A4 Spiral Bound
First Published June 2004

Project Planning and Scheduling Using Primavera® Version 4.1
For Engineering & Construction and Maintenance & Turnaround
ISBN 1-921059-00-1 A4 Paperback
ISBN 1-921059-01-X A4 Spiral Bound
First Published January 05

Project Planning and Scheduling Using Primavera® Version 4.1
For IT Project Office and New Product Development
ISBN 1-921059-02-8 A4 Paperback
ISBN 1-921059-03-6 A4 Spiral Bound
First Published March 05

Project Planning and Scheduling Using Primavera® Contractor Version 4.1
For the Construction Industry
ISBN 1-921059-04-4 A4 Paperback
ISBN 1-921059-05-2 A4 Spiral Bound
First Published January 05

PRINCE2 ™ Planning & Control Using Microsoft® Project
ISBN 1 921059 06 0 B5 Perfect
First Published May 2005

Planning and Control Using Microsoft® Project and PMBOK® Guide Third Edition
ISBN 1-921059-07-9 A4 Spiral
ISBN 1-921059-08-7 B5 Perfect
To Be Published August 2005

1 INTRODUCTION

1.1 Purpose

The purpose of this book is to provide you with a method for planning and controlling projects using Microsoft Project Professional 2003 or Microsoft Project Standard 2003 in a single project environment up to an intermediate level.

The screen shots in this book were captured using Microsoft Project Standard 2003 and Windows XP. Readers using Microsoft Project Professional 2003 will have some additional menu options to those shown in this book, which will operate when their software is connected to Microsoft Project Server software. Microsoft Project Standard 2003 will not operate with Microsoft Project Server and has fewer functions than Microsoft Project Professional 2003 operating with Microsoft Project Server.

At the end of this book, you should be able to:

- Understand the steps required to create a project plan
- Set up the software
- Define calendars
- Add tasks
- Organize tasks
- Format the display
- Add logic and constraints
- Use Tables, Views and Filters
- Print reports
- Record and track progress
- Customize the project options
- Create and assign resources
- Understand the impact of task types and effort driven tasks
- Status projects that contain resources
- Understand the different techniques for scheduling

The book does not cover every aspect of Microsoft Project 2003, but it does cover the main features required to create and status a project schedule. It should provide you with a solid grounding, which will enable you to go on and learn the other features of the software from experimenting with the software, using the help files and reviewing other literature.

This book has been written to minimize superfluous text, allowing the user to locate and understand the information contained in the book as quickly as possible. It does NOT cover functions of little value to common project scheduling requirements. If at any time you are unable to understand a topic in this book, it is suggested that you use the Microsoft Project 2003 Help menu to gain a further understanding of the subject.

1.2 Required Background Knowledge

This book does not teach you how to use computers or to manage projects. The book is intended to teach you how to use Microsoft Project 2003 in a project environment. Therefore, to be able to follow this book you should have the following background knowledge:

- The ability to use a personal computer and understand the fundamentals of the operating system
- Experience using application software such as Microsoft Office which would have given you exposure to Windows menu systems and typical Windows functions such as copy and paste
- An understanding of how projects are managed, such as the phases and processes that take place over the lifetime of a project.

1.3 Purpose of Planning

The ultimate purpose of planning is to build a model that allows you to predict which tasks and resources are critical to the timely completion of the project. Strategies may then be implemented to ensure that these tasks and resources are managed properly, thus ensuring that the project will be delivered both **On Time** and **Within Budget**.

Planning aims to:

- Optimize time
- Evaluate different methods
- Optimize the use of resources
- Provide early warning of potential problems
- Enable you to take proactive and not reactive action
- Identify risks
- Set priorities

Planning helps to avoid the delayed or untimely completion of a project and thus avoid:

- Increased project costs or reduction in scope and/or quality
- Additional changeover and/or operation costs
- Extensions of time claims
- Loss of your client's revenue
- Contractual disputes and associated resolution costs
- The loss of reputation of those involved in a project
- Loss of a facility or asset in the event of a total project failure

1.4 Project Planning Metrics

There are three components that you may measure and control using planning and scheduling software:

- Time
- Effort (resources)
- Cost

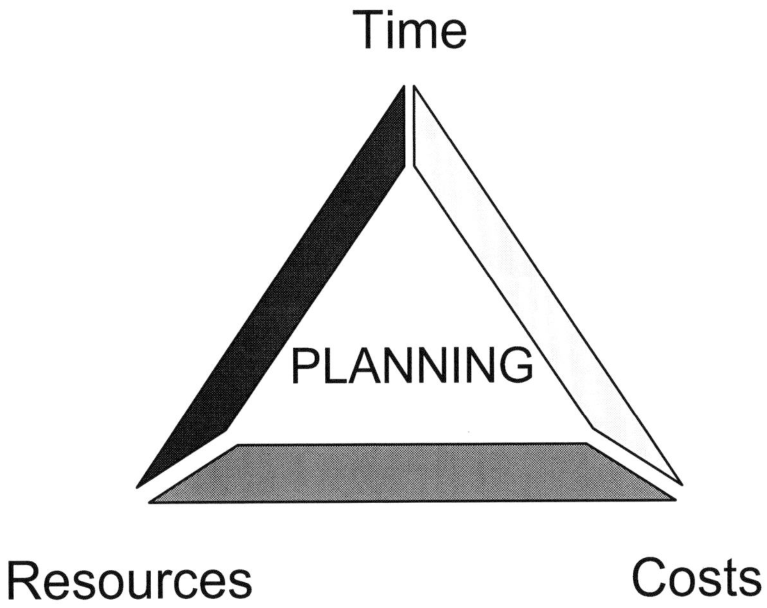

A change in any one of these components normally results in a change in one or both of the other two.

Other project management functions that are not traditionally managed with planning and scheduling software but may have components reflected in the schedule include:

- Document management and control,
- Quality Management,
- Contract Management,
- Issue Management,
- Risk Management,
- Industrial Relations, and
- Accounting.

The development of Enterprise Project Management systems has brought about more of these functions being included in project planning and scheduling software.

1.5 Planning Cycle

The planning cycle is an integral part of managing a project. A software package such as Microsoft Project 2003 makes this task much easier.

When the original plan is agreed to, the **Baseline** or **Target** is set. The **Baseline** is a record of the original plan. The **Baseline** dates may be recorded in Microsoft Project 2003 in data fields titled **Baseline Start** and **Baseline Finish**.

After project planning has ended and project execution has begun, the actual progress is monitored, recorded and compared to the **Baseline** dates.

The progress is then reported and evaluated.

The plan may be changed by adding or deleting tasks and adjusting Remaining Durations or Resources. A revised plan is then published as progress continues.

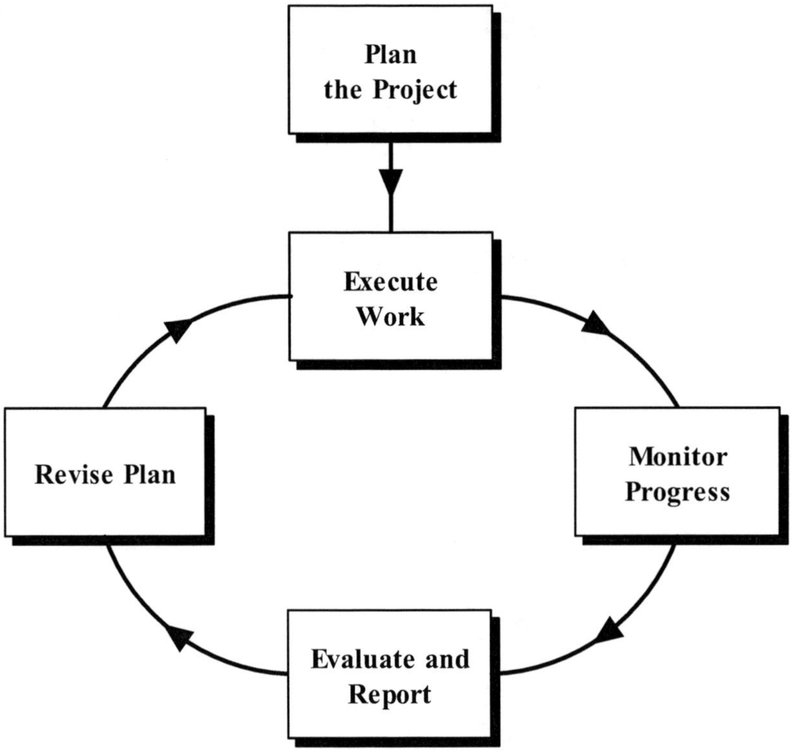

2 CREATING A PROJECT PLAN

The aim of this chapter is to give you an understanding of what a plan is and some practical guidance on how your schedule may be created and statused during the life of a project.

2.1 Understanding Planning and Scheduling Software

A project is essentially a set of unique operations or tasks to be completed in logical order to achieve a defined outcome by a definitive end time. A schedule is an attempt to model these tasks, their durations and their relationships to other tasks. These tasks take time to accomplish and may employ resources such as people, materials, equipment, and money that may have limited availability.

Planning and scheduling software allows the user to:

- Break a project down into tasks that are entered into the software as tasks
- Assign tasks their durations, predecessors, successors and constraints and then calculate the start and finish dates of all the tasks
- Assign resources and costs to the tasks and calculate the planned project resource requirements and project costs,
- Monitor the actual progress of tasks against the original plan and revise the plan when required
- Monitor the consumption of resources and re-estimate the resources required to finish the project.

There are four modes or levels in which planning and scheduling software may be used.

	Planning	**Tracking**
Without Resources	**LEVEL 1** Planning without resources	**LEVEL 2** Tracking progress without resources
With Resources	**LEVEL 3** Planning with resources	**LEVEL 4** Tracking progress with resources

As the level increases, the amount of information required to maintain the schedule will increase. More importantly, your skill and knowledge in using the software will also need to increase. This book is designed to take you from Level 1 through Level 4.

2.2 Understanding Your Project

Before you start the process of creating a project plan, it is important to have an understanding of the project and how it will be executed. On large, complex projects, this information is usually available from the following types of documents:

- Project scope
- Functional specification
- Requirements baseline
- Contract documentation
- Plans and drawings
- Project execution plan
- Contracting and purchasing plan
- Equipment lists
- Installation plan
- Testing plan

It is important to gain a good understanding of the project process before starting to plan your project. You should also understand what level of reporting is required. Providing too little or too much detail will often lead to the schedule being discarded or not be used.

There are three processes required to create or maintain a plan at each of the four levels:

- Collecting the relevant project data
- Entering and manipulating the data in software
- Distributing the plan, reviewing and revising

The ability of the scheduler to collect the data is as important as the ability to enter and manipulate the information using the software. On larger projects, it may be necessary to write policies and procedures to ensure accurate collection of data from the various people, departments, stakeholders/companies, and sites.

2.3 Level 1 – Planning Without Resources

This is the simplest mode of planning.

2.3.1 Creating Projects

To create the project, you will require the following information:

- Project Name
- The Project Start Date (and perhaps the Finish Date)

It would be nice to know other important information such as:

- Client name
- Other project information such as location, project number and stakeholders.

2.3.2 Defining the Calendars

Before you start entering tasks into your schedule it is advisable to set up the calendars. These are used to model the working time for each task in the project. For example, a six-day calendar is created for those tasks that will be worked for six days a week. The calendar should include any public holidays and any other exceptions to available working days such as planned days off.

2.3.3 Defining the Project Breakdown Structure

A project breakdown structure (PBS) is a way of categorizing the activities of a project into numerous codes that relate to the project. The codes act as tags or attributes of each activity.

During or after the tasks are added to the schedule, they are assigned their PBS so that they may be grouped, summarized, and filtered in or out of the display.

The principal method of assigning a PBS to your project with Microsoft Project 2003 is using a function entitled, "Outlining."

Before creating a project, you should design your PBS by asking the following questions:

- Which phases are involved in the project? (E.g. Design, Procure, Install and Test)
- Which disciplines are participating? (E.g. Civil, Mechanical and Electrical)
- Which departments are involved in the project? (E.g. Sales, Procurement and Installation)
- What work is expected to be contracted out and which contractors are used?
- How many sites or areas are there in the project?

Use the responses to these and other similar questions to create the PBS.

2.3.4 Adding Tasks

Tasks must be defined before they are entered into the schedule. It is important that you consider the following factors carefully:

- What is the scope of the task? (What is included and excluded?)
- How long the task is going to take?
- Who is going to perform it?
- What are the deliverables or output for each task?

The project estimate is usually a good place to start looking for a breakdown of the project into tasks, resources and costs. It may even provide an indication of how long the work will take.

Activities may have variable durations depending on the number of resources assigned. You may find that one activity that takes 4 days using 4 workers may take 2 days using 8 workers or 8 days using 2 workers.

Usually project reports are issued on a regular basis such as every week or every month. It is recommended that, if possible, a task should not span more than two reporting periods. That way the tasks should only be **In-Progress** for one report. Of course, it is not practical to do this on long duration activities such as procurement and delivery tasks that may span many reporting periods.

It is also recommended that you have a measurable finish point for each group of tasks. These may be identified in the schedule by **Milestones** and are designated with zero duration. You may issue documentation to officially highlight the end point of one task and the start point of another, thereby adding clarity to the schedule. Examples of typical documents that may be issued for clarity are:

- Issue of a drawing package
- Completion of a specification
- Placing of an order
- Receipt of materials (Delivery logs or tickets or Dockets)
- Competed testing certificates for equipment

2.3.5 Adding the Logic Links

The logic is added to the schedule to provide the order in which the tasks must be undertaken. The logic is designated by indicating the predecessors to or the successors from each activity.

The software will calculate the start and finish dates for each task and the end date of the project based on the start date of the project, the logic amongst the tasks, and durations of the tasks.

It is good practice to create a **Closed Network** with the logic. In a **Closed Network**, all tasks have one or more predecessors and one or more successors except:

- The project start milestone or first task which has no predecessors, and
- The finish milestone or finish task which has no successors.

The project's logic must not loop back on itself. Looping would occur if the logic were stated that A preceded B, B preceded C, and C preceded A. That's not a logical project situation and will cause an error comment to be generated by the software during network calculations.

Thus, when the logic is correctly applied, a delay to a task will delay all its successor tasks and delay the project end date when there is insufficient spare slippage time to accommodate the delay. This spare time is normally called **Float** but Microsoft Project 2003 uses the term **Slack**.

2.3.6 Constraints

To correctly model the impact of events outside the logical sequence, you may use constraints. For example a Start No Earlier Than constraint would be imposed on an event such as the availability of a facility. Constraints should be cross-referenced to the supporting documentation such as Milestone Dates from contract documentation.

2.3.7 Scheduling the Project

The software calculates the shortest time in which the project may be completed. Unstatused tasks without logic or a constraint will be scheduled to start at the Project Start Date.

Scheduling the project will also identify the **Critical Path(s)**. The Critical Path is the chain(s) of tasks that takes the longest time to accomplish. This chain defines the Earliest Finish date of the project. The calculated completion date depends on the critical tasks starting and finishing on time. If any of them are delayed, the whole project will be delayed.

Tasks that may be delayed without affecting the project end date have **Float**.

Total Float is the amount of time a task may be delayed without delaying the project end date. The delay of a task with a positive Total Float value may delay other tasks with positive Total Float but will not delay the end date of the project unless the delay is greater than the float. The delay of any task with a zero Total Float value (and is, therefore, on the **Critical Path)** will delay other subsequent tasks with zero Total Float and extend the end date of the project.

Free Float is the amount of time a task may be delayed without delaying the start date of any of its immediate successor tasks.

2.3.8 Formatting the Display – Views, Tables and Filters

There are tools to manipulate and display the tasks to suit the project reporting requirements. These functions are covered in the **FILTERS** and the **VIEWS, TABLES AND DETAILS** chapters.

2.3.9 Printing and Reports

There are software features that allow you to present the information in a clear and concise manner to communicate the requirements to all project members. These functions are covered in the **PRINTING AND REPORTS** chapter.

2.3.10 Issuing the Plan

All members of the project team should review the project plan in an attempt to:

- Optimize the process and methods employed, and
- Gain consensus among team members as to the project's logic, durations, and PBS.

Correspondence should be used to communicate expectations of team members while providing each with the opportunity to contribute to the schedule and further improve the outcome.

2.4 Level 2 – Monitoring Progress Without Resources

2.4.1 Setting the Baseline

The optimized and agreed-to plan is used as a baseline for measuring progress and monitoring change. The software can record the baseline dates of each activity for comparison against actual progress during the life of the project. These planned dates are stored in the **Baseline Date** fields.

2.4.2 Tracking Progress

The schedule should be **Statused** (updated or progressed) on a regular basis and progress is recorded at a point in time.

Whatever the frequency chosen for statusing, you will have to collect the following task information in order to status a schedule:

- Actual Start Dates of tasks that have begun, whether they were planned to start or not,
- Percentage Complete and Remaining Duration or Expected Finish date for started, but incomplete tasks,
- Actual Finish Dates for completed tasks and
- Any revisions to tasks that have not started

The schedule may be statused after this information has been collected, and then the recorded progress is compared to the **Baseline Dates**.

At this point, it may be necessary to further optimize the schedule to meet the required end date by discussing the schedule with the appropriate project team members. The date as of when progress is reported is commonly known as the **Data Date** or **Status Date** or **Update Date**. The data date is **NOT** the date that the report is printed out but rather the date that reflects when the status information was gathered.

2.5 Level 3 – Scheduling With Resources

2.5.1 Creating and Using Resources

First, establish a resource pool by entering all the project resources required on the project into a table in the software. You then assign the required quantity of each resource to the tasks.

A resource in planning and scheduling software may represent an individual person, a skill or trade, individual pieces of equipment, fleets of equipment, a team or crew, material, space or funds. Each resource may have a quantity and an associated cost.

Entering a cost rate for each resource enables you to conduct a resource cost analysis such as comparing the cost of supplementing overloaded resources against the cost of extending the project deadline.

Costs may also be assigned to tasks without the use of resources by using the Fixed Cost function.

Time-phased cash flows and budgets may be produced from this resource/cost data.

2.5.2 Task Types

Tasks may be assigned a **Type**, which affect how resources are calculated. There are additional software features that enable the user to more accurately model real-life situations. These features are covered in the **CREATING RESOURCES** chapters.

2.6 Level 4 – Monitoring Progress of a Resourced Schedule

2.6.1 Statusing Projects with Resources

When you status a project with resources, you will need to collect some additional information:

- The quantities and/or costs spent to-date per task for each resource, and
- The quantities and/or costs required per resource to complete each task.

You may then status a resourced schedule with this data.

2.6.2 Tools and Techniques for Scheduling

At this point, the book covers some additional software functions making creating and editing schedules simpler.

2.7 The Balance Between the Number of Activities and Resources

On large or complex schedules, you need to maintain a balance between the number of activities and the number of resources that are planned and tracked. As a general rule, the more activities a schedule has, the fewer resources should be created and assigned to tasks.

When you have a schedule with a large number of tasks and a large number of resources assigned to each task, you may end up in a situation where you and members of the project team are unable to understand the schedule and you are unable to maintain it.

Instead of assigning individual resources such as people by name, consider using Skills or Trades, and on very large project use Crews or Teams.

This technique is not so important when you are using a schedule for estimating the direct cost of a project (by assigning costs to the resources) or if you will not be using the schedule to track a project's progress (such as a schedule that is used to support written proposals).

It is more important to minimize the number of resources in large schedules that will be updated regularly, since updating every resource assigned to each task at each schedule update is very time consuming. At this point in time you would be acting as a timekeeper updating resources and not a scheduler looking after the future of the project.

3 CREATING PROJECTS AND SETTING UP THE SOFTWARE

Ensure to check the internet for the latest updates as some of the critical functions such as outlining did not function correctly in the original release of Microsoft Project 2003.

There are three principal methods of creating a new project:

- Start with a blank project, or
- Use a template that contains default data and formats, or
- Open an old project and save it with a new file name.

Before creating a project file it is important to understand the file types that Microsoft Project 2003 will open and save.

3.1 File Types

Microsoft Project is compatible and will operate with the following file types:

- **Project (*.mpp)**. This is the default file format that Microsoft Project 2003, Microsoft Project 2002 and Microsoft Project 2000 uses to create and save files. This is a different format than the ***.mpp** file created by Microsoft Project 98.
- **Microsoft Project 98 (*.mpp)**. This is the format created by Microsoft Project 98.
  - Microsoft Project 98 will not open or save a **Project (*.mpp)** file, created by Microsoft Project 2000, 2002 and 2003.
  - Microsoft Project 2000, 2002 & 2003 will open and save to a **Microsoft Project 98 (*.mpp)** file.
- **MPX (*.mpx).** This is a text format data file created by Microsoft Project 98 and earlier versions of Microsoft Project.
  - This format may be opened by Microsoft Project 2000, 2002 and 2003 but is not created by Microsoft Project 2000, 2002 and 2003.
  - MPX is a format that may be imported and exported by many other project scheduling software packages.
  - Some third-party software will convert mpx files to and from Microsoft Project 2003 mpp format files. You may search the Internet for the latest available products.
- **Template (*.mpt)**. This format is used for creating project templates.
- **Project Database (*.mpd)**. This is a Microsoft Project database format that may be used for exporting data and is intended to replace the mpx format.
- **Microsoft Access Database (*.mdb)**. This is the Microsoft Access format.
- Data may be saved to (and imported from) files in the following additional formats using **File/Save**, **File/Save As** and **File/Open**:
  - Excel (*.xls)
  - Excel **Pivot Table** (save only)
  - Web page (save only) (*.html; *.htm)
  - Tab delimited text files (*.asc)
  - Comma delimited text files (*.txt)
  - Extensible Markup Language format (*.xml)

3.2 Starting Microsoft Project

When opening Microsoft Project 2003, you will be presented with a blank project that you may start working with immediately. The **Startup Task** pane shown in the picture below with a heading **Getting Started** may be displayed on the left-hand side of the screen; this may be closed by clicking on the icon ☒ as shown below, and you may start work immediately.

Close the **Startup Task** pane by clicking here

Getting Started menu

Getting Started Pane

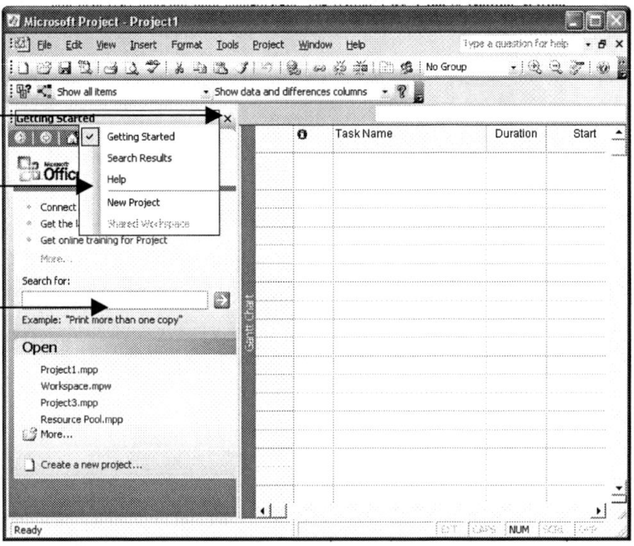

To prevent the pane titled **Getting Started** from being displayed every time Microsoft Project is opened select **Tools**, **Options…**, select the **General** tab and uncheck the **Show Startup Task pane** box.

After closing the **Getting Started** pane your screen may look like the picture below showing a pane on the left hand side titled **Tasks**, this is the called the **Project Guide**. This guide may be used to assist in the creation of project schedules.

Close the **Project Guide** by clicking here:

To prevent the **Project Guide** pane from being displayed every time Microsoft Project is opened select **Tools**, **Options…**, select the **Interface** tab and uncheck the **Display Project Guide** box.

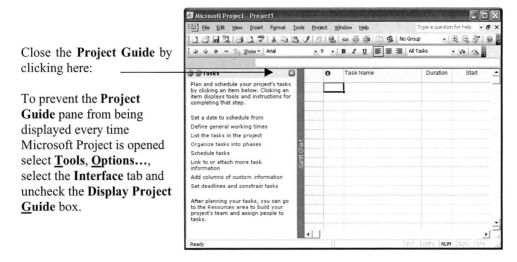

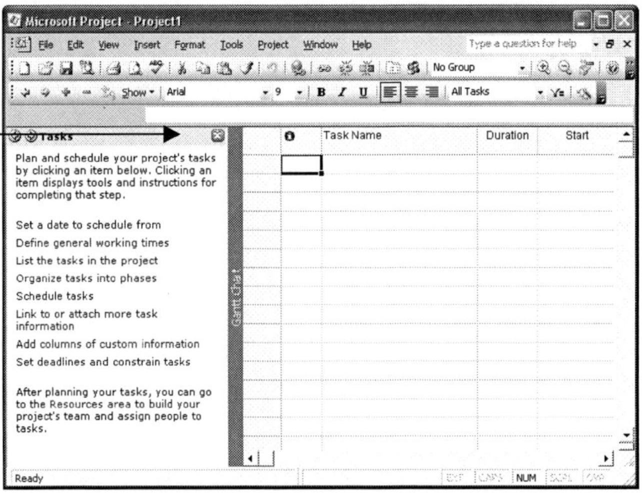

3.3 Creating a Blank Project

A blank project may be created from the **New Project** pane which is displayed by:

- Keying in **Ctrl+N**, or
- Clicking on the **New** tool bar icon, or
- You may also select **File**, **New**, or
- Selecting **New Project** from the **Startup Task** drop down menu, see picture on the previous page.

Select **Blank Project** and a new blank project will be created.

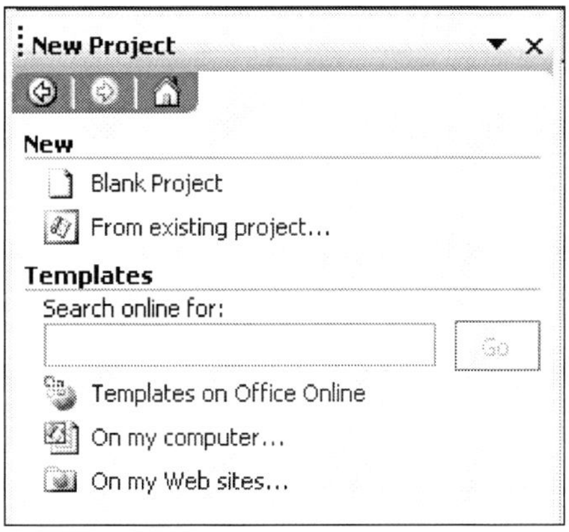

At this point the Project **Start date** is normally set in the **Project Information** form, select **Project**, **Project Information…** to open this form:

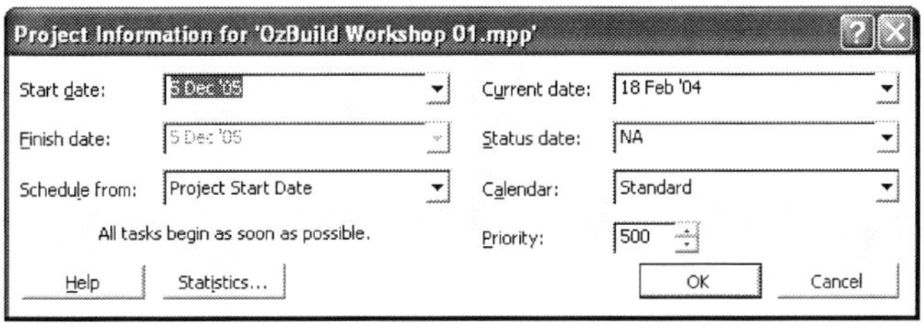

- When **Schedule from:** is set to the **Project Start Date**, which is the usual method of scheduling projects:
  - ➢ Enter the **Start date:** – This is the date before which no task will be scheduled to start.
  - ➢ The **Finish date:** – This is a calculated date and is the date of the completion of all tasks.
- When **Schedule from:** is set to the **Project Finish Date**:
  - ➢ All new tasks are set with a constraint of **As Late As Possible**, and
  - ➢ Therefore, all new tasks are scheduled before the **Project Finish Date** not after the start date.
- **Current Date:** – This field defaults to **today's date**; it represents the date today and may be changed at any time. This date has no effect on most calculations.
- **Status Date:** – This is an optional field used when statusing a project. This topic is covered in the **TRACKING PROGRESS** chapter.
- **Calendar:** – This is the project **Base** calendar that is used to calculate the durations of all tasks unless they have:
  - ➢ A resource with an edited resource calendar, or
  - ➢ Assigned with a different task calendar.
- **Priority:** – This is the project priority when sharing resources over a number of projects. 1000 is highest priority and 0 the lowest.

- Click on the Statistics... button to open the **Project Statistics** form, which outlines statistical information about the project.
- The **Project Information** form may be set to be displayed when a new project is created by selecting **Tools**, **Options**, **General** tab and checking the **Prompt for project info for new projects** check box.

A new blank project copies default values such as the Standard Calendar from the **Global.mpt** file. The **Global.mpt** file may be edited using the **Tools**, **Organizer…** utility.

The default Microsoft Project 2003 blank project has a Standard calendar based on 5 days per week without any holidays, which will not suit many projects. It is recommended that the **Global.mpt** Standard calendar be replaced with a project calendar that has been edited to represent your local public holidays using the **Tools**, **Organizer…** utility.

3.4 Opening an Existing Project

Another method of creating a new project is to open an existing project, saving with a new name and then modifying it. To open an existing project display the **Open** form by selecting:

- **File**, **Open**, or
- **Ctl+O**, or
- Click on the New toolbar icon

Then select the file you want to open.

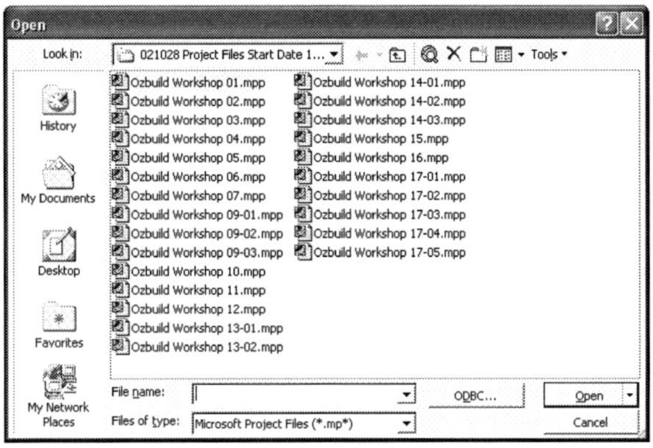

The startup **Task Pane** may be used to open an existing project:

- Select **File**, **New** and this will always display the startup **Task Pane** menu,
  - Then select the **Choose project…** item under the **New from existing project** heading, or
  - Select **More projects…** item under the **Open a project** heading.

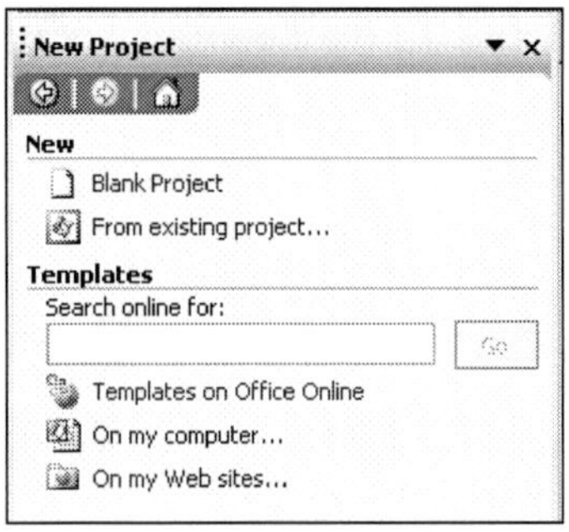

Then use **File**, **Save As** to save the file under a different name:

- Enter a new Project Name,
- Select in which **Current Folder** you want to save the project, and
- Click on Save to save the new project.

You may now alter the contents of this existing plan to reflect the scope of your new project.

3.5 Creating a New Project from a Template

Project templates allow organizations to create project models containing default information applicable to the organization and, in particular, a calendar with the local public holidays. It is normal for organizations to create their own templates to save time when creating a new project.

To create a new project from a template:

- Select **File**, **New** to open the startup **Task Pane**, there are three options for template locations:
  - **Templates on Office Online**, this will take you to a Microsoft web page through Microsoft Explorer.
  - **On my computer…**, this will allow you to open templates on your computer and is covered in the next paragraph.
  - **On my Web sites…**, this opens an Explorer style window where web site addresses may be recorded and templates from these sites used to create projects and templates saved to the sites.
- Click on the **On my computer…**, item under the **New from template** heading to open the **Templates** form to select a Microsoft template. After you have created your own template these will be available from the **General** tab.

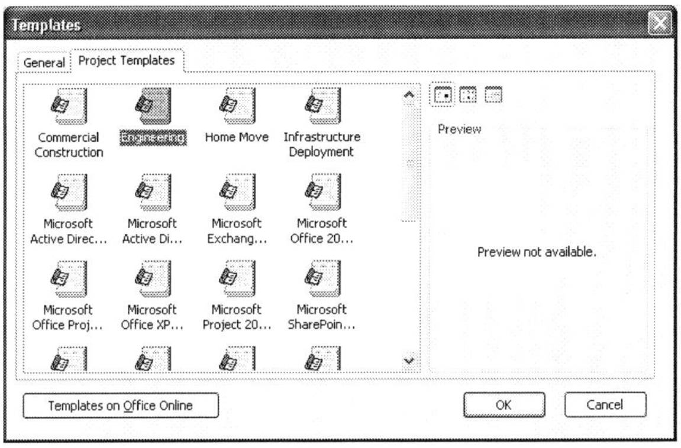

- Select the required template from the form by double-clicking on the template icon or selecting the template by clicking on it and clicking on the OK button.

At this point you would normally set the Project **Start date** in the **Project Information** form, select **Project**, **Project Information…** to open this form.

3.6 Creating a Project Template

To save time when you create new projects, you should create your own templates to suit the different types of projects your organization undertakes. Create a template by saving a project, with or without tasks, in **Template, (*.mpt)** format. This template will be available when you select **File**, **New**.

This directory where user templates are saved may be changed by selecting **Tools**, **Options…**, **Save** tab.

Templates may be deleted by right-clicking to open a menu.

3.7 Saving Additional Project Information

Often additional information about a project is required to be saved with the project such as location, client and type of project. This data may be saved in the **File**, **Properties** form:

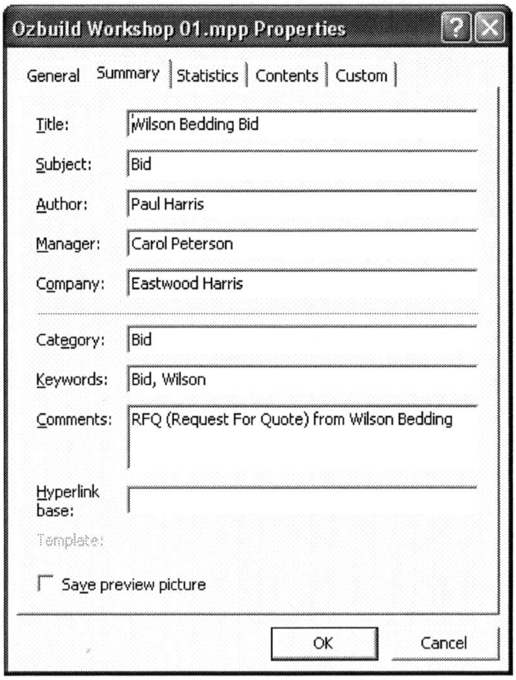

- **Hyperlink base:** This allows you to enter the path to a file or web page.
- **Save preview picture**. This saves a thumbnail sketch, which may be seen when viewing files in Windows Explorer.

WORKSHOP 1

Creating our Project

Preamble

You are an employee of OzBuild Pty Ltd and are responsible for planning the Bid preparation required to ensure that a response to an RFQ (Request For Quote) from Wilson Bedding is submitted on time.

While short-listed, you have been advised that the RFQ will not be available prior to 5 December 2005.

Note: The date format will be displayed according to a combination of your system default settings and the Microsoft Project 2002 Options settings. You may adjust your date format under the system Control Panel, Regional and Language Options and the Microsoft Project 2002 settings in the Options form, which is covered in the **OPTIONS** chapter.

Assignment

1. Create a new project and set the **Start date:** to **Mon 05 Dec 05** with the **Project**, **Project Information…** form but do not edit the **Current date**. Press the **OK** button to save the input data.

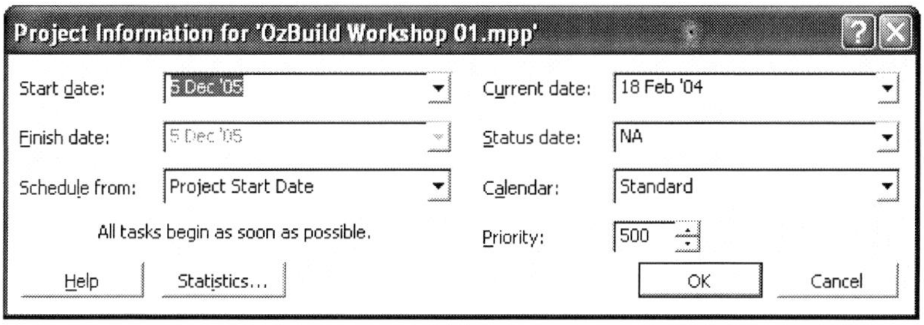

continued over…

2. Add the following project information in the **File, Properties** form.

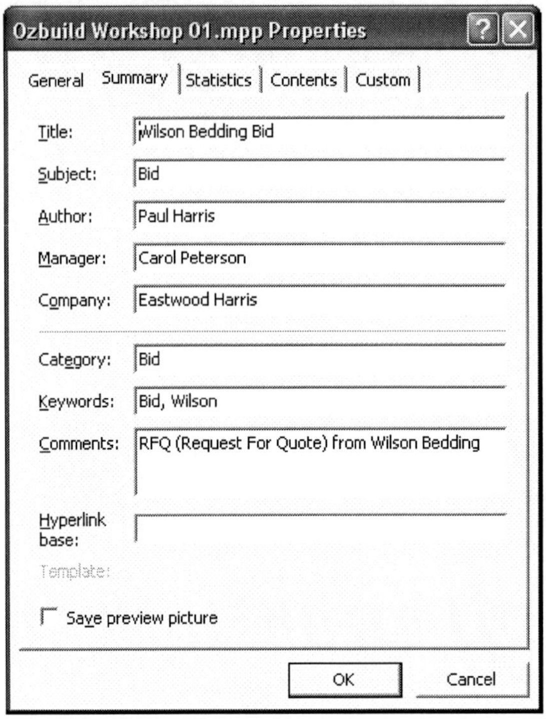

3. **Save** your project as **OzBuild Bid**.

4 NAVIGATING AROUND THE SCREEN

4.1 Identify the Parts of the Project Screen

After a blank project has been created from a template, the default Microsoft Project 2003 screen will look like this:

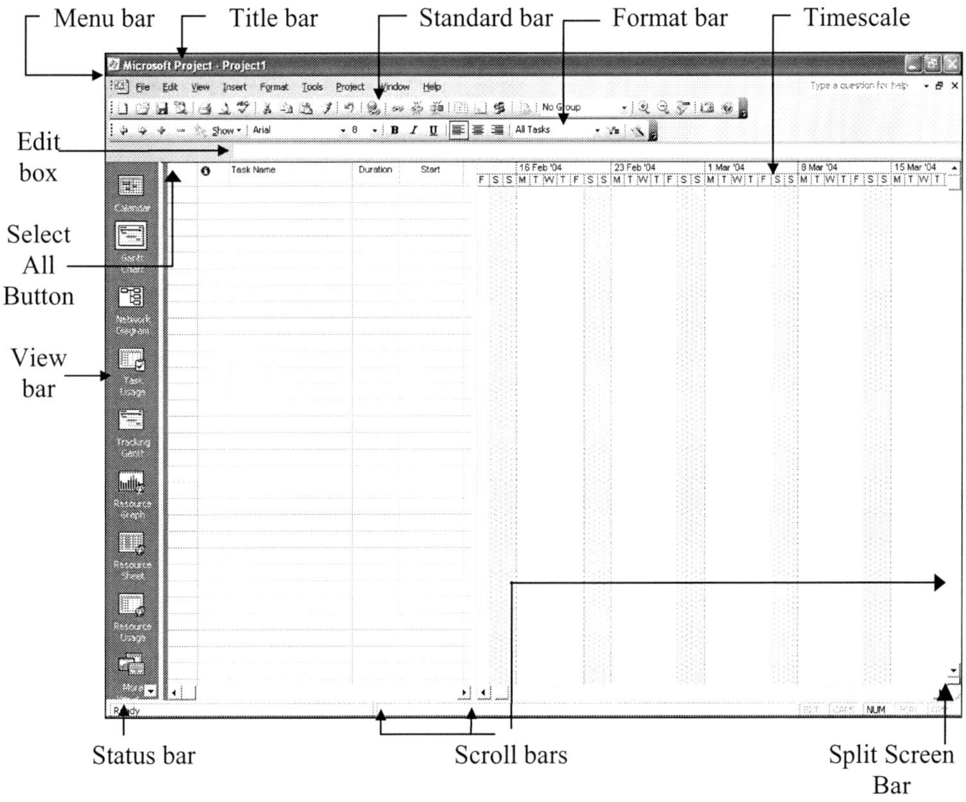

Microsoft Project 2003 has a Windows-style menu system with a typical Windows look-and-feel.

- The project name is displayed after **Microsoft Project** at the top of the left-hand side of the screen.
- The drop-down menus are just below the project name.
- The toolbars are displayed below the menu.
- The left-hand side of the line underneath the toolbars is the **Entry Bar** and **Edit Box**. Any editable data may be edited in the **Edit Box** or directly in the field.

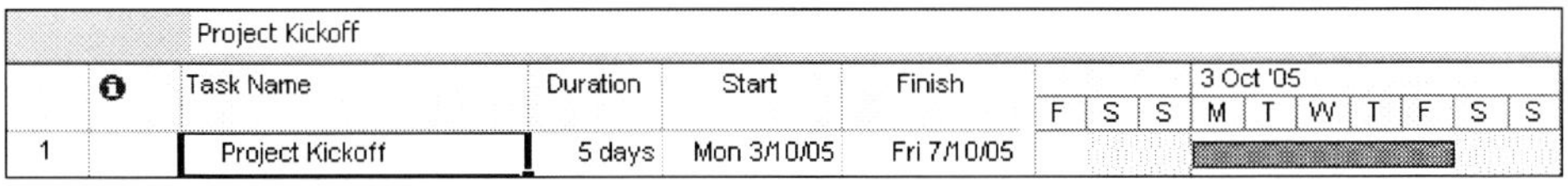

The ✓ and ✗ icons are displayed on the left of the **Edit Box** only when performing editing, and may used for accepting and not accepting data changes that are made from within the edit box.

- The main display has the **Bar Chart** on the right-hand side, with the **Timescale** above, and the **Data Columns** on the left-hand side with their column titles above them. The divider between the two areas may be dragged from side to side by holding down the left mouse button.
- The **View Bar** is designed for the users to select the various chart views and is located on the far left of the screen, this may be hidden or displayed by:
  - Selecting **View**, **View Bar**, or
  - Holding the mouse pointer over the **View Bar**, right-clicking to display a menu, and selecting **View Bar**.
- The horizontal **Scroll Bars** are at the bottom of the screen and the **Status Bar** is below the **Scroll Bars**. The vertical **Scroll Bar** is at the right-hand side of the screen.

4.2 Customizing the Screen

The screen may be customized in a number of ways to suit your preferences. The toolbars and menu bar may be moved around the screen by holding down the right mouse button and dragging them to a new position on the screen.

4.2.1 Toolbars

Toolbars will not be covered in detail but significant productivity improvements may be made by ensuring that functions of frequent use are available on a toolbar.

- There are many built-in toolbars in Microsoft Project 2003. These may be displayed or hidden by:
  - Using the command **View**, **Toolbar** or **Tools**, **Customize**, **Toolbars…** and selecting the **Toolbar** tab, then checking or un-checking the required boxes to display or hide the toolbars.
  - Right-clicking the mouse in the toolbar area to display a Toolbar menu.
- Icons may be added to a bar by selecting **Tools**, **Customize**, **Toolbars…**, **Commands** tab. **Toolbar Icons** may be selected from the dialog box and dragged onto any toolbar.
- Icons may be removed from the toolbars when the **Customize** (Toolbar) form is open by holding down the left mouse button on the icon and dragging them off the toolbar.
- Icons may be reset to default by selecting **Tools**, **Customize**, **Toolbars…**, selecting the **Toolbar** tab and clicking on [Reset...].
- Other toolbar display options are found under **Tools**, **Customize**, **Toolbars…** and then selecting the **Options** tab.

4.2.2 Menu Bar

The **Menu Bar** display options are found under **Tools**, **Customize**, **Toolbars…** and then selecting the **Options** tab.

4.3 Setting up the Options

The basic parameters of the software must be configured so it will operate the way you desire. In order for the software to operate and/or calculate the way you want, some of the defaults must be turned on, or off, or changed. These configuration items may be found under **Tools**, **Options…**.

We will discuss some of the more important options now. All the Options are discussed in the **OPTIONS** chapter. Select **Tools**, **Options…** to display the **Options** form.

Select the **View** tab:

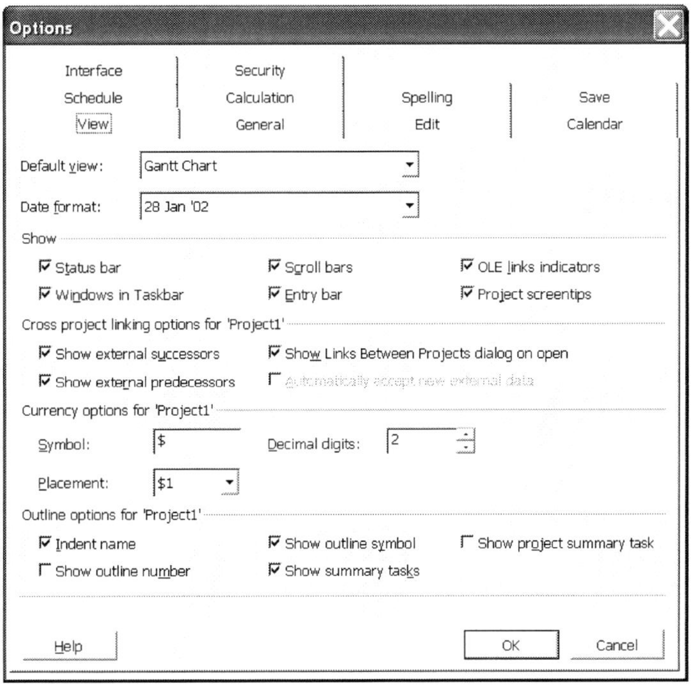

- **Date format:** This is used to select the display style of the dates for all projects. The date format will be displayed according to a combination of your system default settings and the Microsoft Project 2003 Options settings. You may adjust your date format under the system Control Panel, Regional and Language Options and the Microsoft Project 2003 settings in the Options form, which are covered in the **OPTIONS** chapter.

On international projects, to avoid the confusion between the numerical US date style, (mmddyy) and the numerical European date style, (ddmmyy) you should consider adopting the ddmmmyy style, **28 Jan '02** or mmddyy style, **Jan 30 '00**. For example in the United States 020703 is read as 07 Feb 03 and in many other countries as 02 Jul 03.

- Select the Schedule tab:

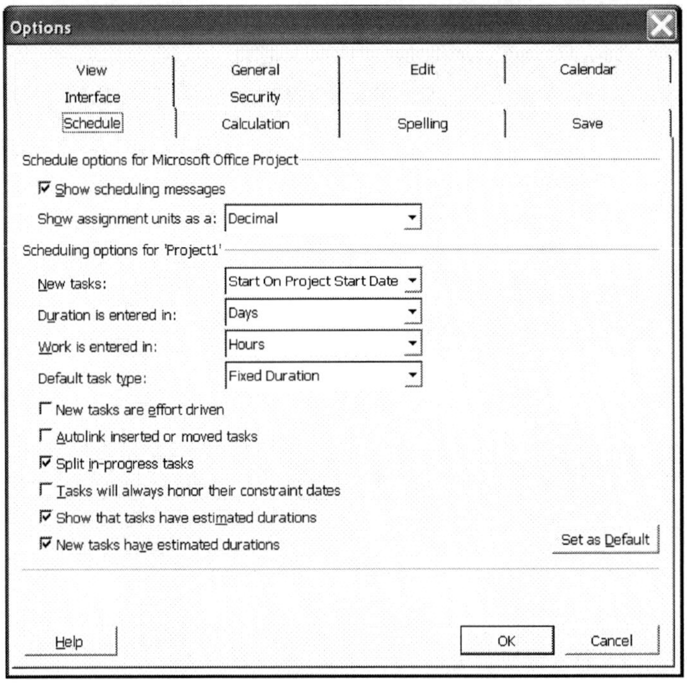

- **Duration is entered in:** – This option specifies the format in which durations are entered via the keyboard. If **Day** is selected as the default, then duration of 2 days is entered as 2 (without the d) and a 2-hour duration should be entered as 2h, and vice versa if **Hours** are selected as **Default type**.

The settings above are the setting normally used by the author.

4.4 Splitting the Screen Views and Details Forms

The screen may be split horizontally into two panes. A different **View** may be displayed in each pane. This is termed **Dual-Pane view**. To open or close the dual-pane view:

- Select **Window**, **Split** or **Window**, **Remove Split**, or
- Grab the horizontal dividing bar at the bottom of the screen (see the picture in paragraph 4.1) by holding down the left mouse button and drag the line to resize the panes.
- Right-click in the right-hand side of the top pane and you will, in most views, be able to display a menu to open or remove the split.
- Double-click the dividing line or dragging it also removes or opens the split window.

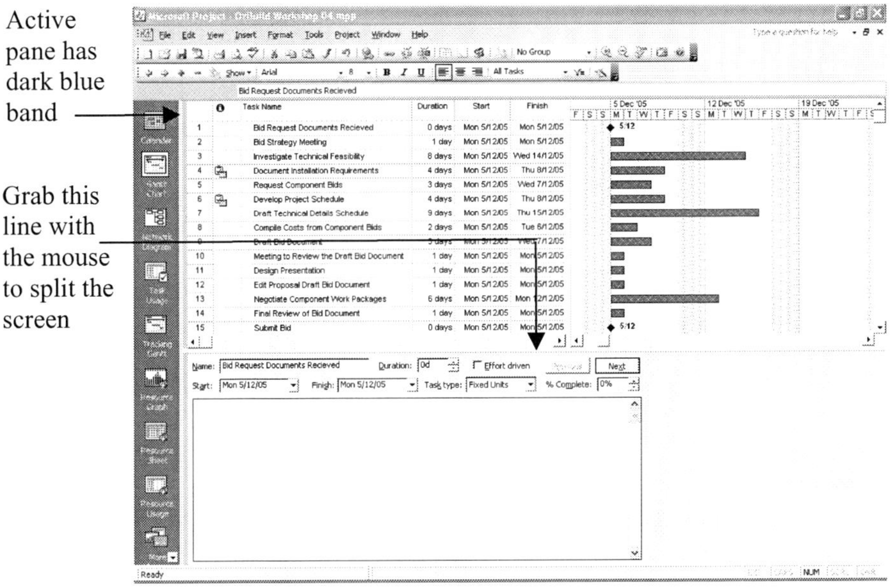
Active pane has dark blue band
Grab this line with the mouse to split the screen

A pane needs to be **Active** before menu items pertaining to that pane become available. A dark blue band (with the standard Windows color scheme) on the left-hand side is displayed in the **Active Pane**.

The menu options will often change when different **Views** are selected in a pane.

A **Pane** is made **Active** just by:

- Clicking anywhere in the pane, or
- Pressing **F6** to swap active panes.

Some **Views** displayed in **Panes** have further options for displaying data. These are titled **Details** forms. The **Details** forms may be selected, when available, by:

- Making the pane active, then
- Selecting:
  - **Format**, **Details**, or
  - Right-clicking in the right-hand side of the screen and clicking the required form.

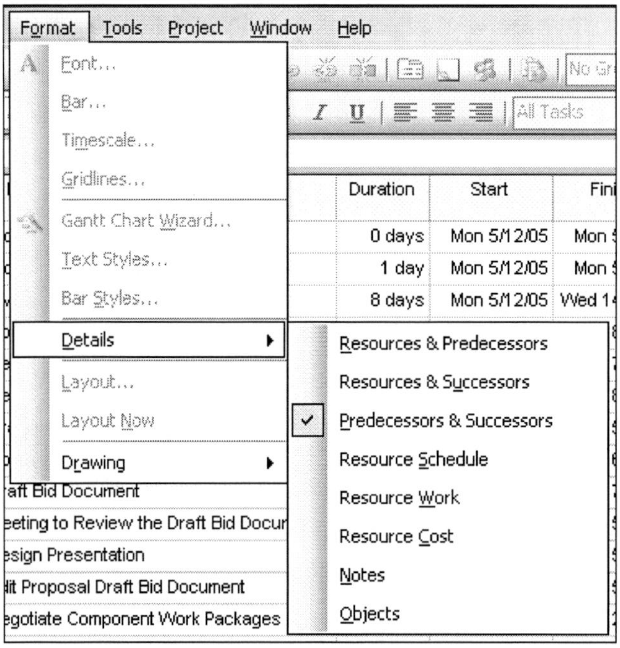

4.5 Right-clicking with the Mouse

It is very important that you become used to using the right-click function of the mouse as this is often a quicker way of operating the software than using the menus. The right-click will normally display a menu, which is often different depending on the displayed View and which pane is the Active Pane. It is advised that you experiment with each view to become familiar with the menus.

WORKSHOP 2

Setting your Project Options

Preamble

For control purposes it is expected that all tasks will be entered in days.

Assignment

1. Open the **Project Options** form and click on each tab and familiarize yourself with the forms. Select **Tools, Options...** and set your options as follows:
   - Select **View** tab, set the **Date format** to:
     1. "**ddmmmyy**" i.e. 31 Jan '00, or
     2. "**mmmddyy**" i.e. Jan 31 '00.

     The available date format will depend on your system settings.
   - Select **Schedule** tab, set the **Durations are to be entered in:** option to **Days** and
   - **Default task type** is **Fixed Units**.
   - **UNCHECK** the **New tasks are efforts driven** option, new tasks are not to be Effort Driven.
   - Press **OK** to commit the changes.
2. Hide and display the **Standard** and **Formatting** toolbars using the **View** menu.
3. Experiment by dragging the toolbars around the screen with your mouse.
4. Hide and display the **View** bar.
5. Split the the screen into two panes by right-clicking with the mouse in the right-hand side of the screen, and selecting **Split** from the menu.
6. Activate the lower pane by clicking in it; note the blue bar on the left-hand side of the screen has moved from the top pane to the bottom pane.
7. Activate the upper pane, by clicking in it.
8. Resize the panes by dragging the Split screen bar.
9. Close the split screen double clicking on the horizontal dividing line.
10. Split the screen by right-clicking on the small bar in the bottom right hand corner of the screen.g.
11. Save your **OzBuild Bid** Project.

5 DEFINING CALENDARS

The finish date (and time) of a task is calculated from the start date (and time) plus the duration over the calendar associated with the task. Therefore, a five-day duration task that starts at the start of the workday on a Wednesday, and is associated with a five-day workweek calendar (with Saturday and Sunday as non-work days) will finish at the end of the workday on the following Tuesday.

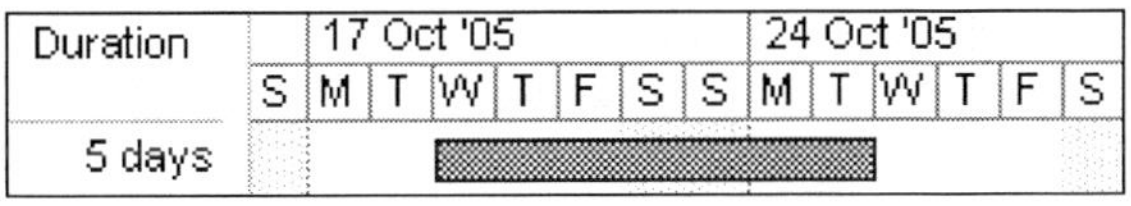

Microsoft Project 2003 is supplied with three calendars, which are termed **Base Calendar**s:

- **Standard** – This calendar is 5 days per week, 8 hours per day.
- **24 Hour** – This calendar is 7 days per week and 24 hours per day.
- **Night shift** – This calendar is 7 days per week and 8 hours per day during the night.

You may create new or edit existing Base Calendars to reflect your project requirements, such as adding holidays or additional workdays or adjusting work times. For example, some tasks may have a 5-day per week calendar and some may have a 7-day per week calendar.

A **Base Calendar** is assigned to each project. Microsoft Project 2003 uses the term **Project Calendar** to describe this calendar. By default all un-resourced tasks use the **Project Calendar** to calculate the end date of the task.

This chapter will cover the following topics:

Topic	Menu Command
• Assigning a base calendar to a project	**Project**, **Project Information…**
• Editing a calendar's working days	**Tools**, **Change Working Time…**
• Creating a new calendar	**Tools**, **Change Working Time…**, **New…**
• Renaming an existing calendar	**Tools**, **Organizer…**, **Calendars** tab, **Rename…**
• Deleting a calendar	**Tools**, **Organizer…**, **Calendars** tab, **Delete**
• Copying a calendar to Global.mpt for use in future projects	**Tools**, **Organizer…**, **Calendars** tab, **Copy** >>
• Copying calendars between projects	**Tools**, **Organizer…**, **Calendars** tab, **Copy** >>

5.1 Assigning a Calendar to a Project

A new project is assigned by default the **Standard** calendar as the **Project Calendar** when it is created. The **Project Calendar** is changed using the **Project Information** form, which is opened by selecting:

- **Project**, **Project Information…**, and
- The alternative calendar from the **Calendar:** drop-down box:

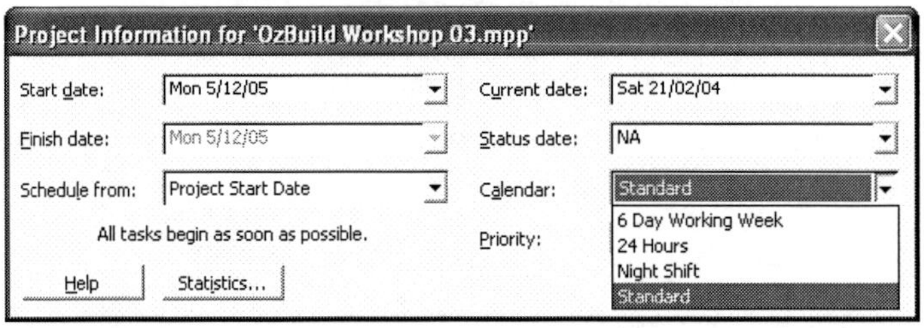

The **Project Calendar** is identified in the **Change Working Time** form as the calendar with **(Project Calendar)** written after the name. This form is covered in the next paragraph.

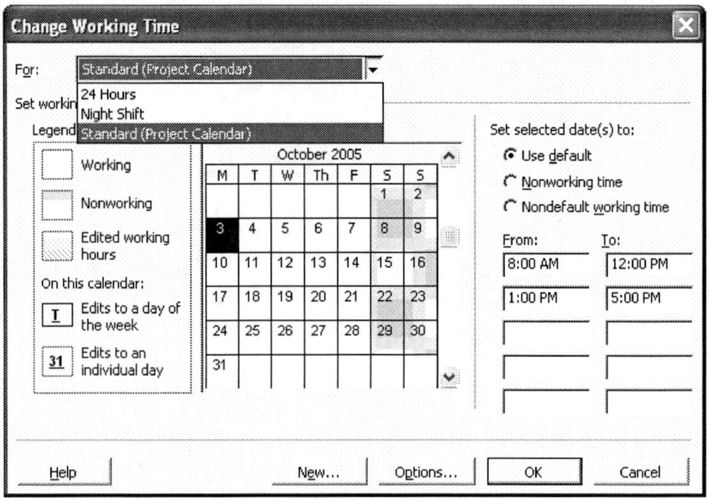

When a **Project Calendar** is changed or edited, the end date of all tasks assigned with the **Project Calendar** will be recalculated based on the new calendar. This may make a considerable difference to your project schedule dates.

5.2 Editing Calendar Working Days

To edit a calendar, select **Tools**, **Change Working Time…** to open the **Change Working Time** form:

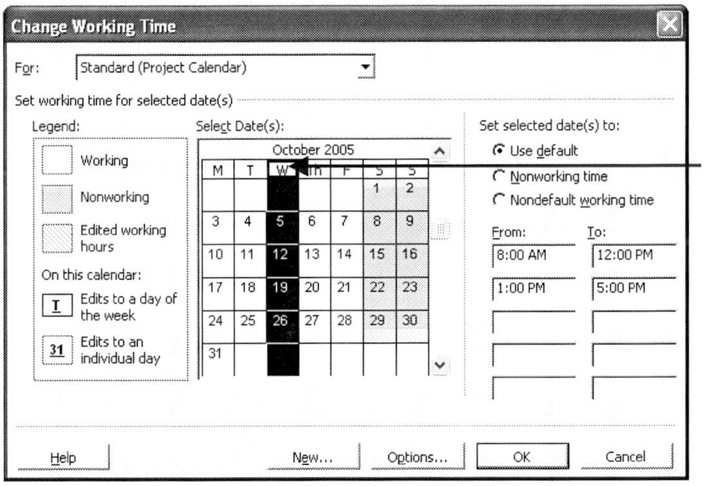

Click here to select a column of days.

- To make **Nonwork Days** into **Work Days**, highlight the day(s) you want to edit by:
  - ➢ Click on an individual day, or
  - ➢ Ctrl-click to select multiple days, or
  - ➢ Click and drag to select multiple days, or
  - ➢ Click on a column or columns of days by clicking the day of the week box, which is located below the month and year.
  - ➢ Then click on the **Nonworking time** radio button to make these days NonWorking.
- To make **Work Days** into **Nonwork Days**, highlight the day(s) you want to edit as described in the paragraph above and then click on the **Nondefault working time** radio button to make these days working days.

5.3 Adjusting Working Hours

To adjust the working hours of one or more days:

- Open the **Change Working Time** form,
- Highlight the days you want to edit, and
- Edit the working hours in the **From:** and **To:** section on the right-hand side of the screen.

The form below shows how three 7-hour shifts are entered with one shift (the first and fourth entries in the **From:** & **To:** mini-table) spanning midnight and therefore having two entries:

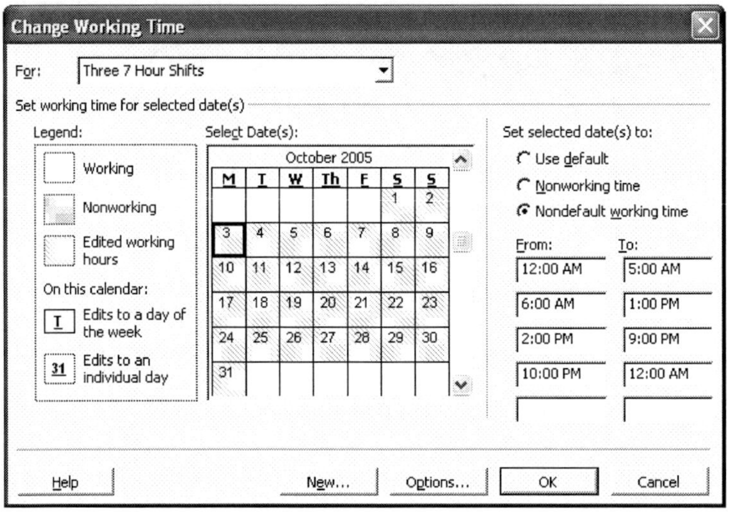

5.4 Creating a New Calendar

Select **Tools, Change Working Time…**, to open the **Change Working Time** form and click on the  button to open the **Create New Base Calendar** form:

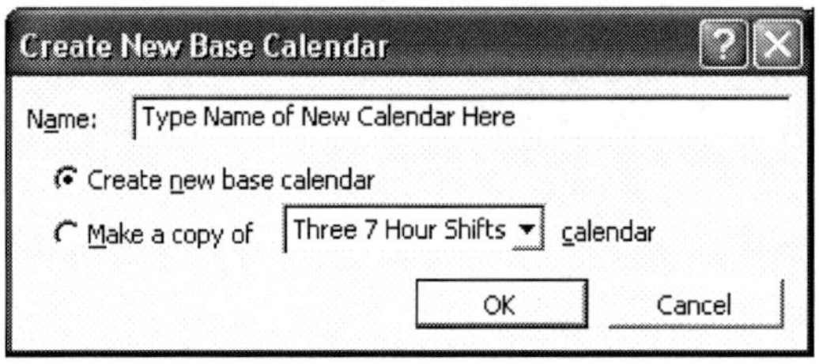

- To create a new calendar, click on the **Create new base calendar** radio button which will be a copy of the Global.mpt base calendar, type in the new calendar name in the **Name:** box and click on the 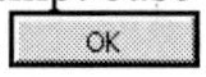 button to create it, or
- To copy an existing calendar, click on the **Make a copy of** radio button and select the calendar you want to copy from the drop-down box and type in the new calendar name.

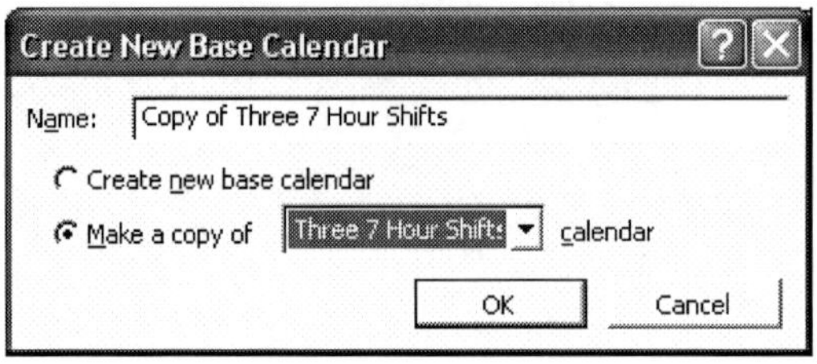

5.5 Renaming a Calendar

To rename a calendar:

- Select **Tools**, **Organizer…**, and select the **Calendars** tab.
- Highlight the calendar you want to rename and click on the Rename... button to open the **Rename** form and type in the new name.

5.6 Deleting a Calendar

To delete a calendar:

- Select **Tools**, **Organizer…**, and select the **Calendars** tab.
- Highlight the calendar you want to delete and click on the Delete... button.

5.7 Copying a Base Calendar to Global.mpt for use in Future Projects

Global.mpt is the default project template and all new projects copy their default settings from the Global.mpt. Once you have set up a calendar with all the holidays for your location or business, it is suggested that you copy it to the Global.mpt. This calendar will then be available in all new schedules.

- Select **Tools**, **Organizer…**, and select the **Calendars** tab.
- Select the **Standard** calendar from your project and click on the << Copy button.

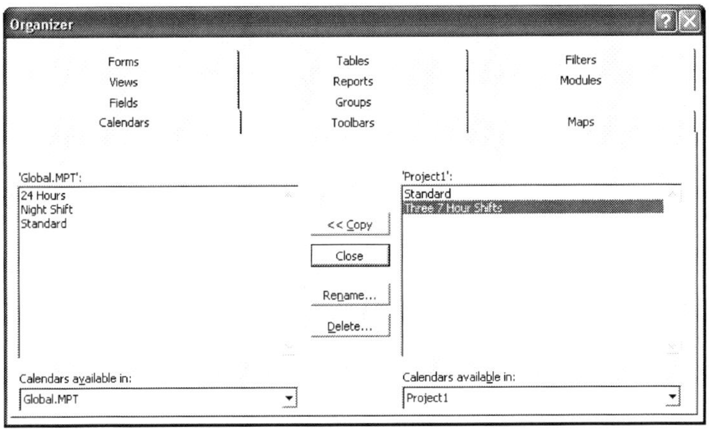

5.8 Copying Calendars Between Projects

To copy a calendar between projects:

- Open both projects,
- Select **Tools**, **Organizer…**, and select the **Calendars** tab,
- From the drop-down boxes at the bottom of the tab under Calendars available in: select the projects that you want to copy to and from, and
- Select the calendar you want to copy and click on the [<< Copy] button.

5.9 Resource Calendars

Individual Resources are allocated a Base Calendar when they are created. The resource Base Calendar may be changed and edited to reflect the availability of the resource. This feature is covered in the **CREATING RESOURCES** chapter.

5.10 Selecting Dates

Microsoft Project 2003 has a function to enable the user to quickly scroll though days, months and years when editing dates from a column. A calendar form may be displayed by clicking on a date cell with the mouse pointer:

- To change the day, click on the required day.
- To change month either:
  - ➢ Scroll a month at a time by clicking on the arrows on the top left hand or top right hand side of the form, or
  - ➢ Click on the month at the top of the form and a drop down list will be displayed to allow any month of the year to be selected.

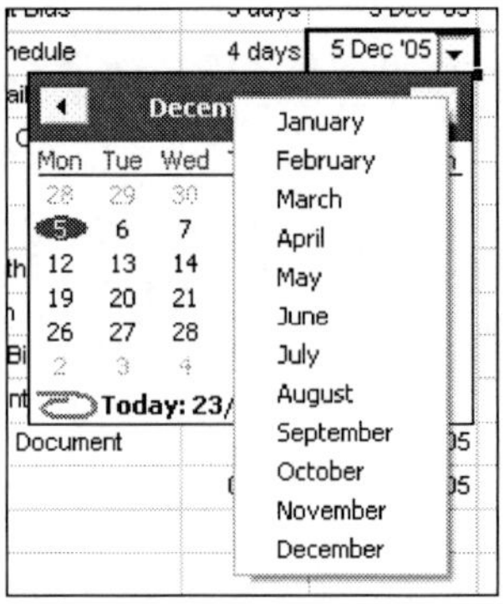

- To change the year, click on the year in the top of the form and scrolling arrows will be displayed allowing scrolling one year at a time.

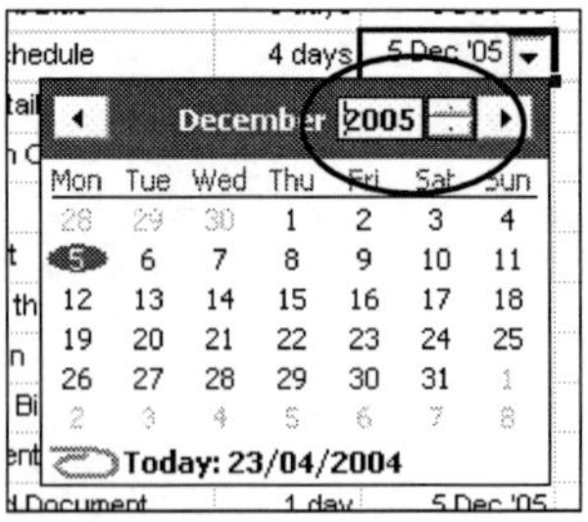

When a date is selected from a column Microsoft Project 2003 will set a constraint without informing the user. This may result in an incorrect constraint being set.

WORKSHOP 3

Maintaining the Calendars

Preamble

The normal working week at OzBuild Pty Ltd is Monday to Friday, 8 hours per day excluding Public Holidays. The installation staff works Monday to Saturday, 8 hours per day.

Assignment

The company observes the following Australian holidays:

	2005	2006	2007	2008
New Years Day	3 January*	2 January*	1 January	1 January
Good Friday	25 March	14 April	6 April	21 March
Easter	28 March	17 April	9 April	24 March
Christmas Day	26 December*	25 December	25 December	25 December
Boxing Day	27 December*	26 December	26 December	26 December

* These holidays occur on a weekend and the dates in the table above have been moved to the next weekday.

Boxing Day is a religious holiday, the day after Christmas, celebrated in many countries.

Assignment

1. Edit the **Standard Calendar** to ensure that only the holidays above in 2005 and 2006 are present by selecting **Tools**, **Change Working Time….**
2. Create a new calendar titled **6 Day Working Week** for the 6-day week by:
   - Copying the **Standard (Project Calendar)**, and then
   - Clicking on the **S** for Saturday in the **Change Working Time** form and then clicking on the **Nondefault working time** radio button to make all Saturdays work days.
3. Save your **OzBuild Bid** project

6 ADDING TASKS

Tasks, or Activities, should be well-defined, measurable pieces of work with a measurable outcome. Task descriptions containing only nouns such as "Specification" have confusing meanings. Does this mean Draft, Review, Approve, Issue, or all of these?

Adequate task descriptions always have a verb-noun structure to them. A more appropriate task description would be "Specification Approved" or "Specification Issued" The limit for task names is 255 characters, but try to keep task descriptions meaningful yet short and concise so they are easier to print.

When tasks are created they may be organized under a Summary task and have logic added to calculate their start and finish dates. Summary or Detailed tasks may be created first and logic added once the relevant tasks have been created.

The creation and sequencing of detailed and summary tasks are discussed in the following chapters:

- Create **Detailed** tasks in this chapter,
- Creating **Summary** tasks in the **ORGANIZING TASKS USING OUTLINING** chapter, and
- Adding the logic in the **ADDING THE DEPENDENCIES** chapter.

This chapter will cover the following topics:

Topic	**Menu Command**
• **Adding New Tasks**	Select a line in the schedule and strike the **Ins** (**Insert Key**) or click on a blank line.
• **Reordering** tasks	Select and drag the task(s) or cut and paste in the required location.
• **Copying** tasks in Microsoft Project 2003	Select the tasks and copy & paste to the required location.
• **Copying** tasks from other Programs	Display the required columns and paste the data.
• **Elapsed** duration tasks	Type "e" after the duration & before the Units, e.g. 5edays.
• **Milestones**	Assign a zero duration, or Check the **Mark task as milestone** box on the **Advanced** tab in the **Task Information** form.
• **Task Information** form	Double-click anywhere on the task line, or Highlight the task line click on the button.
• Assigning **Calendars** to tasks	Set a task calendar in the **Advanced** tab in the **Task Information** form or display the **Task Calendar** column.

6.1 Adding New Tasks

There are several methods of adding a task.

Method 1

The first and easiest is to:

- Click on the first blank line under the title **Task Name** and type the task description.

- The duration of the new task is completed with a default duration, this may have a "?" after to indicate it has been assigned an **Estimated Duration.** Overtype the default duration with the required duration task. The option of assigning and/or displaying a new task with an **Estimated Duration** is controlled in the **Schedule** tab of the **Options** form by checking or un-checking **New tasks have estimated durations**.

	❶	Task Name	Duration	Start	Finish	17 Oct '05								24 Oct '05					
						S	S	M	T	W	T	F	S	S	M	T	W	T	F
1		First Task	1 day?	Wed 19/10/05	Wed 19/10/05			▪											

- Click into the second row and enter a second Task Name. A sequential **Task Number**, starting from "**1**" will be created in the column to the left of the **Task Name**.

	❶	Task Name	Duration	Start	Finish	17 Oct '05								24 Oct '05					
						S	S	M	T	W	T	F	S	S	M	T	W	T	F
1		First Task	5 days	Wed 19/10/05	Tue 25/10/05			▬▬▬▬▬											
2		Second Task	1 day?	Wed 19/10/05	Wed 19/10/05			▪											

The task **Start** and **Finish** dates will be calculated from the **Project Start Date** and the task duration. This information is displayed in the **Start** and **Finish** columns.

Method 2

The second method of adding a task, which may be used for inserting a task between two other tasks is to highlight a task where you want to insert a new task, the highlighted task will be moved down and:

- Select **Insert**, **New Task**, or

- Press the **Insert Key** on the keyboard, or

- Highlight the entire task row by clicking on the Task ID, then right-click and select **New task** from the sub-menu.

	❶	Task Name	Duration	Start	Finish	17 Oct '05								24 Oct '05					
						S	S	M	T	W	T	F	S	S	M	T	W	T	F
1		First Task	5 days	Wed 19/10/05	Tue 25/10/05			▬▬▬▬▬											
2		Inserted Task	3 days	Wed 19/10/05	Fri 21/10/05			▬▬▬											
3		Second Task	2 days	Wed 19/10/05	Thu 20/10/05			▬▬											

Do not type in a Start or Finish date into the **Start** or **Finish** column unless you want to set a constraint, which overrides the logic. Microsoft Project 2003 is programmed to set an Early Constraint when dates are typed into these columns and there is no warning message that these constraints have been set, except when using Windows XP and a Graphical Indicator or Smart Tag that will warn you that constrain has been set.

6.2 Reordering Tasks by Dragging

You may move one or more tasks up or down the schedule by:

- Highlighting any one or more adjacent tasks by using the Task ID column; this will ensure you have selected the whole task and not just cells of the task,
- Moving the cursor to the top line of the selected tasks until it changes to an icon,
- Then left-clicking and holding down the mouse to drag the row up or down. A gray line will indicate where the tasks will be inserted.

The tasks will be renumbered when in their new location.

6.3 Copying Tasks in Microsoft Project 2003

Tasks may also be copied from another project or copied from within the same project using the normal Windows commands, **Copy** and **Paste**, by using the menu commands **Edit**, **Copy Task** and **Edit**, **Paste** or **Ctrl+C** and **Ctrl+V**.

You may also copy one or more adjacent tasks by using the **Ctrl Key** and dragging as follows:

- Select the whole task or tasks, not just a cell,
- Hold down the **Ctrl Key**, and
- Drag the tasks to the location you want to insert them. This will create a copy of the original tasks.

6.4 Copying Tasks from other Programs

Task data may be copied to and from, or updated from other programs such as Excel, by cutting and pasting. The columns and rows in your spreadsheet will need to be formatted in the same way as in your schedule before they may be pasted into your schedule. It is recommended that you first display the data that you want to import or update in your schedule. If you have no tasks then create a dummy task and data, then copy and paste this dummy data into your spreadsheet, update the data and paste it back into the schedule. The data headings are not brought across from the schedule; therefore it is also recommended that you type the headings into the spreadsheet above the imported data, which will assist you in typing the data in the correct columns.

When you copy and paste dates into the schedule, you may find that activities are assigned constraints, which you may not desire. It is recommended that you display the **Indicators** column, which will show an icon if a constraint has been applied.

6.5 Milestones

A Milestone normally has a zero duration and is used to mark the start or finish of a major event. Microsoft Project does allow the user to nominate if a Milestone is a **Start** or **Finish Milestones** as with other products. A Milestone is a Start Milestone when it has no predecessors and is scheduled at the start of a work day and a Finish Milestone when it has predecessors and is schedules at the end of a work day.

Start	Finish	Mon 1 May	Tue 2 May	Wed
		12 AM 6 AM 12 PM 6 PM	12 AM 6 AM 12 PM 6 PM	12 AM
1 May 2006 8:00 AM	1 May 2006 8:00 AM	◆ 1/05		
1 May 2006 8:00 AM	1 May 2006 5:00 PM			
1 May 2006 5:00 PM	1 May 2006 5:00 PM	◆ 1/05		
2 May 2006 8:00 AM	2 May 2006 5:00 PM			
2 May 2006 5:00 PM	2 May 2006 5:00 PM		◆ 2/05	

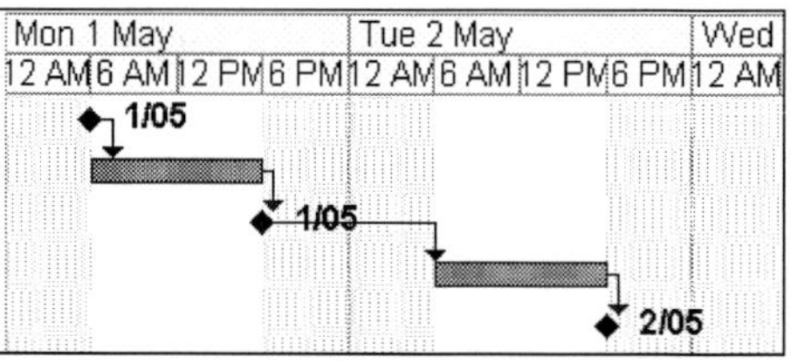

A Task is assigned a zero duration to create a milestone and is then normally displayed in the bar chart with a "◆".

Microsoft Project 2003 also has the ability to display a task with a non-zero duration as a Milestone as follows:

- Highlight the task with a duration that you want to be a milestone,

- Open the **Task Information** form by double-clicking on the task,

- Select the **Advanced** tab, and

- Click the **Mark task as milestone** check box.

	❶	Task Name	Duration	Start	Finish	17 Oct '05	24 Oct '05
						S M T W T F S	S M T W T
1		First Task	5 days	Wed 19/10/05	Tue 25/10/05		
2		Inserted Task	3 days	Wed 19/10/05	Fri 21/10/05		
3		Second Task	2 days	Wed 19/10/05	Thu 20/10/05		
4		Milestone	0 days	Wed 19/10/05	Wed 19/10/05	◆ 19/10	
5		Milestone with Duration	4 days	Wed 19/10/05	Mon 24/10/05	◆ 19/10	

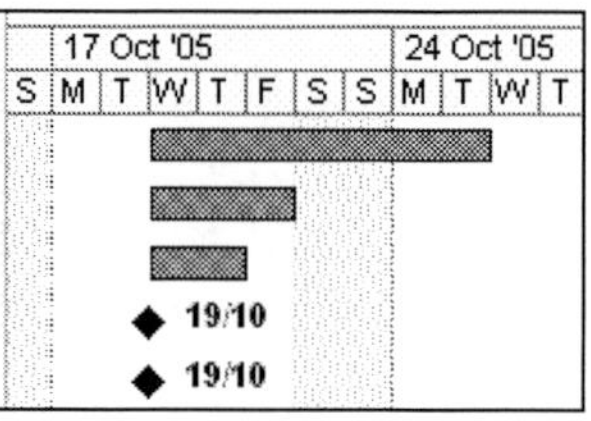

6.6 Elapsed Durations

A task may be assigned an **Elapsed** duration. The task will ignore all calendars and the task will take place 24 hours a day and 7 days per week. A 24-hour, 7-day/week calendar does not need to be created for these tasks. This is useful for tasks such as curing concrete but the Total float will calculate three times longer than a task on an 8 hour a day calendar and this may be misleading.

To enter an elapsed duration, type an "e" between the duration and units. The example below shows the difference between a 7-**Elapsed Day** task and a 7-day task on a **Standard** (5 day/week) calendar.

	❶	Task Name	Duration	17 Oct '05	24 Oct '05
				S M T W T F S	S M T W T F S
1		7 Day task with Elapsed Duration	7 edays		
2		7 Day Task	7 days

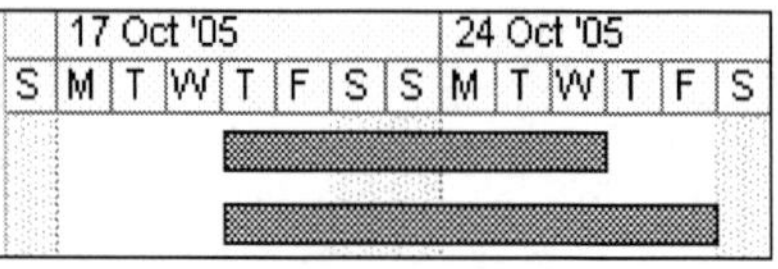

6.7 Task Information Form

The Task Information form may be opened by double-clicking on a task line or by clicking on the  button when the task is highlighted. You are able to make changes to a number of task parameters from this form. Once the form is open, it is not possible to move to another task without closing the form. There are five tabs on the form:

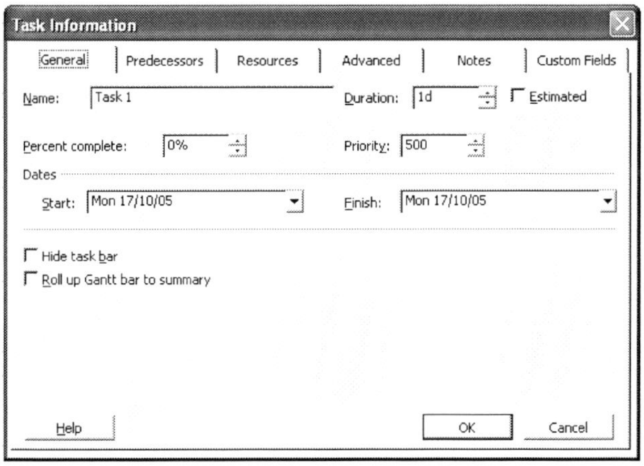

- **General** – This contains the basic information about a task that may be entered and displayed.
  - The **Estimated** check box is used to indicate the task has an **Estimated Duration**. The task duration will be displayed with a "?" which disappears when a duration is entered against a task.
  - **Priority** is used in leveling. 1000 is the highest and 0 the lowest.
  - The **Bar** options are covered in the **Formatting Bars** section.
- **Predecessors** – This is where the task's predecessors are displayed. This is covered in the **ADDING THE DEPENDENCIES** chapter.
- **Resources** – This is where resources may be created, assigned to tasks and assignment information displayed. This topic is covered in the **RESOURCES** chapters.
- **Advanced** – Options are covered in the **CONSTRAINTS** chapter.
- **Notes** – This is where notes about a task may be recorded.
- **Custom Fields** – Any **Custom Fields** that have been customized will be displayed in this tab. **Custom Fields** will not be covered in detail in this book. They are existing yet undefined fields that may be customized by selecting **Tools**, **Customize**, **Fields…** These fields may hold different types of project information such as text, times, values, etc. and are there to make the software more suitable for any given user. There are options on how these fields are summarized at the summary bar level.
- When multiple tasks are selected then the task form may be opened by right clicking and some of the attributes of all the tasks may be edited at the same time.

6.8 Indicators Column

The **Indicators** column will display an icon when a task contains a non-default setting such as a Note, Constraint, or a Task calendar. Placing the mouse over the icon will display information about the task.

6.9 Assigning Calendars to Tasks

Tasks often require a different calendar from the **Project Calendar**. This is assigned in the **Project Information** form. Microsoft Project 2003 allows each task to be assigned a unique calendar. A Task Calendar may be assigned by using the **Task Information** form or displaying the **Task Calendar** column.

6.9.1 Assigning a Calendar Using the Task Information Form

- Select one or more tasks that you want to assign to a different calendar by using Shift-Click or Ctrl-Click.
- Open the **Task Information** form when selecting a single task, by double-clicking on a task.
- Open the **Task Information** form when selecting multiple tasks by:
  - ➢ Select **Project**, **Task Information**, or
  - ➢ **Shift+F2**, or
  - ➢ Right-click and select **Task Information**.
- Then select the Advanced tab:

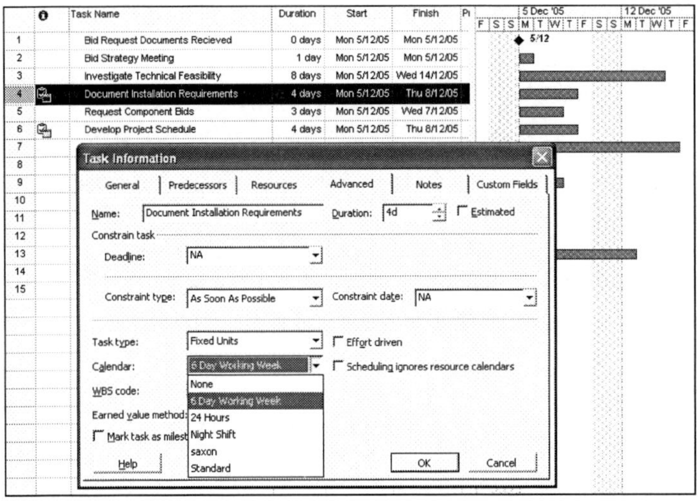

- From the **Calendar:** drop-down box select the calendar you want to assign to the task or tasks.

6.9.2 Assigning a Calendar Using a Column

You may also display the **Task Calendar** column and edit the task calendar from this column. The process of displaying a column is covered in the **FORMATTING THE DISPLAY** chapter. After a calendar has been assigned, an icon will appear in the **Indicators** column as displayed beside the **Bid Strategy Meeting** activity below:

	❶	Task Name	Task Calendar
3		Investigate Technical Feasibility	None
4	[calendar icon]	Document Installation Requirements	6 Day Working Week

WORKSHOP 4

Adding Tasks

Preamble
If you do not have the standard Microsoft Project 2002 settings loaded on your computer or your Global.mpt has been edited then you may not have the same results as displayed in these workshops.

Assignment

1. Use the columns to enter the Name and Original Duration of the tasks as below.
2. Assign the 6-Day Working Week calendar using the **Task Information** form. Double-click on the task to open this form.
3. The task will become a milestone when assigned a zero duration.

Task Name	Duration	Calendar Assignment
Bid Request Documents Received	0 days	
Bid Strategy Meeting	1 day	
Investigate Technical Feasibility	8 days	
Document Installation Requirements	4 days	Assign the 6 Days Working Week calendar
Request Component Bids	3 days	
Develop Project Schedule	4 days	Assign the 6 Days Working Week calendar
Draft Technical Details Schedule	9 days	
Compile Costs from Component Bids	2 days	
Draft Bid Document	3 days	
Meeting to Review the Draft Bid Document	1 day	
Design Presentation	1 day	
Edit Proposal Draft Bid Document	1 day	
Negotiate Component Work Packages	6 days	
Final Review of Bid Document	1 day	
Submit Bid	0 days	

4. Save your **OzBuild Bid** project.
5. Check your answer with the example over the page.

ANSWER TO WORKSHOP 4

6. Your schedule should look like this:

	Information	Task Name	Duration	Start	Finish
1		Bid Request Documents Recieved	0 days	5 Dec '05	5 Dec '05
2		Bid Strategy Meeting	1 day	5 Dec '05	5 Dec '05
3		Investigate Technical Feasibility	8 days	5 Dec '05	14 Dec '05
4	(calendar icon)	Document Installation Requirements	4 days	5 Dec '05	8 Dec '05
5		Request Component Bids	3 days	5 Dec '05	7 Dec '05
6	(calendar icon)	Develop Project Schedule	4 days	5 Dec '05	8 Dec '05
7		Draft Technical Details Schedule	9 days	5 Dec '05	15 Dec '05
8		Compile Costs from Component Bids	2 days	5 Dec '05	6 Dec '05
9		Draft Bid Document	3 days	5 Dec '05	7 Dec '05
10		Meeting to Review the Draft Bid Document	1 day	5 Dec '05	5 Dec '05
11		Design Presentation	1 day	5 Dec '05	5 Dec '05
12		Edit Proposal Draft Bid Document	1 day	5 Dec '05	5 Dec '05
13		Negotiate Component Work Packages	6 days	5 Dec '05	12 Dec '05
14		Final Review of Bid Document	1 day	5 Dec '05	5 Dec '05
15		Submit Bid	0 days	5 Dec '05	5 Dec '05

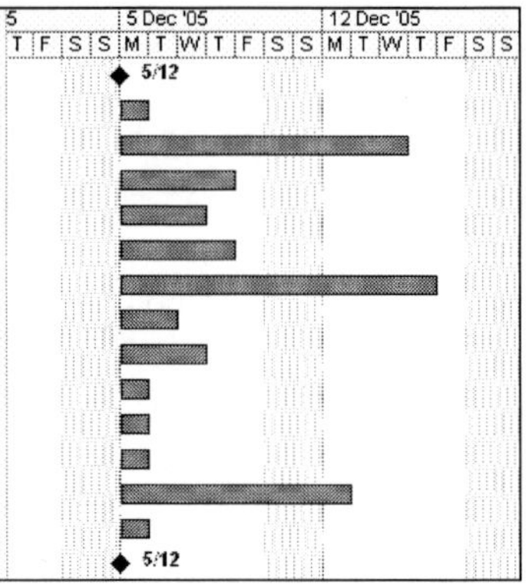

The icon in the Information column on the Left-Hand Side indicates that tasks 4 and 6 have a non-standard calendar which is the 6-Day Working Week calendar set in the last workshop.

7 ORGANIZING TASKS USING OUTLINING

Outlining is used to summarize and group tasks under a hierarchy of **Parent** or **Summary Tasks**. They are used to present different views of your project during planning, scheduling and statusing. These headings are normally based on your project breakdown structure.

Defining the project's breakdown structure can be a major task for project managers. The establishment of templates makes this operation simpler because a standard breakdown is predefined and does not have to be typed in for each new project.

Projects should be broken into manageable areas by using a structure based on a breakdown of the project deliverables, systematic functions, disciplines or areas of work. The Outline structure created in your project should reflect the breakdown of your project.

Microsoft Project 2000 introduced a new feature titled **Grouping**, which is similar to the **Organize** function in Primavera P3 and SureTrak software. This feature allows the grouping of tasks under headings other than the "Outline Structure." Unlike Primavera software, **Grouping** is not the primary method of organizing tasks and is covered in the **GROUPING TASKS, OUTLINE CODES AND WBS** chapter.

7.1 Creating an Outline

To create an **Outline**:

- Insert a new **Summary** task above your proposed **Detailed tasks**:

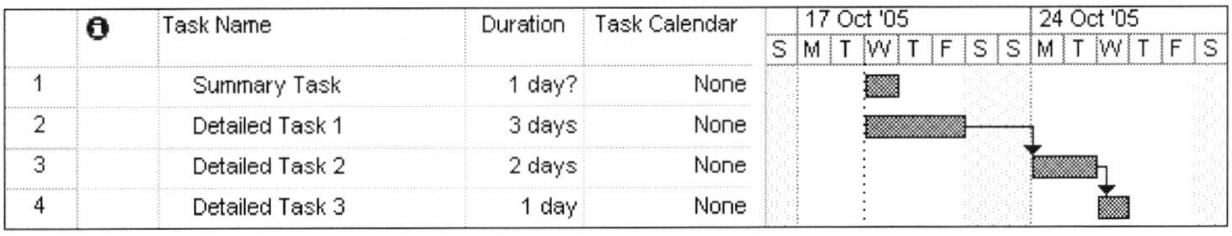

Task Name
Duration
Task Calendar
17 Oct '05
24 Oct '05
S M T W T F S S M T W T F S
1 Summary Task 1 day? None
2 Detailed Task 1 3 days None
3 Detailed Task 2 2 days None
4 Detailed Task 3 1 day None

- Then **Demote** the **Detailed tasks** below **Summary task**. (Demoting is explained in Section 7.2, below.):

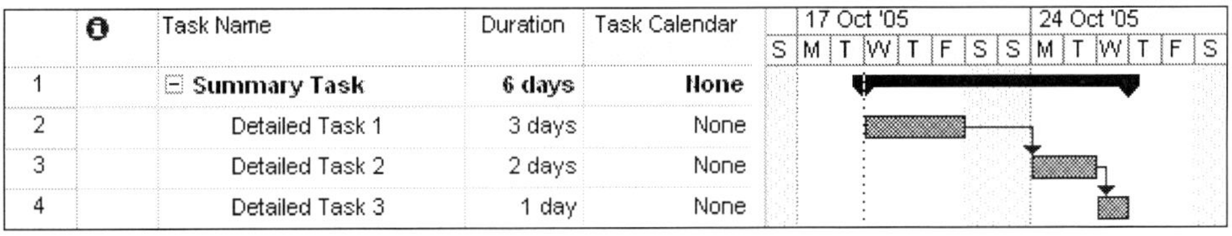

Task Name
Duration
Task Calendar
17 Oct '05
24 Oct '05
S M T W T F S S M T W T F S
1 Summary Task 6 days None
2 Detailed Task 1 3 days None
3 Detailed Task 2 2 days None
4 Detailed Task 3 1 day None

The start and finish dates of the **Summary task** are adopted from the earliest start date and latest finish date of the **Detailed tasks**.

The duration of the **Summary** is calculated from the adopted start and finish dates over the **Summary** task calendar, which is initially the **Project Calendar**.

7.2 Promoting and Demoting Tasks

Demoting or **Indenting** tasks may be achieved in a number of ways. Select the task or tasks you want to **Demote**. Ensure you have selected the whole task and not just some cells. You may use any of the following methods to **Demote a selected task**:

- Click on the Indent  button, or
- Right-click on the **Task Id** column to open the task shortcut menu, click on the 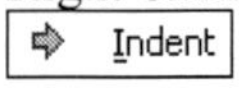

button, or
- Move the mouse until you see a double-headed horizontal arrow in the task name, left-click and drag the task right. A vertical line (see lower of the two pictures below) will appear indicating the outline level you have dragged the task(s) to, or

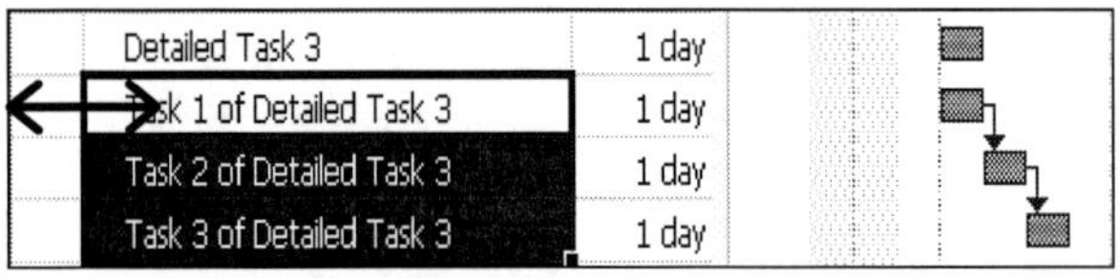

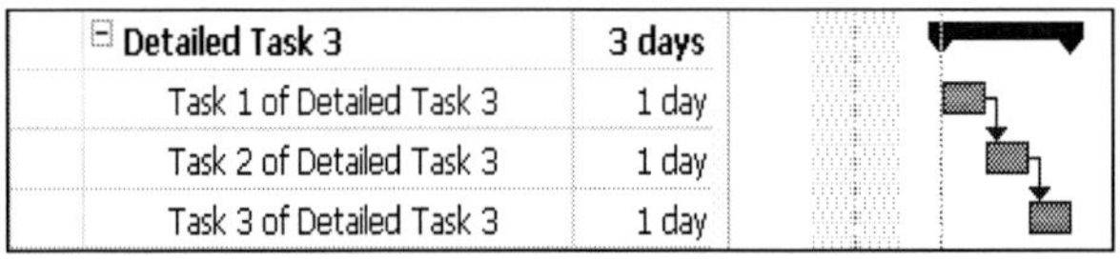

- Hold down the **Alt** and **Shift keys** and press the **Right Arrow Key** on your keyboard.

Demoting or **Outdenting tasks** uses the same principle as promoting tasks. Select the task or tasks you want to demote, ensure you have selected the whole task and not just some cells, then you may:

- Click on the Outdent  button, or
- Right-click on the **Task Id** column to open the task shortcut menu, click on the 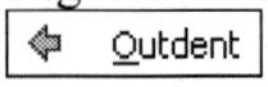

button, or
- Move the mouse until you see a double-headed horizontal arrow in the task name column, left-click and drag the tasks left, or
- Hold down the **Alt** and **Shift keys** together and press the **Left Arrow Key** on your keyboard.

Tasks may be added under a **Detailed** task and demoted to a third level and so on.

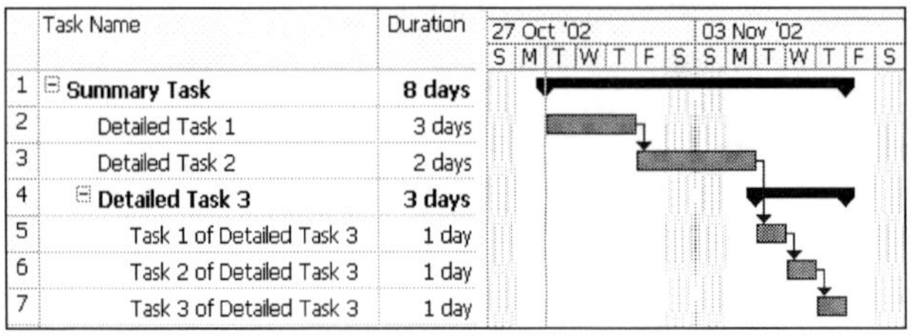

7.3 Summarizing Tasks

Once you have created summary tasks, the detailed tasks may be rolled up or summarized under the summary tasks. Rolled up tasks are symbolized by the + sign to the left of the summarized task description:

- This picture shows **Detailed Task 3** rolled up.

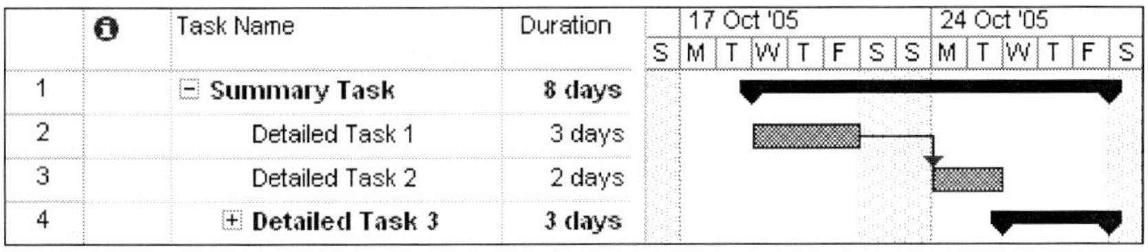

- This picture shows **SUMMARY TASK** rolled up.

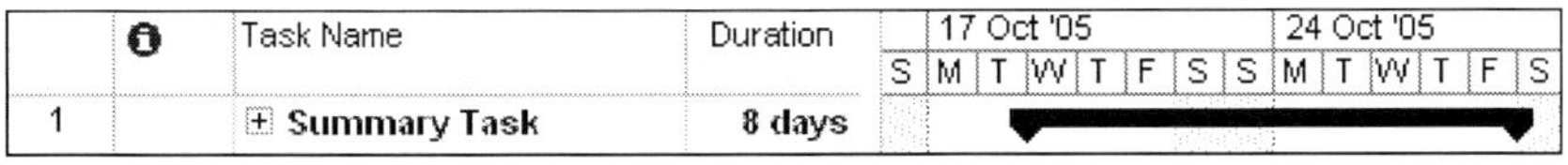

7.3.1 To Rollup Summary Tasks and Show Tasks

The Outline Symbols, ⊞ and ⊟, in front of the tasks may be hidden and displayed from the **Tools**, **Options…**, **View** tab, **Show outline symbol**.

Select the task you want to rollup:

- Click on the ⊟ to the left of the Task Name, or
- Click on the **Hide Subtasks** icon, or
- Double-click on the **Task ID** (not the **Task Name** as this will open the **Task** form).

Displaying rolled-up tasks is similar to rolling them up. To do this, select the task you want to expand. Then:

- Click on the to ⊞ to the left of the Task Name, or
- Click on the **Show Subtasks** icon, or
- Double-click on the **Task ID**.

7.3.2 Roll Up All Tasks to an Outline Level

A schedule may be rolled up to any Outdent Level by selecting the desired Outdent Level from the Show drop-down box on the **Formatting Tool Bar**.

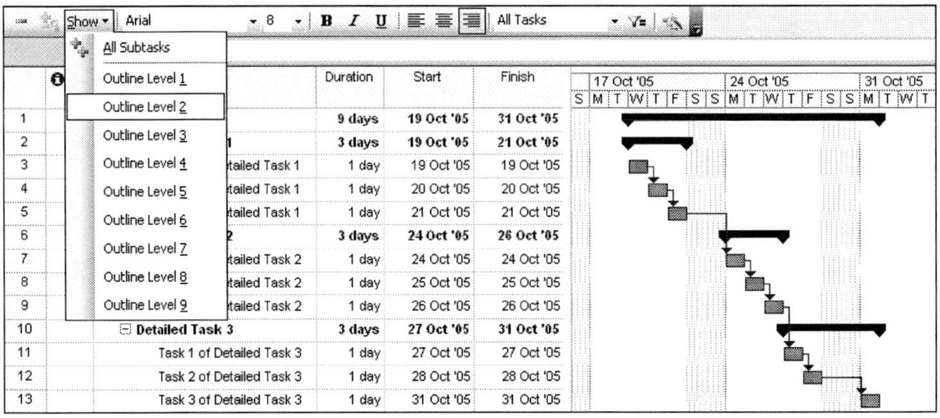

7.3.3 Show All Tasks

Select the All Subtasks option at the top to display all tasks.

7.4 Project Summary Task

A project summary task may be displayed by checking the **Show project summary task** box from the **Tools**, **Options…**, **View** tab. This task spans from the first to the last task in the project and is in effect a built-in Level 1 outline. The description of the Summary Task is the Project Title entered in the **File**, **Properties** form. A project Summary Task is a virtual task and may not have resources, relationships or constraints assigned.

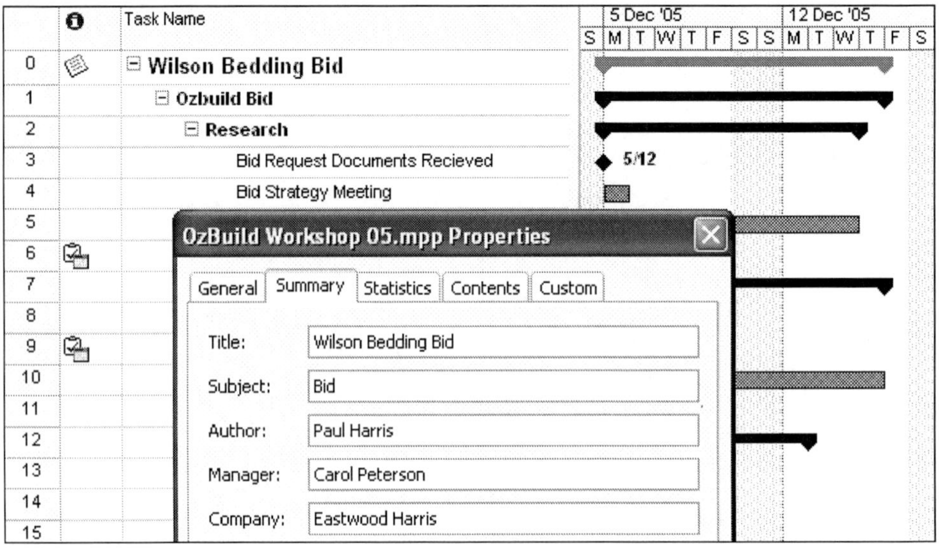

WORKSHOP 5

Maintaining the Task Codes

Preamble

A review of the internal reporting requirements shows that you need to identify the Bid Work Phases:

- Research
- Estimation
- Proposal

Assignment

1. Go to the **View** tab in the **Tools**, **Options…** form and check the option to display the Project summary task, close the form and observe how the Project Summary Task is formatted.
2. Remove the Project summary task as we will create an Outline Level for the Project Summary Task.
3. Create an Outline Level 1 for the whole project entitled "OzBuild Bid" and
4. Create an Outline Level 2 for each of the three phases: Research, Estimation and Proposal. Try using the various alternative methods for indenting and outdenting tasks.
5. Your schedule should look like this:

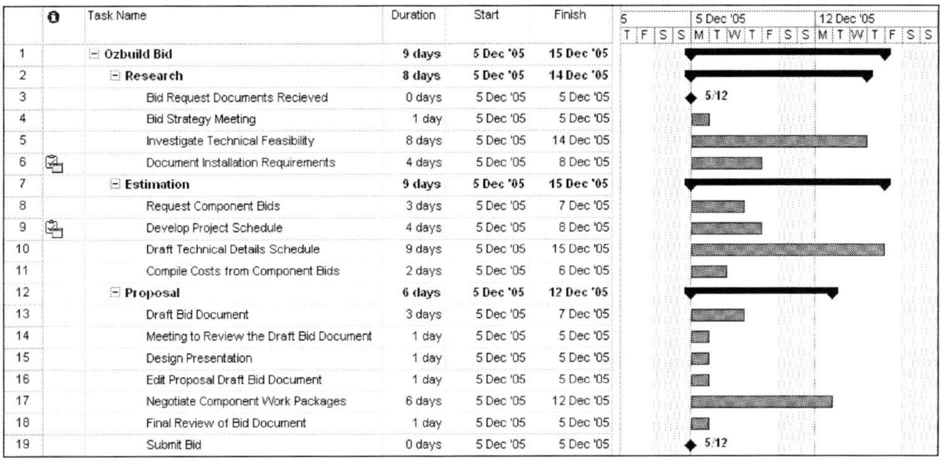

		Task Name	Duration	Start	Finish
1		**Ozbuild Bid**	**9 days**	**5 Dec '05**	**15 Dec '05**
2		**Research**	**8 days**	**5 Dec '05**	**14 Dec '05**
3		Bid Request Documents Recieved	0 days	5 Dec '05	5 Dec '05
4		Bid Strategy Meeting	1 day	5 Dec '05	5 Dec '05
5		Investigate Technical Feasibility	8 days	5 Dec '05	14 Dec '05
6		Document Installation Requirements	4 days	5 Dec '05	8 Dec '05
7		**Estimation**	**9 days**	**5 Dec '05**	**15 Dec '05**
8		Request Component Bids	3 days	5 Dec '05	7 Dec '05
9		Develop Project Schedule	4 days	5 Dec '05	8 Dec '05
10		Draft Technical Details Schedule	9 days	5 Dec '05	15 Dec '05
11		Compile Costs from Component Bids	2 days	5 Dec '05	6 Dec '05
12		**Proposal**	**6 days**	**5 Dec '05**	**12 Dec '05**
13		Draft Bid Document	3 days	5 Dec '05	7 Dec '05
14		Meeting to Review the Draft Bid Document	1 day	5 Dec '05	5 Dec '05
15		Design Presentation	1 day	5 Dec '05	5 Dec '05
16		Edit Proposal Draft Bid Document	1 day	5 Dec '05	5 Dec '05
17		Negotiate Component Work Packages	6 days	5 Dec '05	12 Dec '05
18		Final Review of Bid Document	1 day	5 Dec '05	5 Dec '05
19		Submit Bid	0 days	5 Dec '05	5 Dec '05

6. Save your **OzBuild Bid** project.

8 FORMATTING THE DISPLAY

This chapter shows you how to set up the on-screen presentation so that the schedule will be easier to read and more consistent. This chapter covers the following display customizing topics:

Topic	Menu Command
• **Table** – formatting the columns of data	**View**, **Table:Entry**, **More Tables...** or Select a column and right-click to insert or delete a column.
• **Formatting Columns**	Double-click on the column title.
• **Inserting Columns**	Highlight a column and strike the **Ins Key**, or Select **Insert**, **Column...**, or Right-click and select **Insert Column...**.
• **Deleting Columns**	Highlight a column and strike the **Delete** key, or Select **Edit**, **Hide Column**, or Right-click and select **Hide Column**.
• **Format Bars**	**Format**, **Bar Styles...** or double-click on a bar.
• **Row Height**	Drag with the mouse.
• **Format Text Font**	**Format**, **Font...** to format columns and **Format**, **Text Styles...** to format all other fonts.
• **Timescale**	**Format**, **Timescale...** or Double-click on the timescale.
• **Gridlines**	**Format**, **Gridlines...**.
• **Relationship Lines**	**Format**, **Layout...**.

The formatting is applied to the current **View** and is automatically saved as part of the View when another View is selected. Views are covered in the **VIEWS, TABLE AND DETAILS** chapter.

8.1 Formatting the Columns

There are two methods of formatting the columns:

- Using **Table** function, this is where you set up the data columns in the way you want to see the information on the screen and in printouts. You may edit, create and delete **Tables** and select which one is used display the data.
- Inserting and/or deleting columns of data without using the **Tables** function.

8.1.1 Formatting Columns Using the Table Function

- Select **View**, **Table:Entry** and select from the list of predefined **Tables** you want to display:

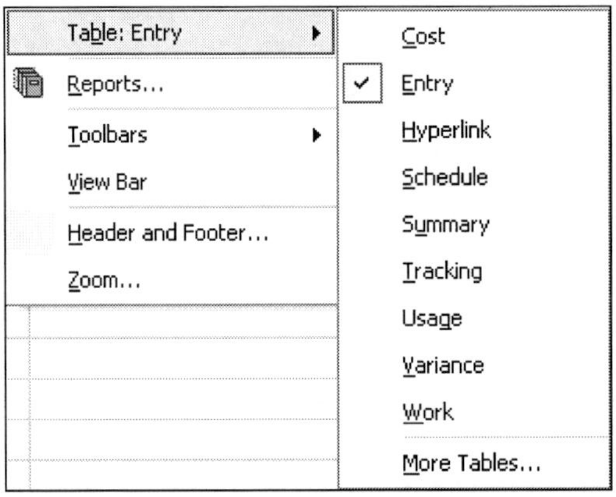

- Select **View**, **Table:**, **More Tables…** to open the **More Tables** form:

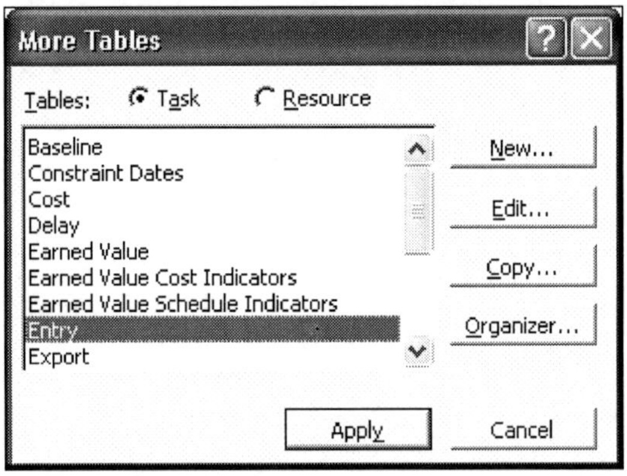

- 

– To create a new Table.
- 

– To edit the highlighted Table.
- 

– To copy the highlighted Table.
- 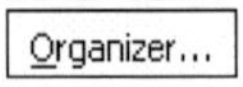

– Opens the **Organizer** form which enables you to copy a Table from one opened project to another or to the Global Project.
- 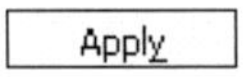

– Applies the selected Table making it visible on the screen.

- When you select New..., Edit... or Copy... you will be presented with the **Table Definition** form:

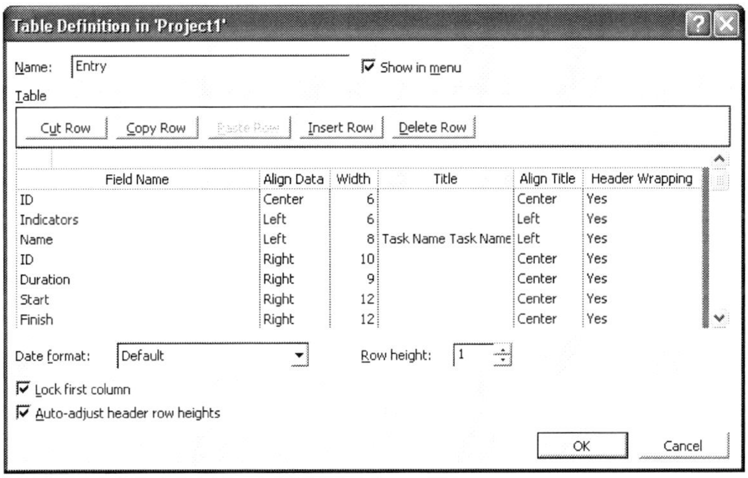

  - Click on the **Show in menu** box to display the Table in the menu found under **View**, **Table:**.
  - The columns of data will be displayed on screen from left to right in the same order as the rows in the form.
  - Highlight a row and then you may use the Cut Row, Copy Row, Paste Row, Insert Row and Delete Row buttons.
  - The data to be displayed may be selected from the drop-down box in the **Field Name** column.
  - **Align Data** and **Width** are used for formatting the data in the columns.
  - The Microsoft Project 2003 **Field Name** may be replaced by typing your own title in the **Title** box.
  - The **Date format:** drop-down box is used to change the format for this table only.
- **Row Height:** sets the default height of all the rows in this table. A row height may be changed by dragging the cell boundary line once a task has been created.
  - **Lock first column** prevents the first column from scrolling and is useful when the first column contains the Task Name.
  - **OK** takes you back to the **More Tables** form where you may click on the **Apply** button to commit the new or edited Table.

If you want to save a table for use in all your new projects then copy the table to the **Global.mpt** template using **Tools**, **Organizer…** and select the **Tables** tab.

You may also copy a **Table** to another project or rename a **Table** using **Tools**, **Organizer…** and selecting the **Tables** tab.

8.1.2 Formatting Columns

When you use this function, you are editing the **Table** currently in use and your changes are saved when you select a different table. Double-click on a column description to open the **Column Definition** form where you may edit the selected column in a similar way to the **Table Definition** form.

The IME Mode button allows the customization of the **Input Mode Editor** for some fields. This feature will only be displayed if it is installed with the Eastern Asian operating systems.

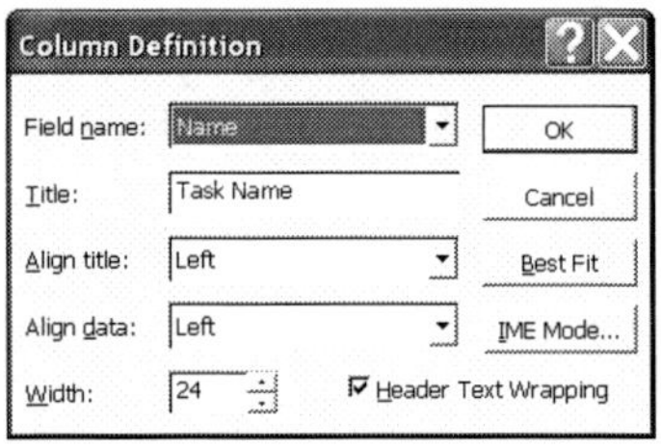

8.1.3 Deleting and Inserting Columns

Insert a column by clicking on the column title where you require the new column. This will highlight the column. To insert a new column:

- Select **Insert**, **Column…**, or
- Hit the **Ins** Key, or
- Right-click and select **Insert Column….**

Delete a column by highlighting the column or by clicking on the title. Then:

- Select **Edit**, **Hide Column**, or
- Hit the **Delete** key, or
- Right-click and select **Hide Column**.

1. There will be no confirmation of deleting a column, but you are allowed to undo the deletion.
2. The term “hide” means “delete” and the column is removed from the Table in this project.

8.1.4 Adjusting the Width of Columns

You may adjust the width of the column either manually or automatically. For manual adjustment, move the mouse pointer to the nearest vertical line of the column. A icon will then appear and enable the column to be adjusted. For automatic adjustment, once again position the mouse pointer to the nearest vertical line of the column, and double right-click the mouse. The column width will automatically adjust to the best fit.

8.1.5 Moving Columns

Columns in a Table may be moved by clicking on the column header, the mouse pointer will change to a and the column may be dragged to a new location

8.1.6 Adding a Space Between the Value and Label

Select **Tools**, **Options…**, **Edit** and check on the **Add space before label** check box to add a space between the value and label in date columns.

8.2 Formatting the Bar Style

The bars in the Gantt Chart may be formatted to suit your requirements for display. Microsoft Project 2003 has the option to:

- Format all the Task Bars, or
- Format one or more specific Task Bars.

8.2.1 Formatting All Task Bars

To format all the bars you must open the Bar Style form by selecting:

- **Format**, **Bar Styles…**, or
- Double-click anywhere in the Gantt Chart area, but not on an existing bar, as this will open the **Format Bar** form for formatting an individual bar.

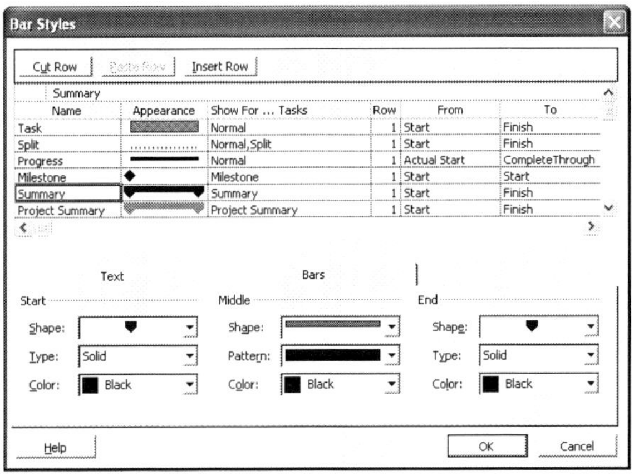

The following notes are the main points for using this function. Detailed information is available in the help facility by searching for "Bar styles dialog box".

- Each bar listed in the table will be displayed on the bar chart.
- Bars may be deleted with the Cut Row button, pasted using Paste Row button and new bars inserted using the Insert Row.
- The **Name** is the title you may assign to the bar and is displayed in the printout legend. To hide the bar on the legend precede the **Name** with an *.
- The appearance of each bar is edited in the lower half of the form. The bar's start point, middle and end point may have their color, shape, pattern, etc. formatted.
- **Show For … Tasks** allows you to select which tasks are displayed. More than one task type may be displayed by separating each type with a ",". Bar types not required are prefixed with "**Not**" For example Normal,Rolled Up,Split,Not Summary bar would not display a bar for a summary task. Should you leave this cell blank then all task types will be displayed in this format.
- The bars may be placed on one of four rows numbered from 1 to 4, top to bottom. If multiple bars are placed on the same row, the bar at the top of the list will be drawn first and the ones lower down the list will be drawn over the top.

- **From** and **To** allow you to establish where the bars start and finish. The picture below shows how to format Total Float and Negative Float. Unlike other planning and scheduling software, the Negative Float is drawn from the Start Date of a task and not the Finish Date and therefore a separate bar is required for Negative and Positive Float.

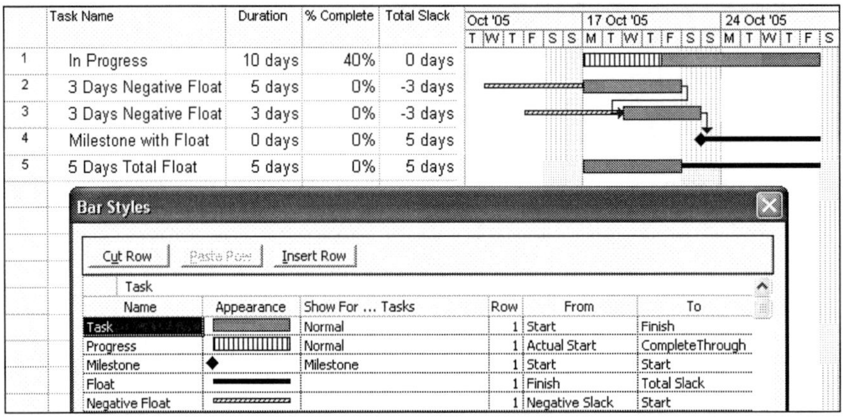

- The **Text** tab allows you to place text inside or around the bar:

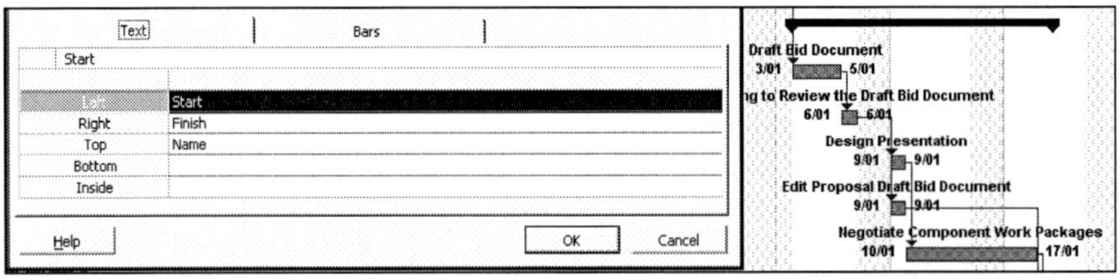

- It is not possible to format the font in this **Bar Styles** form. Select **Format**, **Text Styles…** to format the bar text font.

If you want to show Critical and Non-critical tasks, then the bars should be formatted as shown below, with particular attention paid to the **Show For … Tasks** column Non-critical Tasks are formatted as **Normal, Non-critical** and Critical Tasks as **Normal**, **Critical**.

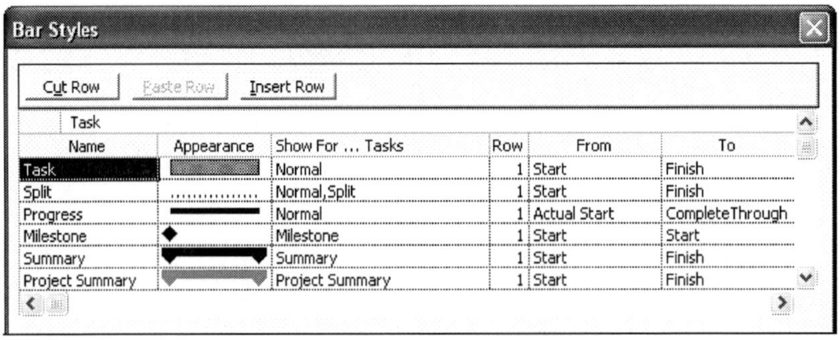

There is also a **Gantt Chart Wizard** available by clicking on the icon or selecting **Format**, **Gant Chart Wizard…**, but this wizard will overwrite any formatting you may have created. This is a straightforward method of formatting your bars.

8.2.2 Layout Form – Format Bars Options

Select **Format**, **Layout…** to open the **Layout** form. This form has some additional bar formatting options for users to customize the appearance of the Gantt Chart.

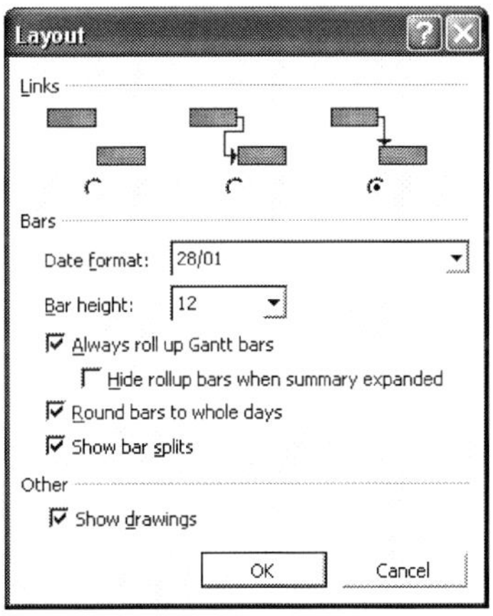

- **Date format:** sets the format for dates displayed on bars only. Dates are displayed on bars using **Format**, **Bar…** or **Format**, **Bar Styles…**.
- **Bar height:** sets the height of all the bars. Individual bars may be assigned different heights by selecting a bar shape in the styles form.
- **Always roll up Gantt bars** and **Hide rollup bars when summary expanded** works as follows:
  - Tasks before rollup:

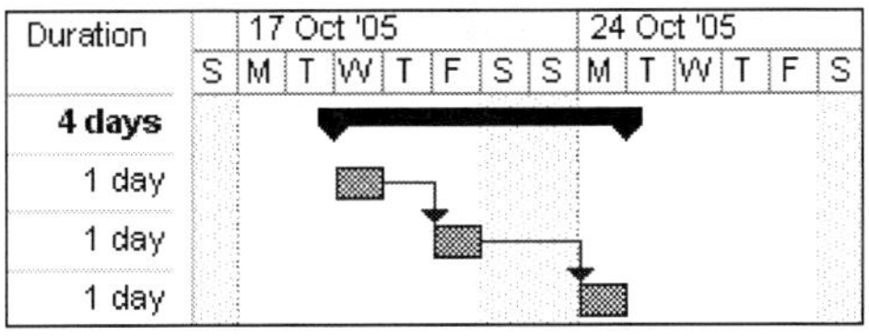

  - With **Always roll up Gantt bars** checked and **Hide rollup bars when summary expanded** unchecked:

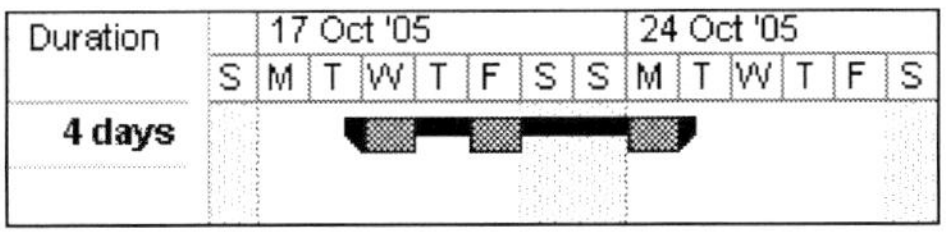

  - With **Always roll up Gantt bars** and **Hide rollup bars when summary expanded** checked:

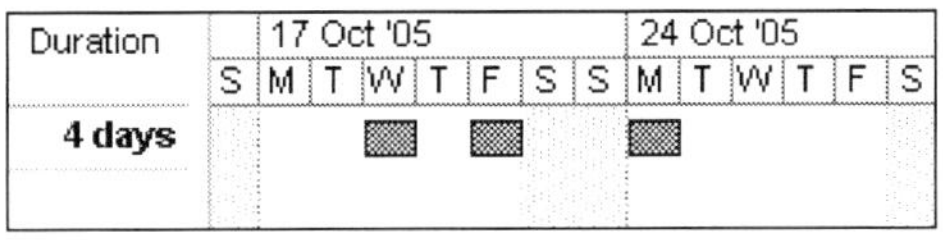

  - An individual bar may be rolled up to a summary task using the **Roll up Gantt bar to summary** option in the **Task Information** form.

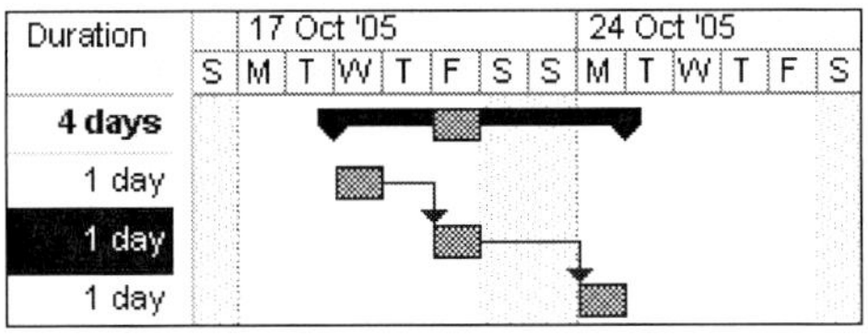

- Round bars to whole days:
  - ➢ When this option is unchecked, the length of the task will be shown in proportion to the total number of hours worked per day over the 24-hour time span. For example, an 8-hour working duration bar is shown below:

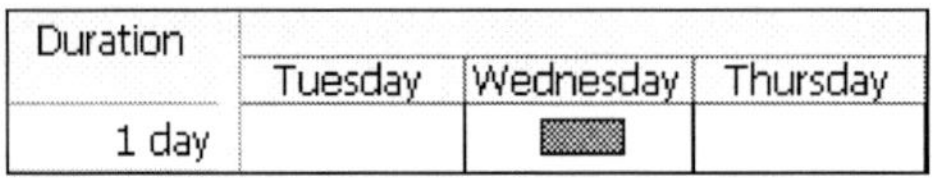

  - ➢ When this option is checked, the task bar will be displayed and spanned over the whole day irrespective of working time:

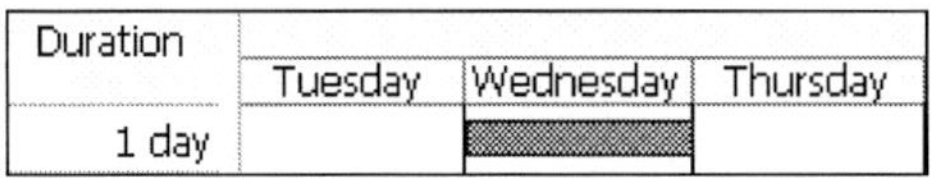

- **Splitting Tasks**
  - ➢ An unstarted task may be split using the Split Task icon on the Task Bar, then highlighting the bar to be split and dragging the section of the bar with the mouse to the location where it is planned to conduct the work. The picture below shows Task 10 – Draft Technical Details Schedule being split. Splits may be removed using the same process.

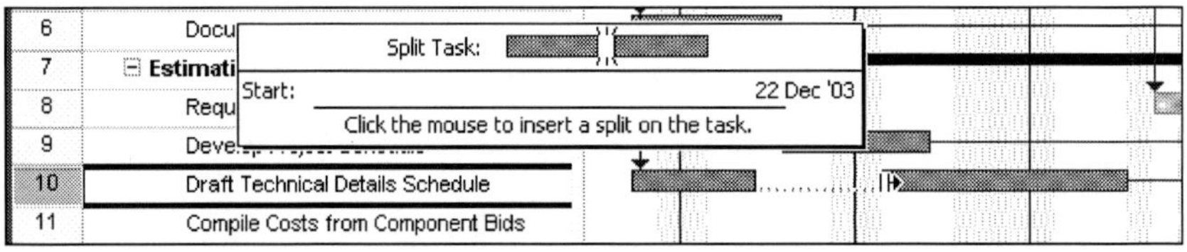

  - ➢ In progress tasks may be split manually by dragging the incomplete portion to the right or a split may be created automatically by commencing a task before its predecessor is complete. Splitting of in progress tasks is covered in both the **TRACKING PROGRESS** and **OPTIONS** chapters. The two pictures below are of the same task, first with the option checked and then unchecked:
- **Show bars splits**
  - ➢ When checked, the activity bar will display splits:

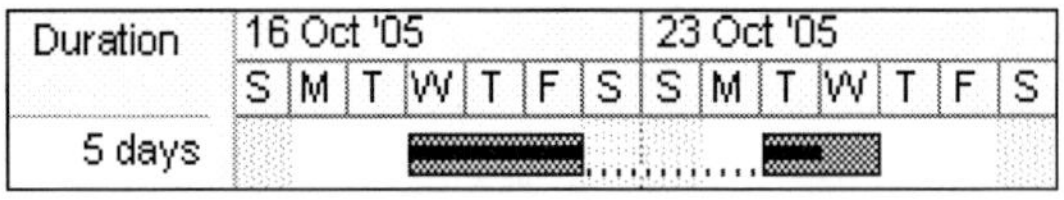

  - ➢ When unchecked, the activity bar will not display splits:

  - ➢ Tasks will only split however when the option in **Tools**, **Options…**, **Schedule** tab, **Spilt in-progress tasks** is checked.

8.2.3 Format One or More Specific Task Bars

Individual bars may be formatted to make them look different from other bars.

To format one or more bars:

- Select:
  - One task bar by clicking on it, or
  - Multiple bars by Ctrl-clicking each bar or left-clicking and dragging with the mouse, then
- Open the **Format Bar** form by:
  - Selecting **Format**, **Bar…**, or
  - Moving the mouse over a bar in the bar chart until the mouse changes to a ✥ and double-clicking. When more than one bar has been selected you will need to hold the **Ctrl Key** down when double-clicking on a bar.
- Then select the required bar formatting options from the **Format Bar** form:

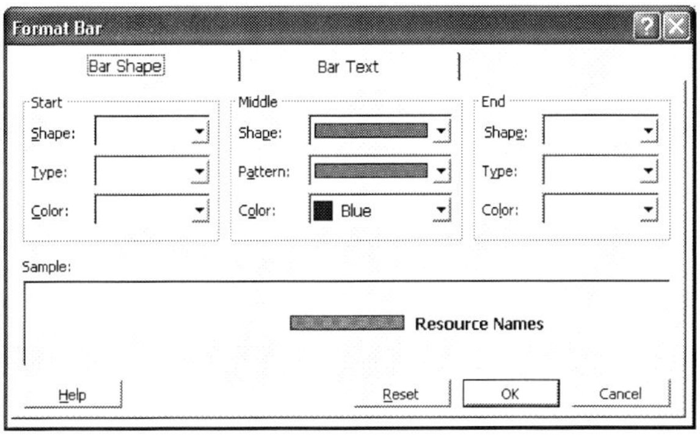

- The Reset is used to restore the default bar formatting to selected bars.

The bar shape may be formatted and text information added in the same way as formatting all the bars described earlier in this chapter.

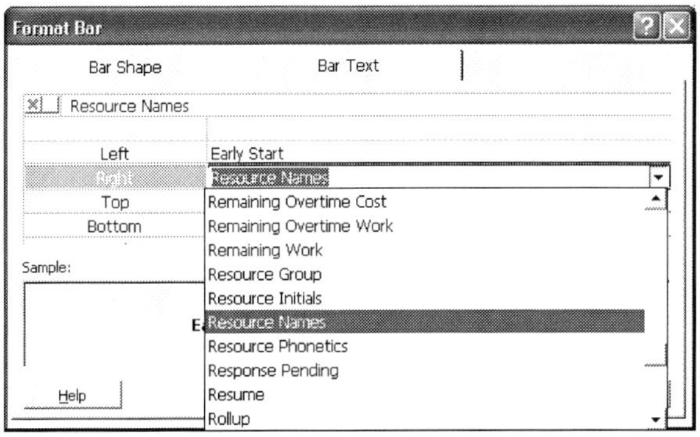

- Select **Format**, **Text styles…** to format the text font.
- A Tasks Bar may be hidden by checking the **Hide task bar** option in the **Task Information** form.

8.3 Row Height

Row heights may be adjusted to display text that would otherwise be truncated by a narrow column.

The row height may be set in the **Table Definition** form by select **View**, **Table:**, **More Tables…**, from this view select the table you wish to edit the row height in and click on the [Edit...] button. Once the **Table Definition** form is open select the row height from the drop down box next to **Row height:**.

The row height of one or more columns may also be adjusted in a similar way to adjusting row heights in Excel, by clicking on the row and dragging with the mouse:

- Highlight one or more rows that need adjusting by dragging or Ctrl-clicking. If all the rows are to be adjusted, then click on the **Select All** button above row number 1, to highlight all the tasks.

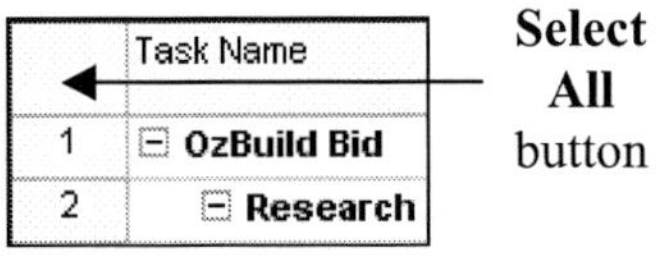

- Then move the mouse pointer to the Left-Hand Side of a horizontal row divider line. The pointer will change to a double-headed arrow ⇕. Click and hold with the left mouse button and drag the row or rows to the required height.

8.4 Format Fonts

8.4.1 Format Font Command

The **Format**, **Font…** function allows you to format any selected text in rows or columns:

- Select all the rows by clicking on the **Select All** button, box above row number 1, or
- Select one or more rows or columns by Ctrl-clicking or dragging, then
- Select **Format**, **Font…** to open the **Font** form:

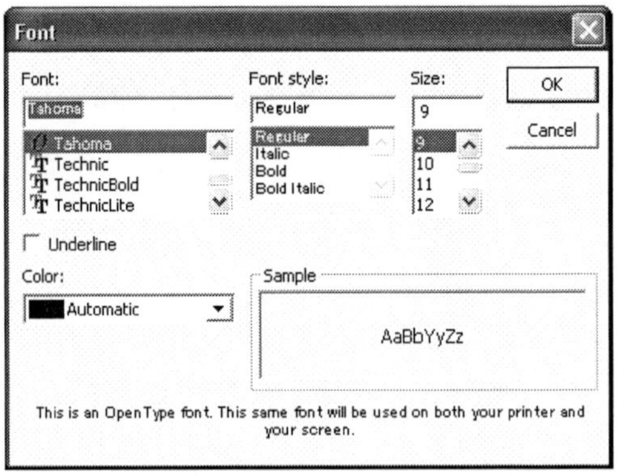

- You may select the font style, font size, and color of the text from the **Font** dialog box.

Once formatted, the selected font style and type will be applied to the data column or task row in any table.

8.4.2 Format Text Style

The **Format**, **Text Styles…** command opens the **Text Styles** form and allows you to select a text type from the **Item to Change:** drop-down box and apply formatting to the selected text style:

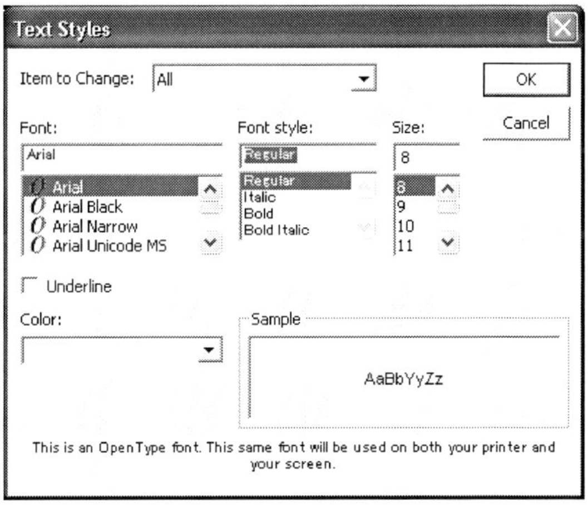

Text may formatted by using any of the styles listed below:

- **All**: This is all text including columns and rows,
- **Noncritical**, **Critical**, **Milestone**, **Summary**, **Project Summary**, **Marked**, **Highlighted** and **External tasks**,
- **Row** and **Column** titles,
- **Top, Middle** and **Bottom Timescale Tiers**, and
- **Bar Text** left, right, below, above and inside.

8.5 Format Colors

Colors are formatted in a number of forms and there is no single form for formatting all colors:

- **Non-working time** colors in the Gantt Chart are formatted in the **Timescale** form, double-click on the timescale.
- **Text** colors are formatted in the **Text Styles** and **Font** forms, found under the **Format** command.
- **Gridline** colors are formatted in the **Gridlines** form, also found under the **Format** command.
- **Hyperlink** colors are formatted under **Tools**, **Options…**, **Edit**.
- **Timescale** colors are formatted with the system color scheme used in the **Start**, **Settings**, **Control Panel**, **Display** option.
- The **Logic Lines**, also known as **Dependencies**, **Relationships**, or **Links**, inherit their color from the predecessor's bar color.

8.6 Format Timescale

8.6.1 Format Timescale Command

The **Timescale** form provides a number of options for the display of the timescale, which is located above the Bar Chart, and the shading of **Non-working** time. This function has been enhanced with the release of Microsoft Project 2002, which had a Major and Minor scale only.

To open the **Timescale** form:

- Double-click on the timescale, or
- Select **Format**, **Timescale…**.

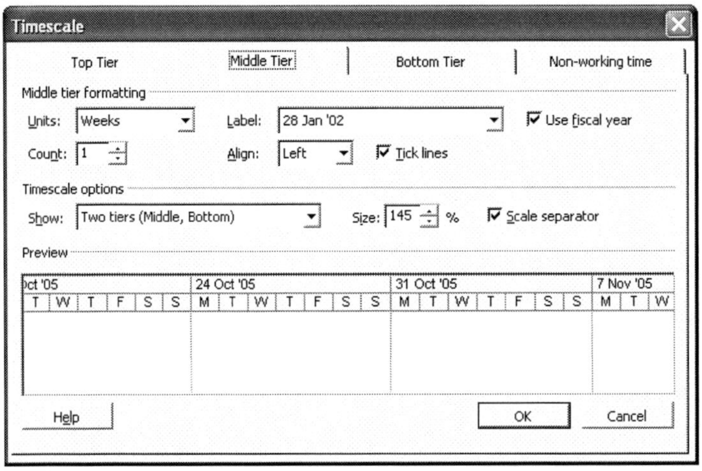

There are many options here, which are intuitive and will not be described in detail.

Top Tier, Middle Tier and Bottom Tier Tab

- These three time scales may have different scales. These are often set at "weeks and days" or "months and weeks." By default, the Top Tier time scale has been disabled. You may enable the three tiers together by selecting Three Tiers (Top, Middle, Bottom) from the **Timescale options**, **Show:**.
- The **Label** will affect how much space the timescale will occupy, so the selection of a long label will result in longer Task bars.
- **Tick lines** and **Scale separator** hide and display the lines between the text.
- **Size:** controls the horizontal scale of the timescale and in association with the **Label:** are the two main tools for scaling the horizontal axis in the Gantt Chart.
- If you choose to use the **Use fiscal year** function to display the financial year, then you will need to select the **Tools**, **Options…**, **Calendar** tab to choose the month on which the fiscal year starts.
- Should you wish to say number the time periods say **Week 1**, **Week2**, etc, there are a number of sequential numbering options for the time periods available at the bottom of the label list.

Non-working Time Tab

The **Non-working time** tab allows you to format how the non-working time is displayed. You may select only one calendar. The non-working time may be presented as shading behind the bars, in front of the bars or hidden.

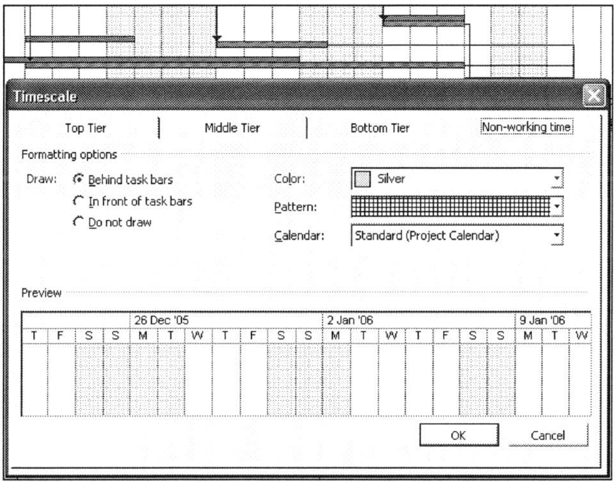

8.6.2 View Zoom

The **View**, **Zoom…** function or clicking on the and icons is used to adjust the horizontal scale of the Gantt Chart only. It does not work like most other Windows products, which scale the whole work area.

The **View**, **Zoom…** function is not a temporary change and overwrites any customized timescale settings.

8.6.3 Format Timescale Font

To format the Timescale font, select **Format**, **Text styles…** to open the **Text Styles** form:

The timescale fonts may be formatted separately by selecting the appropriate line item under **Item to Change:**.

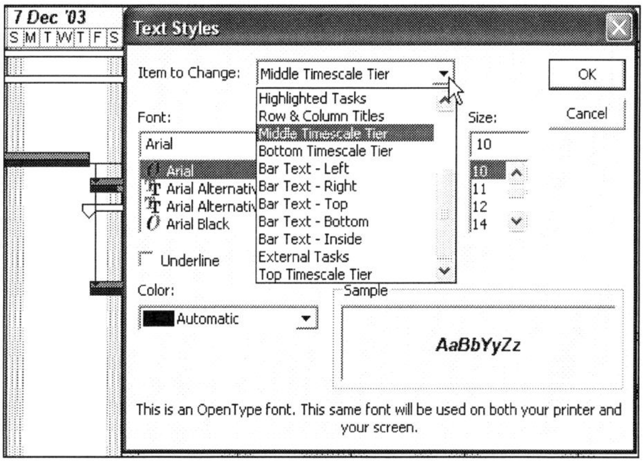

8.7 Format Gridlines

Gridlines are important to help divide the visual presentation with dividing lines on the Bar Chart. This example shows **Middle Tier Gridlines** every week and **Bottom Tier Gridlines** every day.

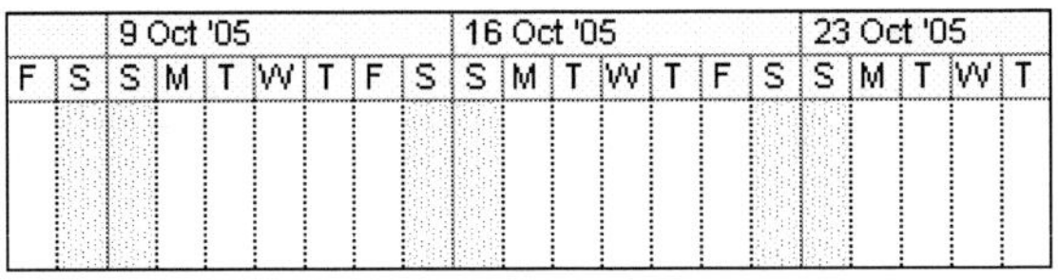

To format the Grid Line select **Format**, **Gridlines…** to open the Gridlines form:

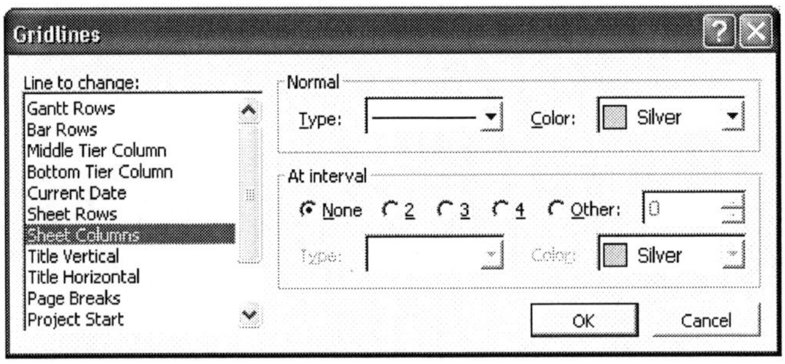

- Select the gridline from the dropdown box under **Line to change:**.
- Select color and type from under **Normal**.
- Some gridlines may be set to occur at intervals under the **At interval** option.

Some of the titles for the gridlines are not intuitive, so some interpretation is given below:

- For Data Column and Row dividing lines, use **Sheet Rows** and **Sheet Columns**.
- For Timescale and Column Titles, use **Title Horizontal** and **Title Vertical**.
- Gantt Chart area, including lines for Project Start and Finish Date, Current and Status Date, are clearly described.
- Page Breaks will only display manually-inserted breaks. A page will only break if the **Manual page breaks** check box in the **Print** form is checked.

Many laser printers will not print light gray lines clearly, so it is often better to use dark gray or black Sight Lines for better output.

8.8 Format Links, Dependencies, Relationships, or Logic Lines

The Links, also known as Dependencies, Relationships, or Logic Lines, may be displayed or hidden by using the **Layout** form.

- Select **Format**, **Layout…** to open the **Layout** form and click on one of the three radio buttons under **Links** to select the style you require:

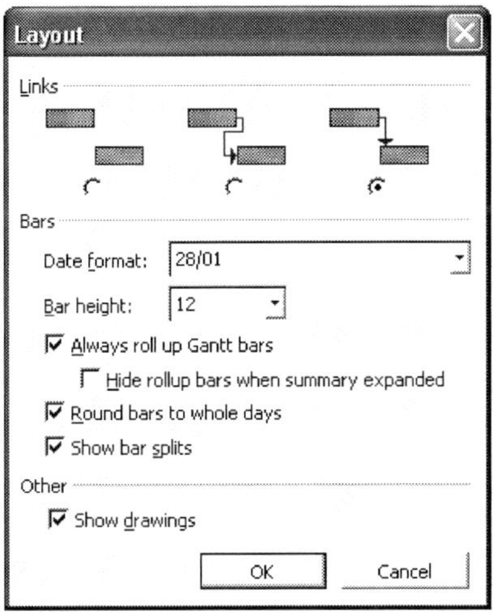

- ➢ The color of the **Link** is inherited from the color of the predecessor task.
- ➢ To display critical path on the relationship lines you will need to format the bars with a different color. This is often set to red.

i The color of the successors' relationship lines is adopted from the task bar color. Therefore, re-formatting critical bars with the **Format Bar** form will also re-format the color of the successors' relationship lines and they will no longer display the Critical Path color on the Logic Lines. This will effectively mask the critical path and could provide misleading results.

8.9 Customize Fields

Select **Tools**, **Customize Fields…** to open the **Customize Fields** form. This function includes a number of predefined fields for both Task and Resources. This is a significant enhancement of the Microsoft Project 98 Customize Fields function form has two tabs:

- Custom Fields which are lineal fields, and
- Hierarchical Custom Outline Codes fields.

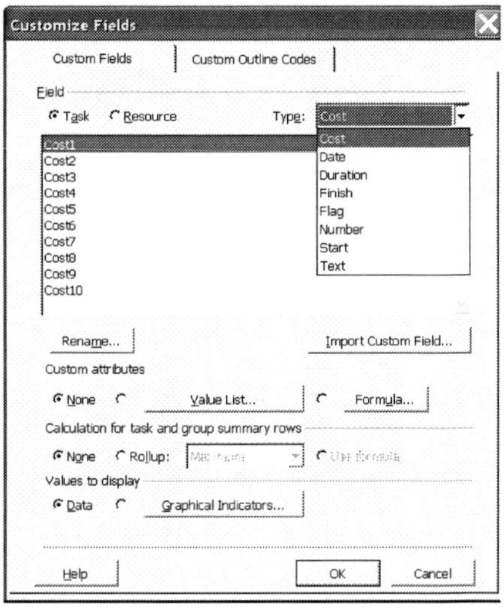

8.9.1 Custom Fields

These fields fall into the following categories:

- Cost
- Date
- Duration
- Finish (date)
- Number
- Start (date)
- Text

Both the title and content of these fields may be edited with options including:

- Renaming the filed name,
- Assigning a Value and Description of the field
- Assigning formulae for the calculation of field value, including Summary Tasks,
- Assigning lists that may be selected for the field value and
- Generating graphical indicators, traffic lights.

People who have used Primavera software will find the formatting options available when using Value and Description restrictive as the description may not be displayed in columns and the value not displayed when Grouping

8.9.2 Custom Outline Codes

The ability to create hierarchical Code Structure with the Custom Outline Codes and their use for reorganizing tasks under other hierarchical project breakdown structures is covered in more detail in **Chapter 16 GROUPING TASKS, OUTLINE AND WBS**.

There are ten Task Custom Outline Codes and ten Resource Outline Codes that may be renamed to suit the project.

- The Task Custom Outline Codes may be used for any project breakdown structure, such as a PRINCE2 Product Break Down Structure, Contract Breakdown Structure, and
- The Task Custom Outline Codes may be used for organizational breakdown structures such as the hierarchy of authority, locations and departments.

WORKSHOP 6

Formatting the Bar Chart

Preamble

Management has received your draft report and requests some changes to the presentation.

Assignment

Format your schedule as follows:

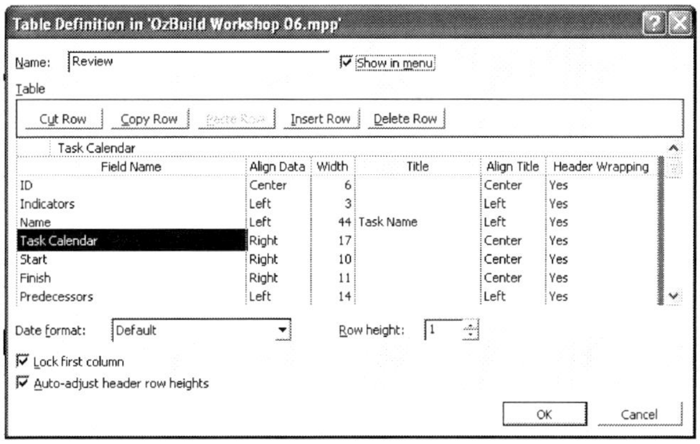

1. Create a new **Table** titled **Review** by copying the **Entry** table.
2. Add **Task Calendar** column between **Duration** and **Start** columns in the **Table Definition** form. Left Align this column
3. Check the **Show in menu** option.
4. Apply the Table.
5. Click on the **Duration** column and right-click to display a menu, then hide this column.
6. Format the **Middle Tier Timescale (Major scale in MSP2000)** with **Units:** of **Months** with **Label:** of **January** and the **Bottom Tier Timescale (Minor scale in MSP2000)** with **Units:** of **Days** with **Label:** of **M,T,W,...** and **Size:** of 120%.
7. Format the Gridlines with **Middle Tier Columns (Major Columns in MSP2000)** to solid black lines and **Bottom Tier Columns (Minor Columns in MSP2000)** to dotted black lines.
8. Remove the date from the milestones. If you do not have dates by your milestones then add the date to the right of the Milestones and then remove them
9. Add the task **Name** to the right of each **Task** bar.
10. Save your **OzBuild Bid** project.
11. Check your result with the example shown on the next page.

ANSWER TO WORKSHOP 6

	Indicators	Task Name	Task Calendar	Start	Finish
1		**Ozbuild Bid**	**None**	**5 Dec '05**	**15 Dec '05**
2		**Research**	**None**	**5 Dec '05**	**14 Dec '05**
3		Bid Request Documents Recieved	None	5 Dec '05	5 Dec '05
4		Bid Strategy Meeting	None	5 Dec '05	5 Dec '05
5		Investigate Technical Feasibility	None	5 Dec '05	14 Dec '05
6		Document Installation Requirements	6 Day Working Week	5 Dec '05	8 Dec '05
7		**Estimation**	**None**	**5 Dec '05**	**15 Dec '05**
8		Request Component Bids	None	5 Dec '05	7 Dec '05
9		Develop Project Schedule	6 Day Working Week	5 Dec '05	8 Dec '05
10		Draft Technical Details Schedule	None	5 Dec '05	15 Dec '05
11		Compile Costs from Component Bids	None	5 Dec '05	6 Dec '05
12		**Proposal**	**None**	**5 Dec '05**	**12 Dec '05**
13		Draft Bid Document	None	5 Dec '05	7 Dec '05
14		Meeting to Review the Draft Bid Document	None	5 Dec '05	5 Dec '05
15		Design Presentation	None	5 Dec '05	5 Dec '05
16		Edit Proposal Draft Bid Document	None	5 Dec '05	5 Dec '05
17		Negotiate Component Work Packages	None	5 Dec '05	12 Dec '05
18		Final Review of Bid Document	None	5 Dec '05	5 Dec '05
19		Submit Bid	None	5 Dec '05	5 Dec '05

9 ADDING TASK DEPENDENCIES

The next phase of a schedule is to add logic to the tasks. There are two types of logic that you may use:

- **Dependencies**, (**Relationships** or **Logic** or **Links**) between tasks, and
- Imposed **Constraints** to task start or finish dates. These are covered in the **CONSTRAINTS** chapter.

Microsoft Project 2003's Help file and other text uses the terms "**Dependencies, Relationships** and **Links"** for Dependencies but does not use the term "**Logic**."

There are a number of methods of adding, editing and deleting task **Dependencies**. We will look at the following techniques in this chapter:

Topic	Notes for creating a SF Dependency
• Graphically in the Gantt, Calendar or Network Diagram Views	Drag the 4-headed mouse pointer ✥ from one task to another to create an FS dependency.
• Through the **Link** and **Unlink** icon on the Standard toolbar	Select the tasks in the order they are to be linked and click on the Link icon.
• By using the Menu command	Select the tasks in the order they are to be linked and select **Edit**, **Link Tasks**, or **Ctrl+F2**.
• By opening the **Task Information** form	Predecessor only may be added and deleted.
• Through the **Predecessor** and **Successor Details** forms	Open the bottom pane, **Windows**, **Split** and then select **Format**, **Details**, **Predecessors and Successors**.
• By editing or deleting a dependency using the **Task Dependency** form	Double-click on a task link (relationship line) in the **Bar Chart** or **Network Diagram** view.
• **Autolink** new inserted tasks and moved tasks	Select **Tools**, **Options…**, **Schedule** tab and check the **Autolink Inserted or Moved Tasks** box.
• By displaying the **Predecessor** or **Successor** column	Edit the relationships in the columns.

Dependencies

Generally, there are two types of dependencies that may be entered into the software:

- **Hard Logic** are dependencies that may not be avoided. For example a specification would have to be prepared before a price is requested from suppliers.
- **Soft Logic**, also referred to as **Sequencing Logic** or **Preferred Logic**, which often may be changed at a later date to reflect planning changes. An example would be determining the order in which specifications have to be prepared.

There is no simple method of documenting which is hard and which is soft logic. A schedule with a large amount of soft logic has the potential of becoming very difficult to maintain when the plan is changed. You will also find that as a project progresses, soft logic converts to hard logic as commitments are made and tasks are started.

Constraints

Constraints are applied to Tasks when relationships do not provide the required result. Typical applications of a constraint are:

- The availability of a site to commence work.
- The supply of information by a client.
- The required finish date of a project.

Constraints are often entered to represent contract dates and may be directly related to contract items.

Constraints are covered in detail in the **CONSTRAINTS** chapter.

9.1 Understanding Dependencies

There are four types of dependencies available in Microsoft Project 2003:

- Finish-to-Start (**FS**) (also known as conventional)
- Start-to-Start (**SS**)
- Start-to-Finish (**SF**)
- Finish-to-Finish (**FF**)

Two other terms you must understand are:

- **Predecessor**, a task that controls the start or finish of another immediate subsequent task.
- **Successor**, a task whose start or finish depends on the start or finish of another immediately preceding task.

The following pictures show how the dependencies appear graphically in the **Gantt Chart** and **Network Diagram** (PERT) views.

The **FS** (or conventional) dependency looks like this:

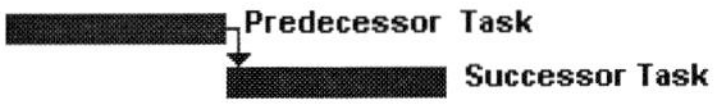

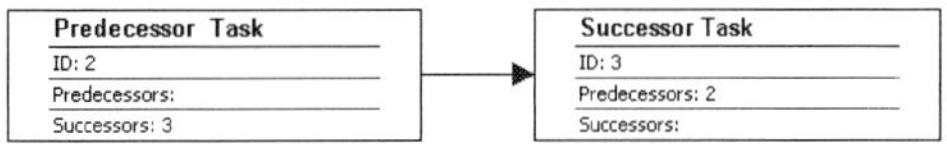

While the **SS** dependency is like this:

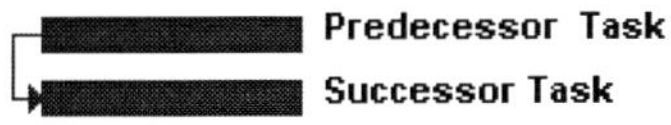

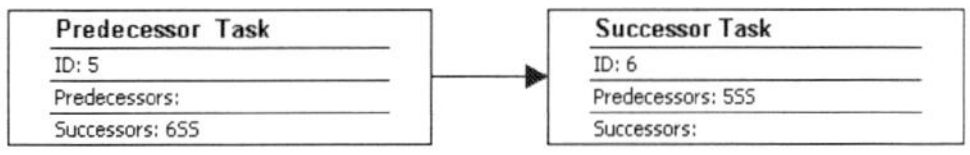

The **SF** dependency looks like:

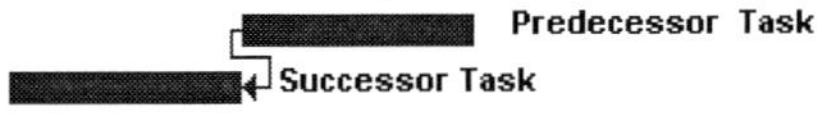

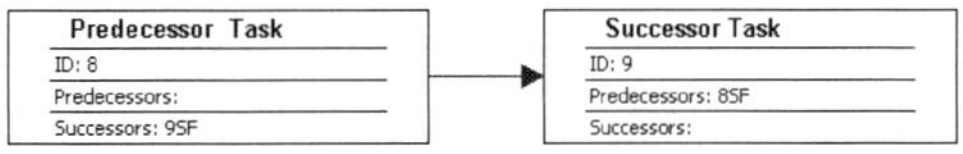

The **FF** dependency would be:

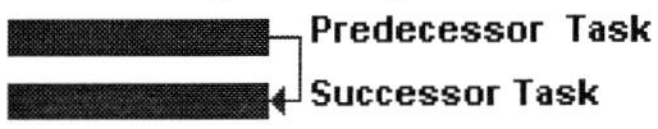

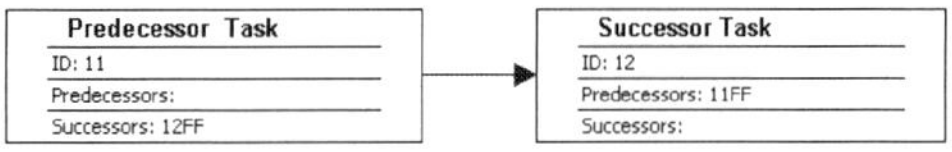

To create a **Closed Network** each task will require a Start predecessor and a Finish successor. Most schedules may be created using only Finish to Start relationships with positive or negative lags. This method ensures a Closed Network is created and the **Critical Path** flows through the activities and not just through the relationships. A delay to the completion of the first activity in the example on the left below will not delay the second or third task and the critical path flows through the relationships and not the tasks, therefore a critical path has not been created. The example on the right has created a true critical path

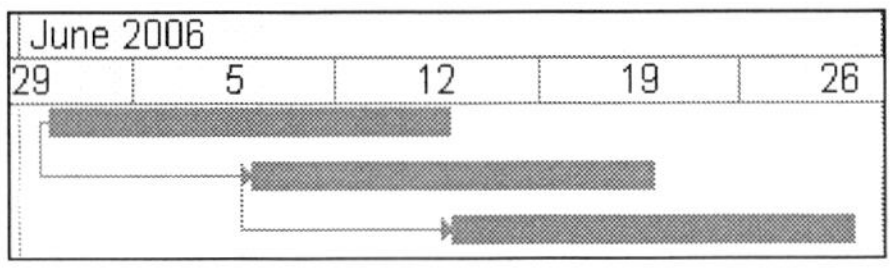

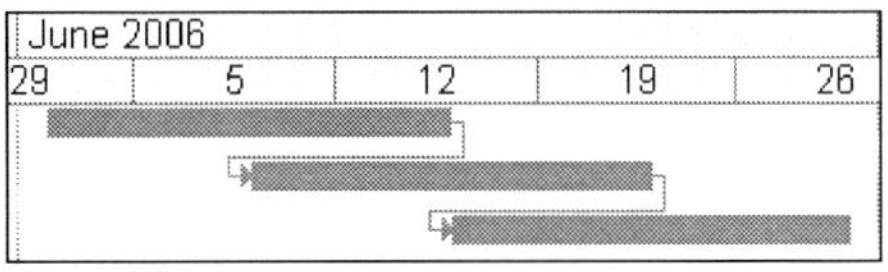

9.2 Understanding Lags and Leads

A **Lag** is a duration that is applied to a dependency to make the successor start or finish earlier or later.

- A successor task will start later when a positive **Lag** is assigned. Therefore, a task requiring a 3-day delay between the finish of one task and start of another will require a positive lag of 3 days.
- Conversely, a lag may be negative (also called a **Lead)** when a new task can be started before the predecessor task is finished.
- **Leads** and **Lags** may be applied to any relationship type including Summary Task relationships.

An example of a **FS** with positive lag

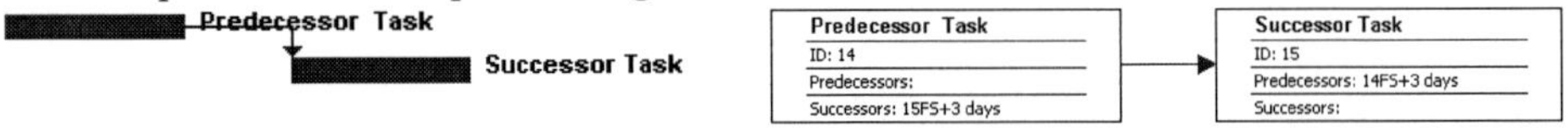

An example of a **FS** with negative lag:

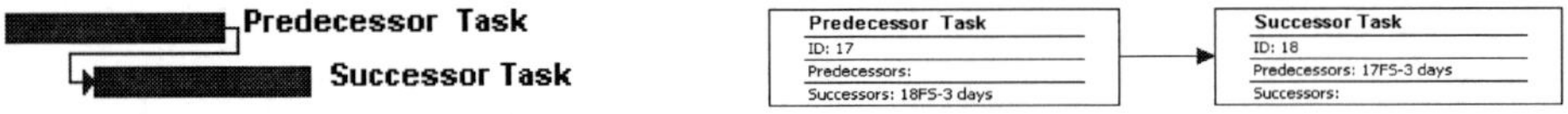

Here are some important points to understand about Lags.

- Lags are normally calculated on the **Project Calendar**. (Primavera P3 and SureTrak software uses the predecessor's calendar.)
- Lags may be assigned **Elapsed** durations, therefore they will be based on a 24-hour, 7-day/week. To enter an elapsed lag type an "e" before the unit, e.g. **5ed**.
- Lags may be expressed in terms of % and in this situation the lag is a percentage of the predecessor's duration. The example below shows a FS dependency +250%; the predecessor is 1day so the Lag is 2.5days.

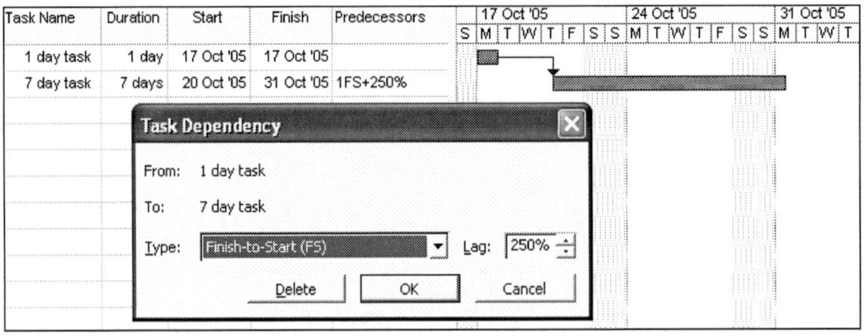

You must be careful when using a lag to allow for delays such as curing concrete when the Base Calendar is not a seven-day calendar. Since this type of activity lapses non-work days, the activity could finish before Microsoft Project 2003's calculated finish date. You may want to use elapsed durations in a lag in this situation.

9.3 Restrictions on Summary Task Dependencies

Dependencies may be made between Summary and Detailed tasks of a different Summary task. Consider the following points when using dependencies at Summary Task level:

- There is a built-in dependency between Summary and Detailed tasks. Detailed tasks may be considered as Start-to-Start successors and Finish-to-Finish predecessors of their Summary Task.
- Summary tasks may only have **FS** and **SS** dependencies; you will receive a warning message when you attempt to enter an illegal dependency.

It is recommended that dependencies be maintained at the detail level. This is particularly important when moving tasks from one summary task to another since the dependencies will still be valid. Again, be aware that there is a function found under **Tools**, **Options…**, **Schedule** tab which will Autolink moved tasks and this function should be turned off if you wish to move a task and keep the existing logic.

9.4 Displaying the Dependencies on the Gantt Chart

The dependencies may be displayed or hidden with the **Layout** form.

- Select **Format**, **Layout…** to open the **Layout** form and click on the radio button under the style you require:

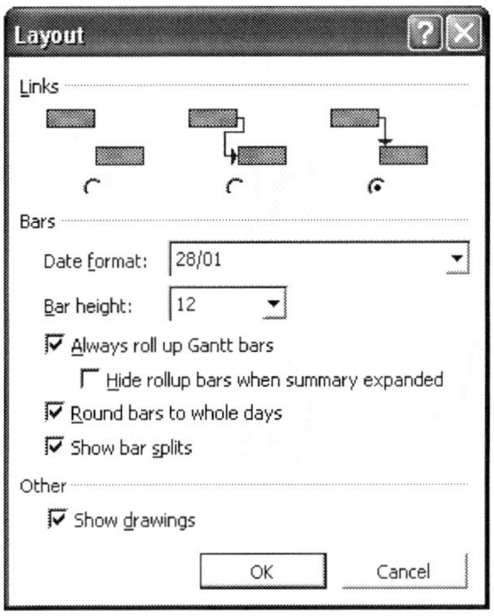

- The color of the dependency line is inherited from the color of the predecessor task.
- To display a critical path on the relationship lines you will need to format the bars as critical.

9.4.1 Graphically Adding a Dependency

You may graphically add a **Finish-to-Start** dependency only by:

- Selecting the **Gantt**, **Calendar** or **Network Diagram** views from the **View Bar** on the right-hand side of the screen or selecting **View**, **Gantt Chart**, **Calendar** or **Network Diagram**, then:

- Move the mouse pointer over a task until the mouse pointer changes to a ✥, left-click and drag to the successor task. The cursor will change to a shape during this operation.

9.4.2 Using the Link and Unlink Icon on the Standard Toolbar

The **Link Tasks** icon on the toolbar may be used for linking tasks with a Start-to-Finish dependency:

- Highlight one of more tasks using Ctrl-left-click to select one task at a time and Shift and left-click to select a contiguous group of tasks.

- Then click the **Link Tasks** icon on the toolbar and the tasks will be linked with Start-to-Finish dependencies in the order that they were selected.

To remove a dependency, select the tasks and click on the **Unlink Tasks** icon.

9.4.3 Linking Using the Menu Command

The menu may be used for linking tasks with a Start-to-Finish dependency:

- Highlight one or more tasks using Ctrl-left-click to select one task at a time or Shift and left-click to select a group of tasks.

- Then select **Edit**, **Link Tasks**, or **Ctrl+F2** and the tasks will be linked with Start-to-Finish dependencies in the order that they were selected.

Dependencies may be removed using the menu using **Edit**, **Unlink Tasks**.

Primavera P3 and SureTrak software links tasks that have been highlighted from top to bottom. Microsoft Project 2003 links the tasks in the order they are highlighted.

9.4.4 Adding and Deleting Predecessors with the Task Information Form

The **Task Information** form may be used for adding and deleting predecessors only.

- Double-click on a task to open the **Task Information** form,
- Select the **Predecessor** tab,
- To select the predecessor, you may either:
  - ➢ Type in the Predecessor Task ID in the first line under ID, or
  - ➢ Use the drop-down box under task name:

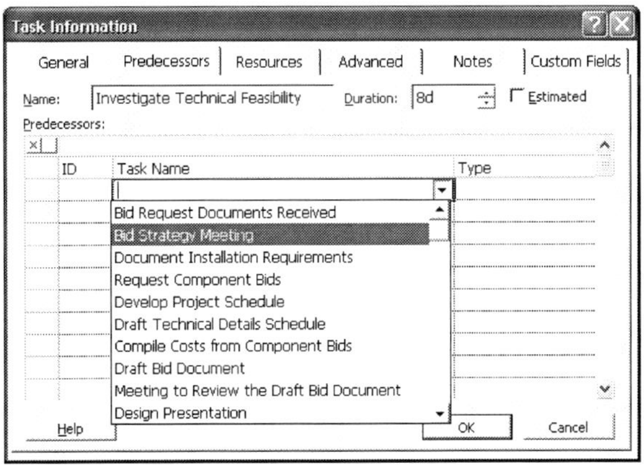

- Now enter the Relationship Type from the **Type** drop-down list and the lag, if required, from the **Lag** drop-down list.

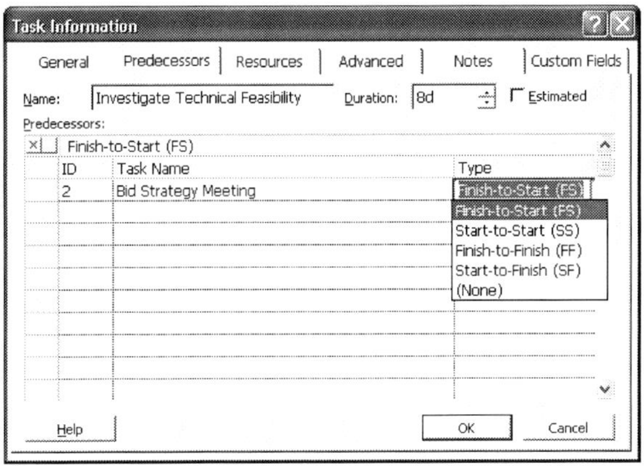

- To enter another relationship click on the next line.
- You are not able to scroll up or down to another task while the **Task Information** form is open.

To complete your operation, either:

- Press the **Enter Key** or click on the OK button to commit the changes, or
- Click on the **Esc Key** or click on the Cancel button to abort any changes.

9.4.5 Predecessor and Successor Details Forms

The Predecessor and Successor Details form may be displayed by:

- Opening the bottom pane by selecting **Window**, **Split**,
- Then make the lower pane active by clicking anywhere in the lower pane. The bar on the left-hand side of the lower pane will turn blue when it is active,
- Display the **Task Details Form**, **Task Entry** or **Task Form**,
- Then select **Format**, **Details**, **Predecessors and Successors** to display the **Predecessors and Successors Detail** form:

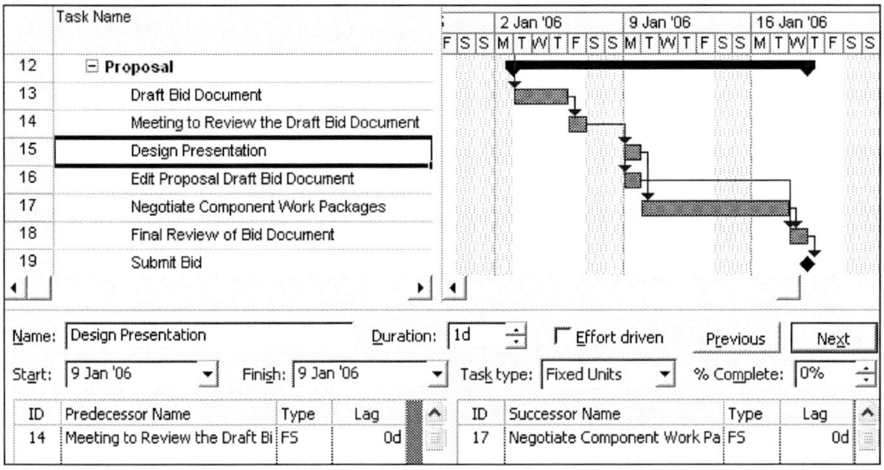

Predecessors and successors may be added using the same method as the **Task Information** form.

Double Clicking on a Predecessor or Successor in any of these forms will display the Predecessor or Successor **Task Information** form, allowing the dates, constraints etc of related activities to be examined.

9.4.6 Editing or Deleting Dependencies Using the Task Dependency Form

To use the **Task Dependency** form, a logic link between tasks must already exist. To open the **Task Dependency** form, double-click on a task link (relationship line) in the Bar Chart or Network Diagram.

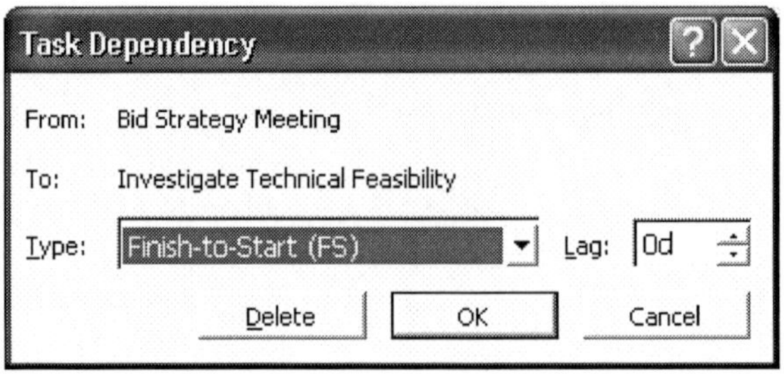

A Link may only be edited or deleted from this form.

9.4.7 Autolink New Inserted Tasks or Moved Tasks

This function automatically creates predecessors to tasks above it and successors below it when a task is moved or inserted. This option may be activated by selecting **Tools**, **Options...**, **Schedule** tab and check the **Autolink inserted or moved tasks** box. When activated you must ensure that:

- You have selected the whole task by clicking on the task ID and therefore highlighting all the columns before you drag the task to a new location. Otherwise, you will only move the cell contents.

When the task is moved, **Autolink New Tasks or Moved** tasks will change the existing predecessor and successor logic without warning. This function potentially makes substantial changes to your project logic and may affect the overall project duration. It is suggested that the option is **NEVER** switched on as dragging an activity to a new location may completely change the logic of a schedule.

9.4.8 Editing Relationships Using the Predecessor or Successor Columns

The **Predecessor** or **Successor** column may be displayed and edited following the example below.

		Task Name	Predecessors	Successors
1		Predessor		2SS+3 days,3
2		Successor 1	1SS+3 days	3
3		Successor 2	1,2	

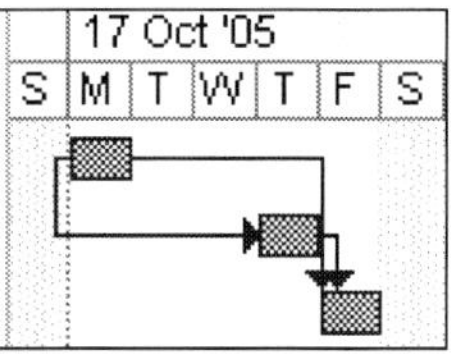

9.4.9 Viewing Relationships Using the Unique ID Predecessor or Unique ID Successor Columns

Each task is assigned a Unique ID when it is created and this number is not used again in the schedule, even if the task is deleted. There are two other columns that may be used to edit and display relationships using the Unique ID:

- The **Unique ID Predecessor**, and
- The **Unique ID Successor**.

9.4.10 Editing Relationships Using WBS Predecessor or WBS Successor Columns

There are two other columns that may be used only to display (and not edit) the **Predecessors** and **Successors**:

- The **WBS Predecessor**, and
- **WBS Successor**.

9.5 Scheduling the Project

Once you have your tasks and the logic in place, Microsoft Project 2003 calculates the tasks' dates/times. More specifically, Microsoft Project 2003 has **Scheduled** the project to calculate the **Early Dates**, **Late Dates** and the **Total Float**. This will allow you to review the **Critical Path** of the project. Microsoft Project 2003 uses the term **Slack** instead of the standard term **Float**. Both terms are used interchangeably throughout this book.

Sometimes it is preferable to prevent the **Automatic Calculation** of your project's start/end dates. To do this, select **Tools**, **Options…**, **Calculation** tab. Click on **Manual**.

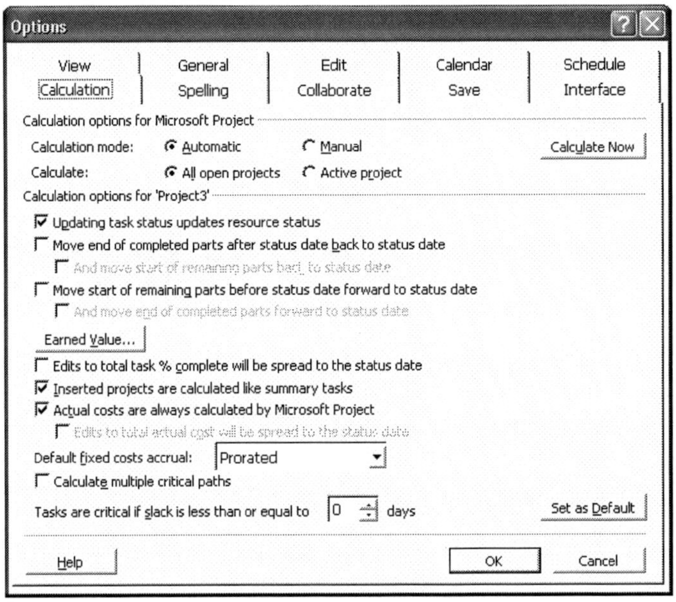

To calculate the schedule with the calculation mode set to manual:

- Press the **F9 Key**, or
- Click on the **Select All** button, top left hand corner of the Gantt Chart view, right-click to open a menu and select **Calculate Project**.

WORKSHOP 7

Adding the Relationships

Preamble

You have determined the logical sequence of tasks, so you may now create the relationships.

Assignment

1. Remove the text from the bars and format the **Middle Tier Timescale** to **Weeks – 31 Jan '01** and **Bottom Tier Timescale** to **Days – M,T,W,… Timescale size** to **70%** and remove the Minor Gridlines.
2. Apply the **Entry** table.
3. Input the logic below using the several of the methods detailed in this chapter.

		Task Name	Predecessors
1		**OzBuild Bid**	
2		**Research**	
3		Bid Request Documents Recieved	
4		Bid Strategy Meeting	3
5		Investigate Technical Feasibility	4
6		Document Installation Requirements	5
7		**Estimation**	
8		Request Component Bids	5
9		Develop Project Schedule	6
10		Draft Technical Details Schedule	5
11		Compile Costs from Component Bids	8
12		**Proposal**	
13		Draft Bid Document	9,10,11
14		Meeting to Review the Draft Bid Document	13
15		Design Presentation	14
16		Edit Proposal Draft Bid Document	14
17		Negotiate Component Work Packages	15
18		Final Review of Bid Document	16,17
19		Submit Bid	18

4. Display the Logic Links using **Format**, **Layout….** (If your links are displayed by default, then hide and then display them again.)
5. Check your results against the diagram on the next page.
6. Save your **OzBuild Bid** project.

ANSWER TO WORKSHOP 7

		Task Name	Duration	Start	Finish	Predecessors
1		**Ozbuild Bid**	**30 days**	**5 Dec '05**	**18 Jan '06**	
2		**Research**	**12 days**	**5 Dec '05**	**20 Dec '05**	
3		Bid Request Documents Recieved	0 days	5 Dec '05	5 Dec '05	
4		Bid Strategy Meeting	1 day	5 Dec '05	5 Dec '05	3
5		Investigate Technical Feasibility	8 days	6 Dec '05	15 Dec '05	4
6		Document Installation Requirements	4 days	16 Dec '05	20 Dec '05	5
7		**Estimation**	**9 days**	**16 Dec '05**	**30 Dec '05**	
8		Request Component Bids	3 days	16 Dec '05	20 Dec '05	5
9		Develop Project Schedule	4 days	21 Dec '05	24 Dec '05	6
10		Draft Technical Details Schedule	9 days	16 Dec '05	30 Dec '05	5
11		Compile Costs from Component Bids	2 days	21 Dec '05	22 Dec '05	8
12		**Proposal**	**12 days**	**3 Jan '06**	**18 Jan '06**	
13		Draft Bid Document	3 days	3 Jan '06	5 Jan '06	9,10,11
14		Meeting to Review the Draft Bid Document	1 day	6 Jan '06	6 Jan '06	13
15		Design Presentation	1 day	9 Jan '06	9 Jan '06	14
16		Edit Proposal Draft Bid Document	1 day	9 Jan '06	9 Jan '06	14
17		Negotiate Component Work Packages	6 days	10 Jan '06	17 Jan '06	15
18		Final Review of Bid Document	1 day	18 Jan '06	18 Jan '06	16,17
19		Submit Bid	0 days	18 Jan '06	18 Jan '06	18

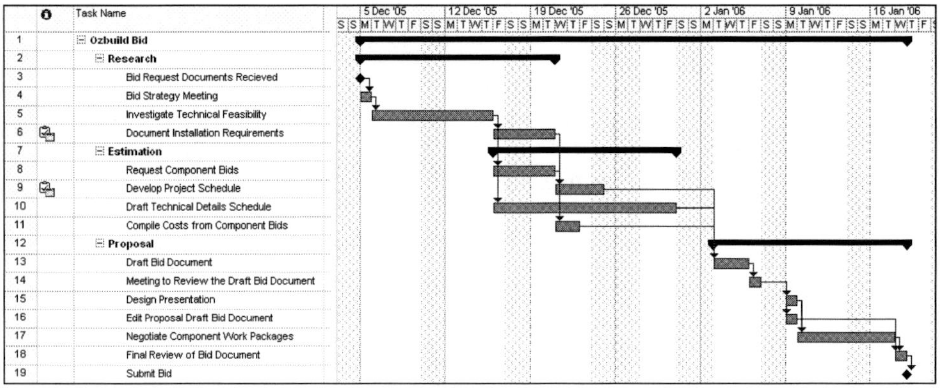

10 NETWORK DIAGRAM VIEW

The **Network Diagram View** is an enhancement of the earlier versions of Microsoft Project **Pert View** and displays tasks as boxes connected by the relationship lines. This chapter will not cover this subject in detail but will introduce the main features.

To view your project in the Networking View:

- Select **View**, **Networking Diagram**, or
- Click on the **Networking Diagram** icon on the **View Bar** on the left-hand side of the screen.

Many features available in the **Gantt Chart View** are also available in the **Network Diagram View**, including:

Topic	Menu Command
• Add new tasks	Click and drag into a blank area. This will add a new Task with a Finish-to-Start relationship, or Use the **Insert Key** or selecting **Insert**, **New Task**. These tasks are added after the highlighted task without a relationship.
• Delete Tasks	Select the task and press the **Delete** key.
• Display the **Task Information** form	Double-click on a Task Box, or Highlight a task and right-click on it and select **Task Information…** for the menu.
• Format Dependencies	Select **Format**, **Layout…**
• Display the **Task Dependency** form	Double-click on a relationship line.
• Display information in the lower pane	Select **Window**, **Split**, or Drag the dividing bar.
• Format Task Boxes	Select **Format**, **Box Styles…**, or Double-click on the outside edge of a box.
• Format an individual Task Box	Select **Format**, **Box…**
• Change the scale of the display	Select **View**, **Zoom…** which displays the **Zoom** form.

10.1 Understanding the Network Diagram View

The following list describes the main display features of the Network Diagram View:

- Summary Tasks, Detail Tasks and Milestones are normally formatted with a different shape. Typical shape examples are provided as follows:
  - ➢ Summary Tasks are Trapezoidal –
  - ➢ Detail Tasks are Rectangular –
  - ➢ The Milestone Task is an elongated diamond –

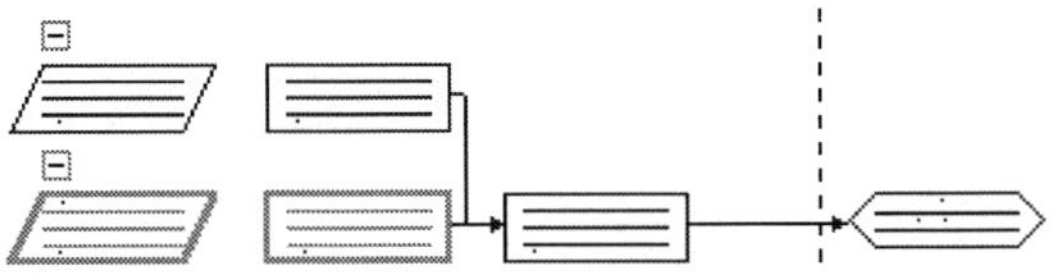

- Summary Tasks are positioned to the left and at the same level or above Detail Tasks.
- Summary Tasks may be rolled up by clicking on the ⊟ above the Task and expanded by clicking on the ⊞ above a rolled-up summary task.
- All relationship lines are drawn as **Finish-to-Start** even when they are not linked as a **Finish-to-Start** relationship. The link type may be displayed on the arrow. See **Link style** later in this chapter.
- Critical Tasks may be formatted to have different borders and backgrounds.
- The contents of the Task Boxes and relationship lines may also be formatted.

10.2 Adding and Deleting Tasks in the Network Diagramming View

- A **New Task** may be created with a **Finish-to-Start** relationship by dragging from the center of a Task Box into a blank part of the screen.
- A **New Task** may be created without a relationship below the tasks position in the Gantt Chart by:
  - ➢ Using the **Insert Key**, or
  - ➢ Selecting **Insert**, **New Task**.

10.3 Adding, Editing and Deleting Dependencies

Dependencies may be added, deleted or edited using the following methods:

- Graphically add a relationship by clicking on the center of one task and dragging to the successor.
- Hold the **Ctrl Key** to select two or more tasks and then use the **Link** function.
- The **Unlink** removes dependencies between selected tasks.
- Open the **Task Information** form by double-clicking on a task and selecting the **Predecessors** tab.
- Double-click on a **Relationship line** to open the **Task Dependency** form and edit or delete a dependency.
- Create a split window by selecting **Window**, **Split.** Display the predecessors and successors in the lower pane by selecting **Format**, **Details**, **Predecessors and Successors**.

10.4 Formatting the Task Boxes

Task Boxes may be formatted from the **Box Styles** form, which is displayed by:

- Selecting **Format**, **Box Styles…**, or
- Double-clicking on the outside edge of a box.

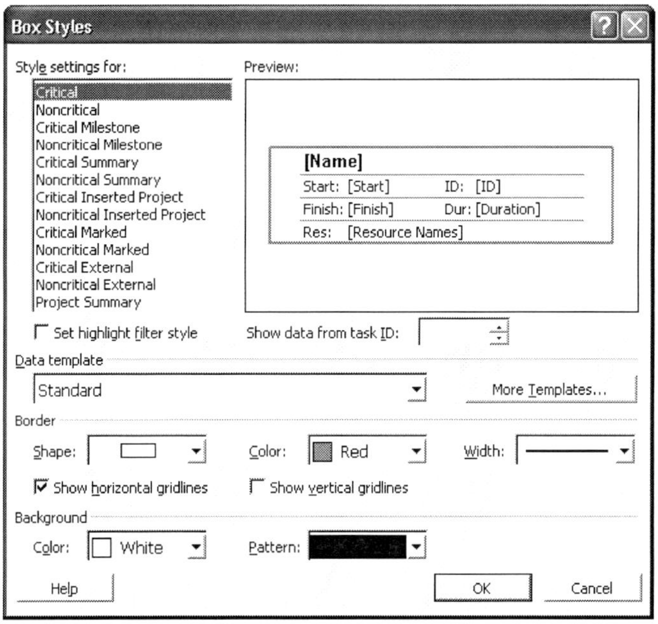

- A new template may be created by clicking on the [More Templates...] button which will open the **Data Templates** form.

10.5 Formatting Individual Boxes

One highlighted **Task Box** may be formatted differently than all others by selecting **Format**, **Box...** to open the **Format Box** form.

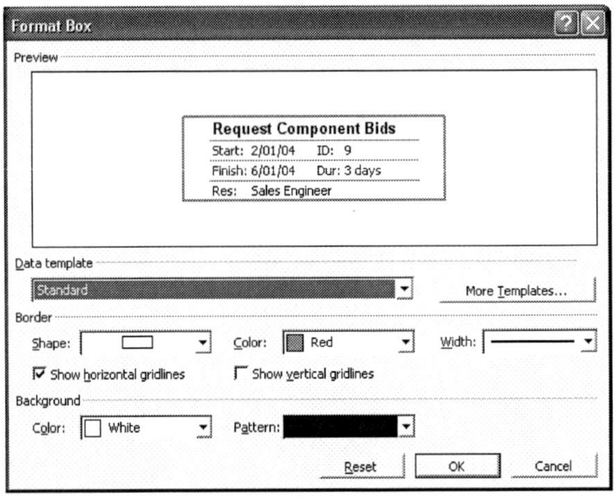

The [Reset] button is used to set the formatting back to default.

10.6 Formatting the Display and Relationship Lines

Most formatting, except formatting the boxes, is set within the **Layout** form. Select **Format**, **Layout...** to open the **Layout** form:

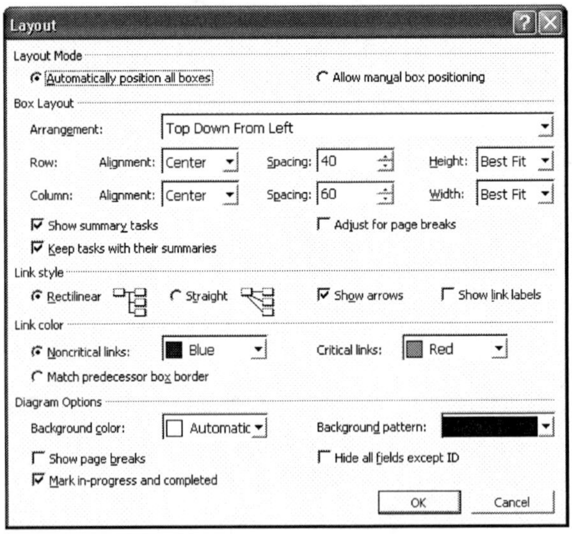

- The **Layout Mode** allows you to drag a Task Box to a new position. When **Allow manual box positioning** is selected, place the mouse over the Task Box. When it changes to a ✥, drag the box to the required location.

- **Box Layout** allows you to specify how the Task Boxes are arranged. This option will only work when **Layout Mode** is set to **Automatically position all boxes**. For example, when "Top Down by Week" is selected, all the Tasks that start in the first week will be placed in the first column. The other options under **Box Layout** should be self-explanatory.

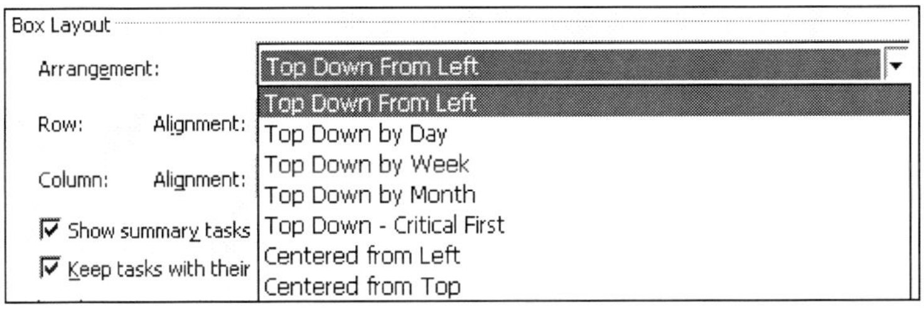

- **Link style** is an option to display relationships. Checking **Show link labels** will place the Relationship Type and Lag on the relationship line and is useful when all your relationships are not Finish-to-Start. See below.

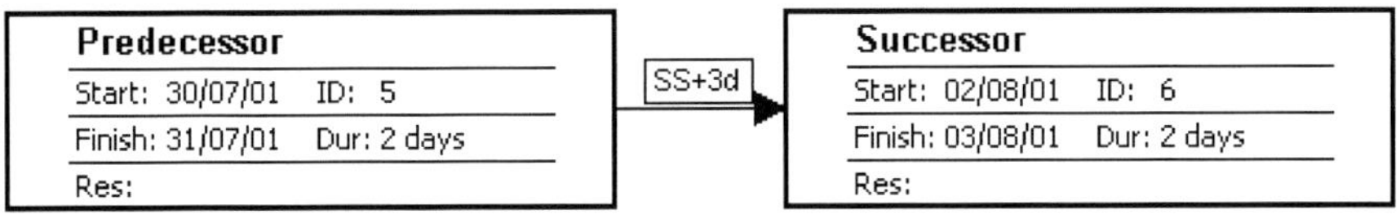

- **Link color** allows you to specify the color of the critical links and other link types. Selecting the option **Match predecessor box border** sets the format of the relationship lines to the predecessor's box border format.

- **Diagram Options** are self-explanatory. The **Hide all fields except ID** will display the Task Boxes, as below. In the example, note that **Show link labels** has also been checked:

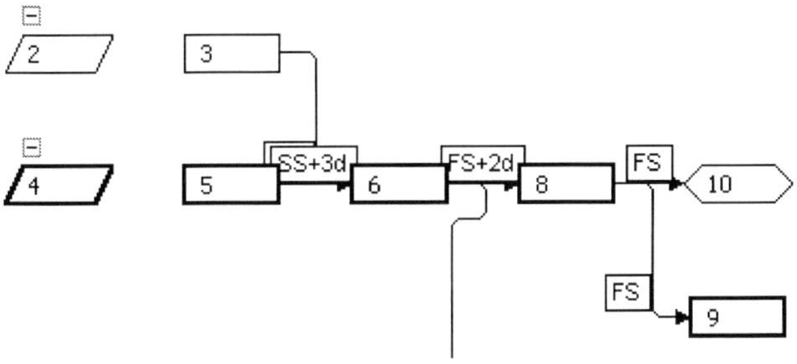

10.7 Early Date, Late Date and Float/Slack Calculations

To help understand the calculation of late and early dates, float and critical path, we will now manually work through an example. The boxes below represent tasks.

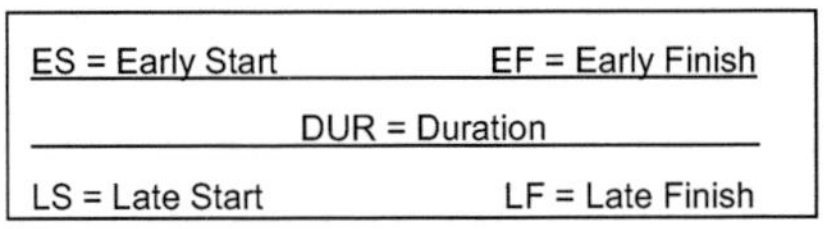

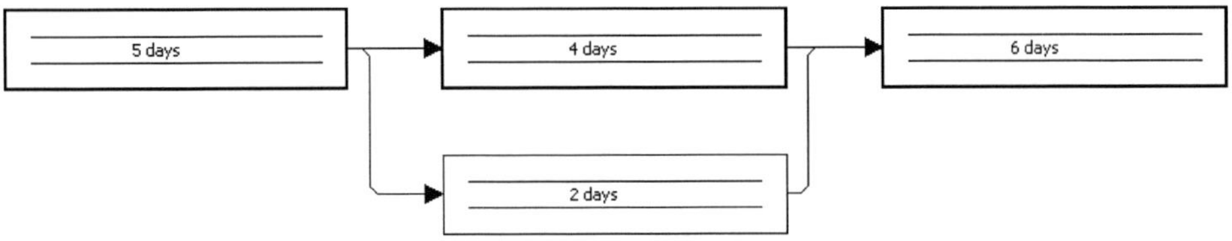

The forward pass calculates the early dates: EF = ES + DUR – 1

Start the calculation from the first task and work forward in time.

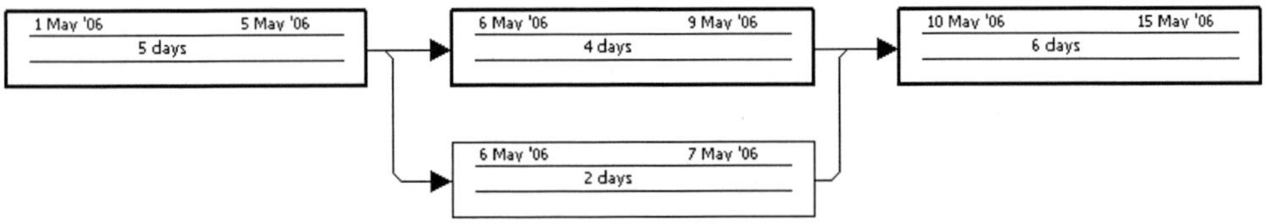

The backward pass calculates the late dates: LS = LF – DUR + 1

Start the calculation at the last task and work backwards in time.

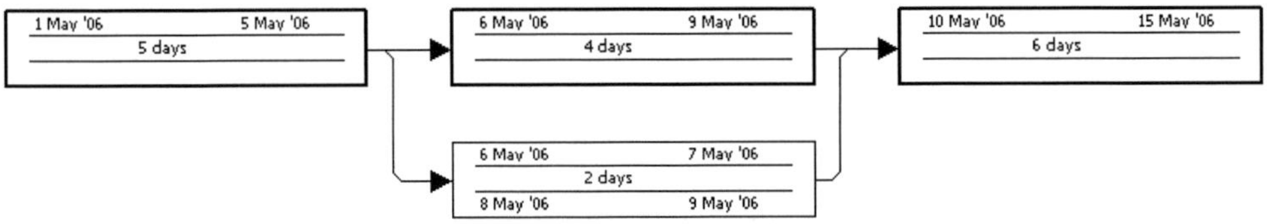

The **Critical Path** is the path where any delay causes a delay in the project and runs through the top row of tasks.

Total Float is the difference between either the **Late Finish** and the **Early Finish** or the difference between the **Late Start** and the **Early Start**. The lower 2 days' task has float of 9 – 7 = 8 – 6 = 2 days. None of the other tasks has float.

A task may not be on the Critical Path and may have more than one predecessor. A **Driving Relationship** is the predecessor that determines the Task Early Start. Microsoft Project 2003 does not identify the difference between **Driving** and **Non-driving Relationships**, which may often make analyzing a schedule difficult.

WORKSHOP 8

Scheduling Calculations

Preamble

We want to practice calculating Early and Late dates with a simple manual exercise.

Assignment

Calculate the early and late dates for the following tasks, assuming a Monday to Friday working week and the first task starts on 1 May 06.

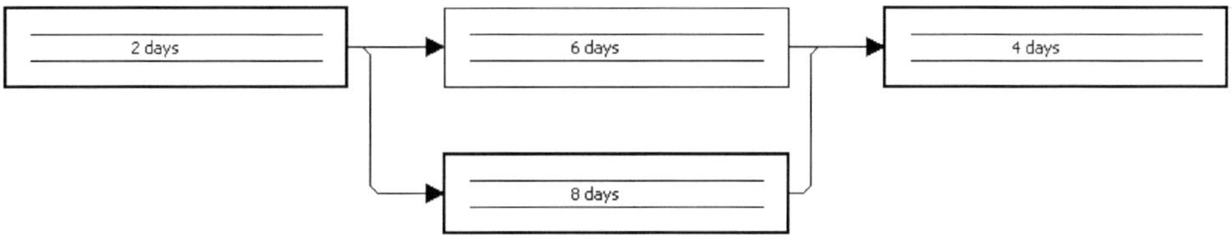

May 2006

M	T	W	Th	F	S	S
1	2	3	4	5	6	7
8	9	10	11	12	13	14
15	16	17	18	19	20	21
22	23	24	25	26	27	28
29	30	31				

ANSWER TO WORKSHOP 8

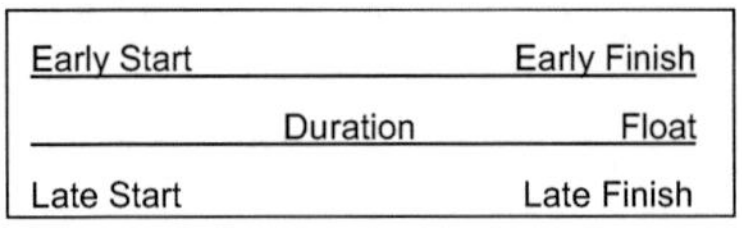

Forward Pass EF = ES + DUR –1

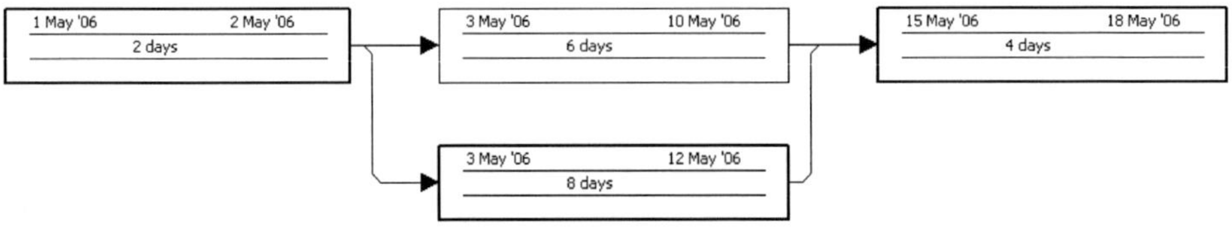

Backward Pass LS = LF – DUR +1

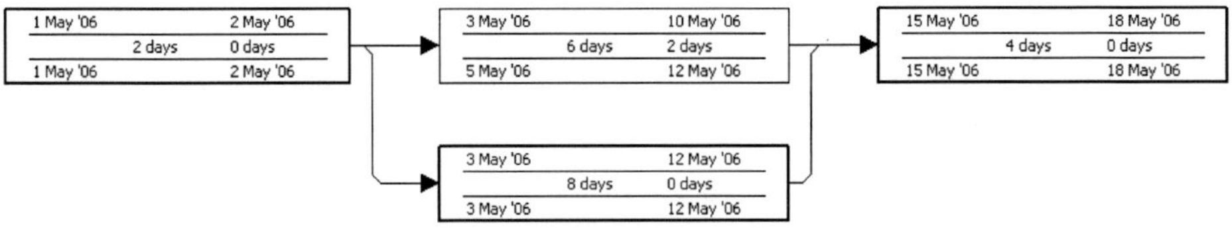

May 2006						
M	T	W	Th	F	S	S
1	2	3	4	5	6	7
8	9	10	11	12	13	14
15	16	17	18	19	20	21
22	23	24	25	26	27	28
29	30	31				

11 CONSTRAINTS

Constraints are used to impose logic on tasks that may not be realistically scheduled with logic links. Microsoft Project 2003 will only allow one constraint against a given task. This chapter will deal with the following constraints in detail:

- **Start No Earlier Than**
- **Finish No Later Than**

These are the minimum number of constraints that are required to effectively schedule a project.

These two constraints may be applied to summary and detailed tasks; all others may only be applied to detailed tasks.

Start No Earlier Than (also known as an "Early Start" constraint) is used when the start date of a task is either known precisely or approximately. Microsoft Project 2003 will not show the task start date prior to this date.

Finish No Later Than (also known as "Late Finish" constraint) is used when the latest finish date is stipulated. Microsoft Project 2003 will not show the task's late finish date after this date.

The following chart summarizes the methods used to assign Constraints to Tasks:

Topic	Notes for Creating a Constraint
• Open the **Task Information** form.	Double-click on the Task and select the **Advanced** tab.
• Display the **Constraint Type** and **Constraint Date** column.	Apply the **Constraint Table** or insert the columns in an existing table.
• Type a date into a date field of the **Task Information** form or the **Details** form or a column.	Type a date in an Early Start box will apply an **Start No Earlier Than** of that date, without warning. A date typed into an Early Finish box will apply a **Finish No Later Than** constraint, without warning.
• Display a **Combination** view displaying the **Task Details** form in the lower pane.	Open the bottom pane, select **View**, **More Views…**, highlight the **Task Details** and click on Apply.

There is an option found under the **Tools**, **Options…**, **Schedule** tab titled **Tasks will always honor their constraint dates**. This function is described in detail in the **OPTIONS** chapter, **Schedule** section. When this option is checked, which is the default, constraints override a logic link. When unchecked, a logic link will override a constraint and a task may be delayed.

Only one constraint may be applied to an activity except when a **Deadline Date** is assigned to an activity. This topic is discussed in this chapter.

A full list of **constraints** available in Microsoft Project 2003 is:

- **As Soon As Possible** This is the default for a new task. A task is scheduled to occur as soon as possible and does not have a Constraint Date.
- **As Late As Possible** A Task will be scheduled to occur as late as possible and does not have any particular Constraint Date. The Early and Late dates have the same date. A task with this constraint has no Total Float and delays the start of all the successor activities.
- **Start No Earlier Than** This constraint sets a date before which the task will not start.
- **Start No Later Than** This constraint sets a date after which the task will not start.
- **Must Start On** This constraint sets a date on which the task will start. Therefore the task has no float. The early start and the late start dates are set to be the same as the Constraint Date.
- **Must Finish On** This constraint sets a date on which the task will finish and therefore has no float. The early finish and the late finish dates are set to be the same as the Constraint Date.
- **Finish No Earlier Than** This sets a date before which the task will not finish.
- **Finish No Later Than** This sets a date after which the task will not finish.
- **Deadline Date** This is similar to applying a **Finish No Later Than** constraint. This offers the opportunity of putting a second constraint on a task.

Earlier Than constraints operate on the **Early Dates** and **Later Than** constraints operate on **Late Dates**. The picture below demonstrates how constraints calculate Total Float (Total Slack) of tasks (without predecessors or successors) against the first task of 10 days' duration:

	Duration	Constraint Date	Constraint Type	Total Slack	Late Start	Late Finish
1	10 days	NA	As Soon As Possible	0 days	18 Oct '05	31 Oct '05
2	3 days	NA	As Late As Possible	0 days	27 Oct '05	31 Oct '05
3	3 days	24 Oct '05	Start No Earlier Than	3 days	27 Oct '05	31 Oct '05
4	3 days	25 Oct '05	Start No Later Than	5 days	25 Oct '05	27 Oct '05
5	3 days	24 Oct '05	Must Start On	0 days	24 Oct '05	26 Oct '05
6	3 days	21 Oct '05	Must Finish On	0 days	19 Oct '05	21 Oct '05
7	3 days	25 Oct '05	Start No Earlier Than	2 days	27 Oct '05	31 Oct '05
8	3 days	26 Oct '05	Finish No Later Than	4 days	24 Oct '05	26 Oct '05

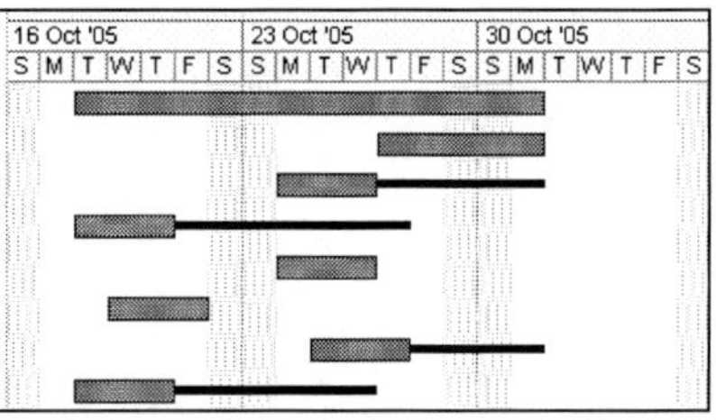

1. An **Expected Finish** is used in Primavera P3 and SureTrak software to calculate the remaining duration of a task. There is no equivalent of this constraint in Microsoft Project 2003.
2. An activity assigned with an **As Late as Possible** constraint in Primavera P3 and SureTrak software will schedule the activity so it absorbs only **Free Float** and will not delay the start of successor activities. In Microsoft Project 2003, a task assigned with an **As Late as Possible** constraint will be delayed to absorb the Total Float and delay all its successor activities, not just the activity with the constraint. Therefore an **As Late as Possible** constraint must be used with care in Microsoft Project.

11.1 Assigning Constraints

11.1.1 Open the Task Information Form

To assign a constraint using the **Task Information** form:

- Double-click on a task to open the form,
- Select the **Advanced** tab,
- Select the **Constraint type:** from the drop-down box,
- Select the **Constraint date:** from the calendar, or type the date in the box, and
- Click on the OK button to accept the constraint.

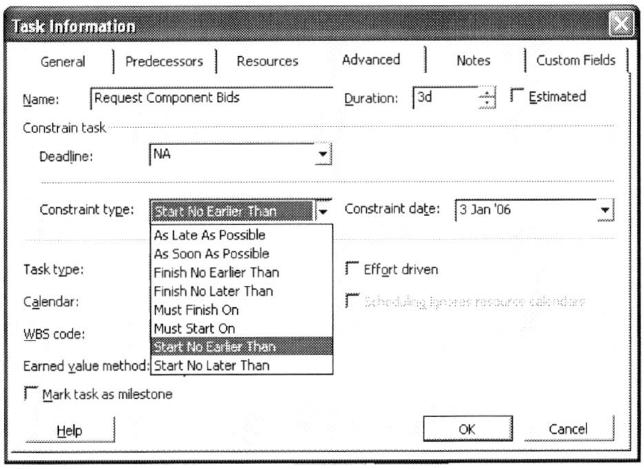

11.1.2 Displaying the Constraint Type and Constraint Date Column

To assign a constraint using the **Constraint Type** and **Constraint Date** columns:

- Either
  - ➢ Insert the **Constraint Type** and **Constraint Date** columns in your existing table by selecting **Insert**, **Column…**, or
  - ➢ Select the **Constraint Dates** table by selecting **View**, **Table:**, **More Tables…**.
- Assign your **Constraint Type** and **Constraint Date** from the drop-down boxes in the columns.

7		⊟ **Estimation**	**14 days**	**As Soon As Possible**	**NA**
8		Request Component Bids	3 days	Start No Earlier Than	3 Jan '06
9		Develop Project Schedule	4 days	As Soon As Possible	NA
10		Draft Technical Details Schedule	9 days	As Late As Possible	NA
11		Compile Costs from Component Bids	2 days	As Soon As Possible	NA
12		⊟ **Proposal**	**10 days**	Finish No Earlier Than	**NA**
13		Draft Bid Document	3 days	Finish No Later Than	NA
14		Meeting to Review the Draft Bid Document	1 day	Must Finish On	NA
15		Design Presentation	1 day	Must Start On	NA
16		Edit Proposal Draft Bid Document	1 day	Start No Earlier Than	NA
				Start No Later Than	

11.1.3 Typing a Date into the Task Information or Details Form

You may assign some constraints from the **Task Information** or **Task Details** form:

- A **Start No Earlier** constraint is assigned by overtyping the Start date, and
- A **Finish No Earlier** constraint is assigned by overtyping the Finish date.

You may overtype or select a date from the drop-down box.

This function provides no warning that a constraint has been set and new users need to be careful when they enter a date into a Start or Finish box.

11.1.4 Display a Combination View Through the Task Details Form

There are a number of forms where constraints may be set, including the **Task Details** form, which is covered in detail in the **VIEWS, TABLES AND DETAILS** chapter. The **Task Details** form may be displayed by:

- Splitting the screen by selecting **Window**, **Split**,
- Make the bottom pane active by clicking in the bottom half anywhere, or
- Select **View**, **More Views...**, and select the **Task Details** form from the drop-down list:

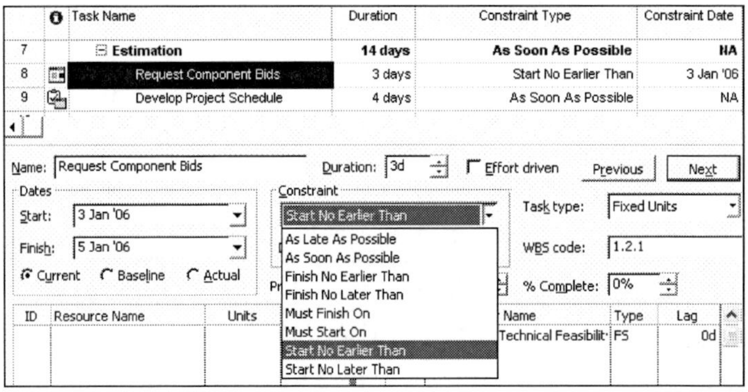

11.2 Deadline Date

Deadline Date was a new Microsoft Project 2000 feature, which allows you to set a date by when a task should be completed. A **Deadline Date** is similar to placing a **Finish No Later Than** constraint on a task and affects the calculation of the **Late Finish** date and float of the activity:

The Deadline Date may be displayed as a column and the display on the bar chart may be formatted as such. This is covered in the **FORMATTING THE DISPLAY** chapter. An Indicator icon is placed in the Indicator column.

11.3 Schedule From Project Finish date

You are able to impose an absolute project finish date by setting the **Schedule from:** option to **Project Finish Date** in the project using the **Project Information** form.

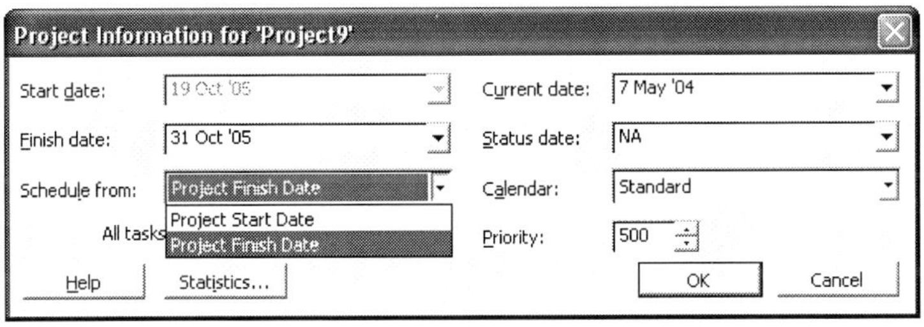

This option may be set after tasks have been added to the schedule. From the time that the project is set to schedule from the Project Finish date all new tasks will be assigned with an **As Late As Possible** (ALAP) constraint as they are created.

Any original tasks which may be set as **As Early As Possible** (AEAP) may either be:

- Reset as ALAP, and calculated with all new tasks as ALAP, or
- Left as AEAP.

When the original tasks are left as AEAP, the tasks will be scheduled as AEAP with a Start No Earlier constraint which is calculated either:

- With no float when their combined durations are greater than the ALAP tasks, or

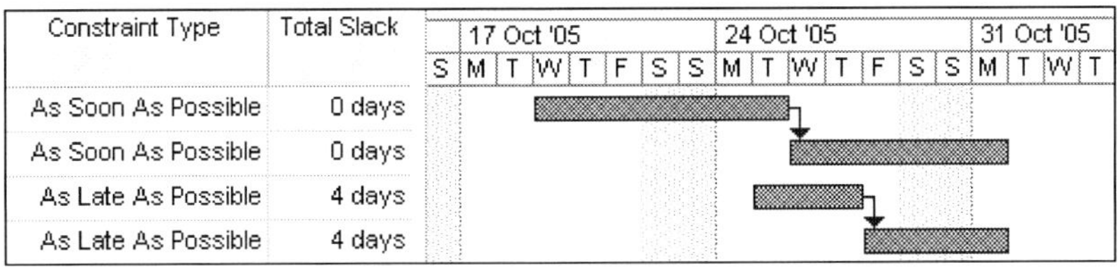

- Will not start earlier than the earliest Early Start of the ALAP tasks when their combined durations are greater than the ALAP tasks.

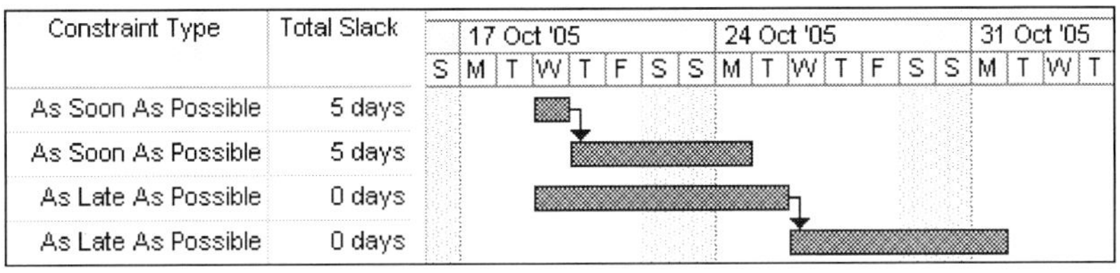

11.4 Task Notes

It is often important to note why constraints have been set. Microsoft Project 2003 has functions that enable you to note information associated with a task, including the reasons associated for establishing a constraint.

The **Task Information** form has a **Note** tab, which has some word processing type formatting functions.

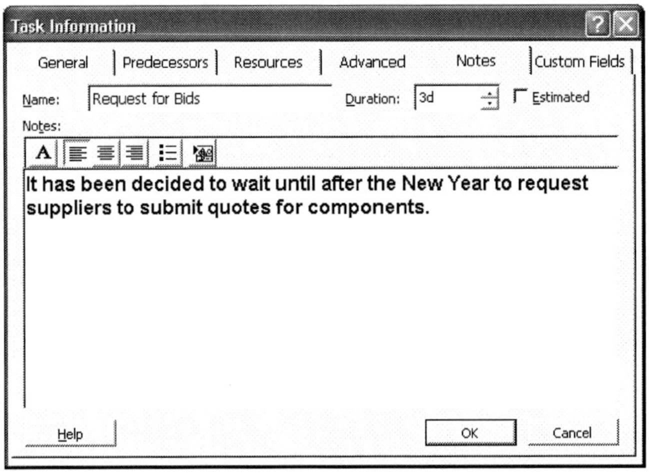

The notes may be displayed by:

- Inserting a column (The formatting of rows is covered in the **FORMATTING THE DISPLAY** chapter.):

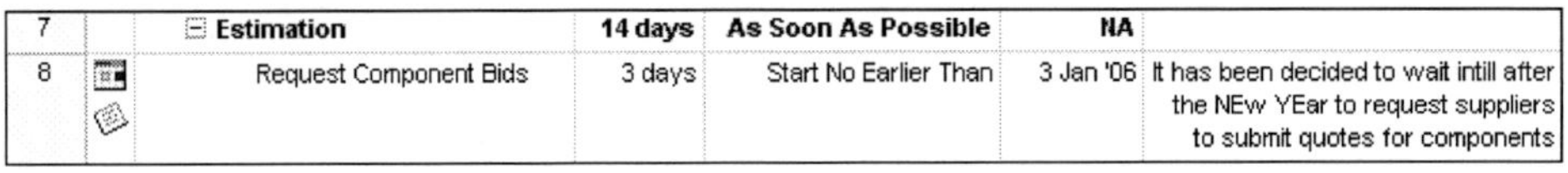

- Displaying on the bar chart (This topic is covered in the **FORMATTING THE DISPLAY** chapter.):

- There is an option for printing task notes in the **Page Setup…**, **View** form.

WORKSHOP 9

Constraints

Preamble

Management has provided further input to your schedule.

Assignment

1. Insert the **Total Slack** column into the **Entry** table between the **Finish** and **Predecessor** column.
2. Run the **Gantt Chart Wizard** found under the **Format** command, and display Critical Path in the Gantt Chart. Do not choose to display resources and dates, and show the link lines between dependent tasks.
3. Observe the calculated finish and the critical path of the project before applying any constraints.
4. The client has said that they require the submission on 23 Jan 06. Apply a **Finish No Later Than** constraint and assign a constraint date of 23 Jan 06 to the **Submit Bid** task and review float. Should you be presented with an error message, read the message carefully and then set the constraint. There should be no change in the Total Float.
5. Due to proximity to Christmas, management has requested we delay the **Request Component Bids** until first thing in the New Year (03 Jan 06). Consensus is that a better response and sharper prices will be obtained after the Christmas rush. Record this in the task notes.
   - To achieve this, set a **Start No Earlier Than** constraint and a constraint date of 03 Jan 06 on the **Request Component Bids** task. Should you be presented with an error message, allow scheduling conflict and set the constraint.
   - Now observe the impact on the critical path and end dates.
6. After review, it is agreed that two days can be deducted from **Negotiate Component Work Packages** task. Change the duration of this task to 4 days. Press **F9** to ensure the schedule is recalculated.
7. Add a **Total Float** bar to all tasks as shown below.

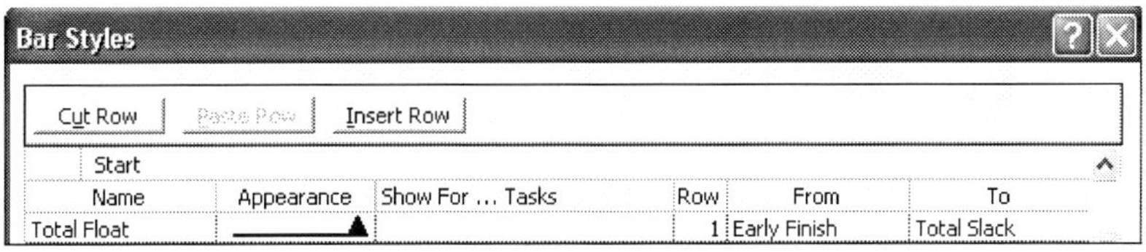

8. Save your **OzBuild Bid** project.
9. Save the project as **OzBuild No Resources** and close it. We will require a copy of this schedule later for Workshop 15.

ANSWER TO WORKSHOP 9

Before delaying the **Request Component Bids** until 3 Jan 06:

	Task Name	Duration	Start	Finish	Total Slack
2	**Research**	**12 days**	**5 Dec '05**	**20 Dec '05**	**0 days**
3	Bid Request Documents Recieved	0 days	5 Dec '05	5 Dec '05	0 days
4	Bid Strategy Meeting	1 day	5 Dec '05	5 Dec '05	0 days
5	Investigate Technical Feasibility	8 days	6 Dec '05	15 Dec '05	0 days
6	Document Installation Requirements	4 days	16 Dec '05	20 Dec '05	4 days
7	**Estimation**	**9 days**	**16 Dec '05**	**30 Dec '05**	**0 days**
8	Request Component Bids	3 days	16 Dec '05	20 Dec '05	4 days
9	Develop Project Schedule	4 days	21 Dec '05	24 Dec '05	4 days
10	Draft Technical Details Schedule	9 days	16 Dec '05	30 Dec '05	0 days
11	Compile Costs from Component Bids	2 days	21 Dec '05	22 Dec '05	4 days
12	**Proposal**	**12 days**	**3 Jan '06**	**18 Jan '06**	**0 days**
13	Draft Bid Document	3 days	3 Jan '06	5 Jan '06	0 days
14	Meeting to Review the Draft Bid Document	1 day	6 Jan '06	6 Jan '06	0 days
15	Design Presentation	1 day	9 Jan '06	9 Jan '06	0 days
16	Edit Proposal Draft Bid Document	1 day	9 Jan '06	9 Jan '06	6 days
17	Negotiate Component Work Packages	6 days	10 Jan '06	17 Jan '06	0 days
18	Final Review of Bid Document	1 day	18 Jan '06	18 Jan '06	0 days
19	Submit Bid	0 days	18 Jan '06	18 Jan '06	0 days

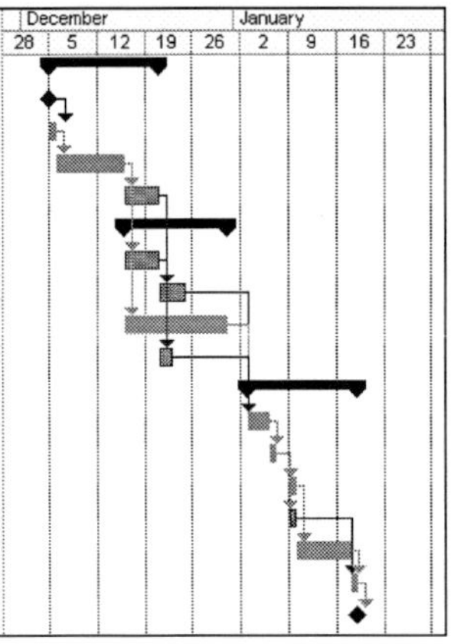

After delaying the **Request Component Bids** until 3 Jan 06:

	Task Name	Duration	Start	Finish	Total Slack
2	**Research**	**12 days**	**5 Dec '05**	**20 Dec '05**	**3 days**
3	Bid Request Documents Recieved	0 days	5 Dec '05	5 Dec '05	3 days
4	Bid Strategy Meeting	1 day	5 Dec '05	5 Dec '05	3 days
5	Investigate Technical Feasibility	8 days	6 Dec '05	15 Dec '05	3 days
6	Document Installation Requirements	4 days	16 Dec '05	20 Dec '05	4 days
7	**Estimation**	**14 days**	**16 Dec '05**	**9 Jan '06**	**-2 days**
8	Request Component Bids	3 days	3 Jan '06	5 Jan '06	-2 days
9	Develop Project Schedule	4 days	21 Dec '05	24 Dec '05	4 days
10	Draft Technical Details Schedule	9 days	16 Dec '05	30 Dec '05	3 days
11	Compile Costs from Component Bids	2 days	6 Jan '06	9 Jan '06	-2 days
12	**Proposal**	**12 days**	**10 Jan '06**	**25 Jan '06**	**-2 days**
13	Draft Bid Document	3 days	10 Jan '06	12 Jan '06	-2 days
14	Meeting to Review the Draft Bid Document	1 day	13 Jan '06	13 Jan '06	-2 days
15	Design Presentation	1 day	16 Jan '06	16 Jan '06	-2 days
16	Edit Proposal Draft Bid Document	1 day	16 Jan '06	16 Jan '06	4 days
17	Negotiate Component Work Packages	6 days	17 Jan '06	24 Jan '06	-2 days
18	Final Review of Bid Document	1 day	25 Jan '06	25 Jan '06	-2 days
19	Submit Bid	0 days	23 Jan '06	23 Jan '06	-2 days

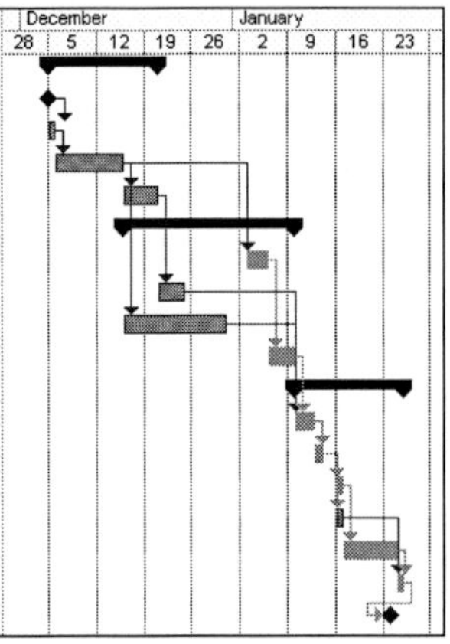

After trimming 2 days from **Negotiate Component Work Packages**:

	Task Name	Duration	Start	Finish	Total Slack
2	**Research**	**12 days**	**5 Dec '05**	**20 Dec '05**	**5 days**
3	Bid Request Documents Recieved	0 days	5 Dec '05	5 Dec '05	5 days
4	Bid Strategy Meeting	1 day	5 Dec '05	5 Dec '05	5 days
5	Investigate Technical Feasibility	8 days	6 Dec '05	15 Dec '05	5 days
6	Document Installation Requirements	4 days	16 Dec '05	20 Dec '05	10 days
7	**Estimation**	**14 days**	**16 Dec '05**	**9 Jan '06**	**0 days**
8	Request Component Bids	3 days	3 Jan '06	5 Jan '06	0 days
9	Develop Project Schedule	4 days	21 Dec '05	24 Dec '05	10 days
10	Draft Technical Details Schedule	9 days	16 Dec '05	30 Dec '05	5 days
11	Compile Costs from Component Bids	2 days	6 Jan '06	9 Jan '06	0 days
12	**Proposal**	**10 days**	**10 Jan '06**	**23 Jan '06**	**0 days**
13	Draft Bid Document	3 days	10 Jan '06	12 Jan '06	0 days
14	Meeting to Review the Draft Bid Document	1 day	13 Jan '06	13 Jan '06	0 days
15	Design Presentation	1 day	16 Jan '06	16 Jan '06	0 days
16	Edit Proposal Draft Bid Document	1 day	16 Jan '06	16 Jan '06	4 days
17	Negotiate Component Work Packages	4 days	17 Jan '06	20 Jan '06	0 days
18	Final Review of Bid Document	1 day	23 Jan '06	23 Jan '06	0 days
19	Submit Bid	0 days	23 Jan '06	23 Jan '06	0 days

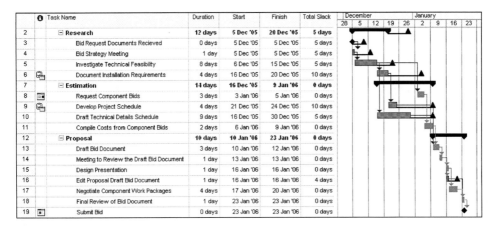

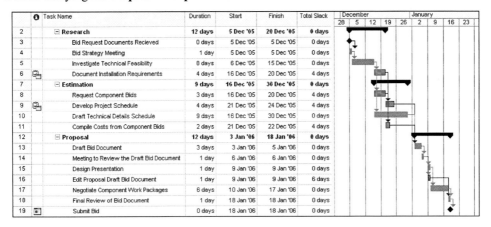

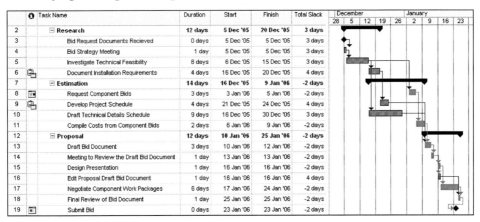

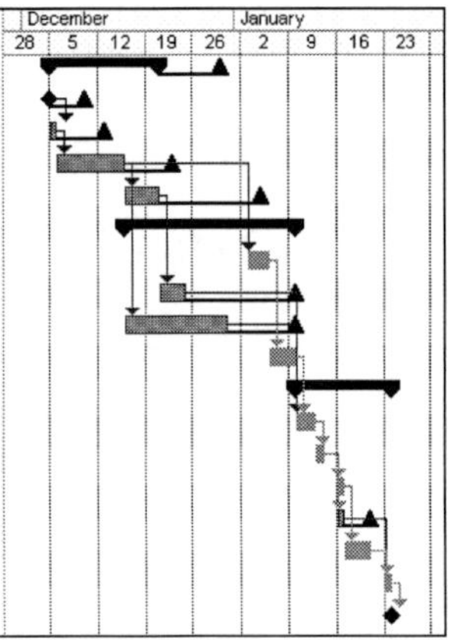

Note: If your Total Float is not calculating as above, press F9 to recalculate.

12 FILTERS

This chapter covers the ability of Microsoft Project 2003 to control which activities are displayed, both on the screen and in printouts, by using **Filters**.

Filters could be created to display activities for Phases, Disciplines, Systems and physical areas of a project.

12.1 Understanding Filters

Microsoft Project 2003 has an ability to display tasks that meet specific criteria. You may want to see only the incomplete Tasks, or the work scheduled for the next couple of months, or the Tasks that are in-progress.

Microsoft Project 2003 defaults to displaying all tasks. It has a number of pre-defined filters available that you may use or edit. You can also create one or more of your own.

A filter may be applied to display or to highlight only those tasks that meet a criteria.

There are two types of filters:

- The first is where you select a **Filter** which exists or has been created on the **More Filters** form.
- The second is to create an **AutoFilter** which is very similar to the Excel **AutoFilter** (Drop-down filter) function.

Topic	Menu Command
• Apply a **Filter**	Select **Project**, **Filter for: All Tasks**, or With the **Formatting** toolbar displayed, select from the drop-down list.
• Create or modify a **Filter**	Select **Project**, **Filter for: All Tasks**, **More Filters…** to open the **More Filters…** form.
• Turn on **AutoFilter**	Select **Project**, **Filter for: All tasks**, **AutoFilter**, or Click on the **AutoFilter** icon on the **Formatting** toolbar.
• Apply an **AutoFilter**	Click on the icon in the column headers.

12.2 Applying a Filter

Filters may be applied using several methods:

Method 1

Select **Project**, **Filter for: All Tasks**, and select the filter required from the menu.

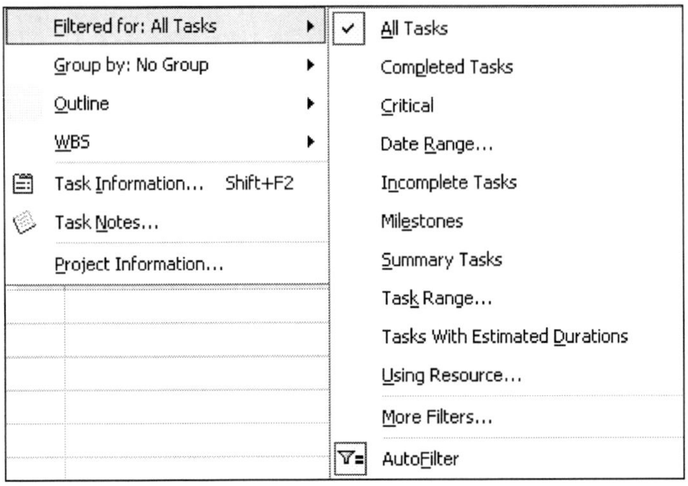

Only filters established in the **Filter Definition** form are displayed. The currently applied filter is always written after **Filtered for:** in the menu. See above.

This may not be a complete list of filters as only filters chosen to be displayed in the menu when they are created will be displayed in this list.

Method 2

Display the **Formatting** toolbar, and select the filter you require from the drop-down list:

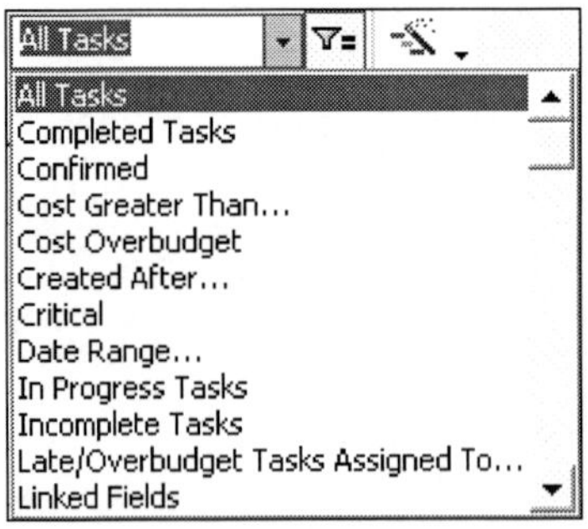

Select **View**, **Toolbars**, **Formatting** to display the formatting toolbar.

This may not be a complete list of filters since only filters chosen to be displayed in the menu when they are created will be displayed in this list.

Method 3

Select **Project**, **Filter for: All Tasks**, **More Filters…** to open the **More Filters…** form:

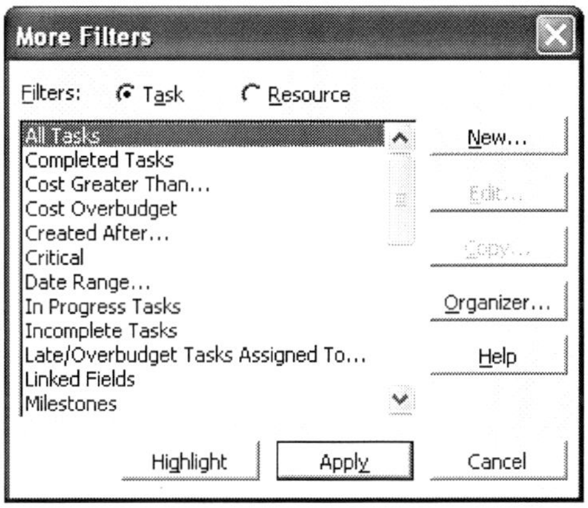

- The two radio buttons at the top of the form allow you to select filters that operate either on a **Task** criteria or on a **Resource** criteria.
  - **Task** criteria will operate on the criteria of most data such as costs, dates, durations, text columns, number columns, outline and WBS.
  - **Resource** criteria will operate on a similar criteria as Tasks but will select the resource value and not the task value.
- Select the required filter from the drop-down list, and
- You have a choice of selecting:
  - 

 to highlight the tasks that meet the criteria, or
  - 

 to display only the tasks that meet the criteria.

12.3 Creating and Modifying Filters

Select **Project**, **Filter for: All Tasks**, **More Filters…** to open the **More Filters** form where you may create or modify a filter:

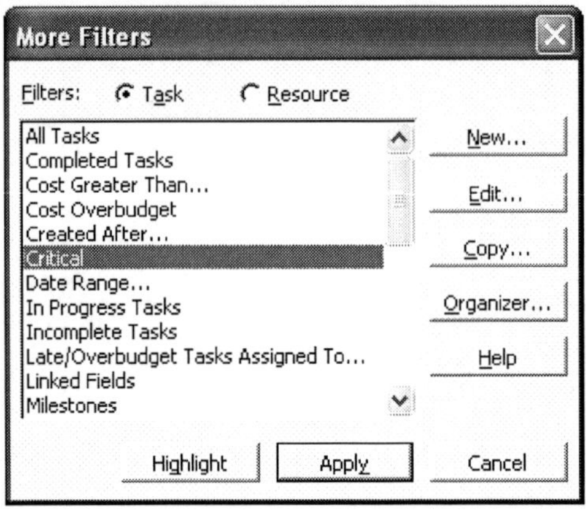

- Click on 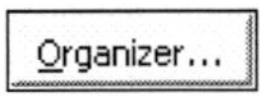 to copy a filter from another file.
- Click on:
  -  to create a new filter, or
  -  to edit an existing filter, or
  -  to copy an existing filter.
- The **Filter Definition** form will be displayed. The one-line example below will only display critical tasks.

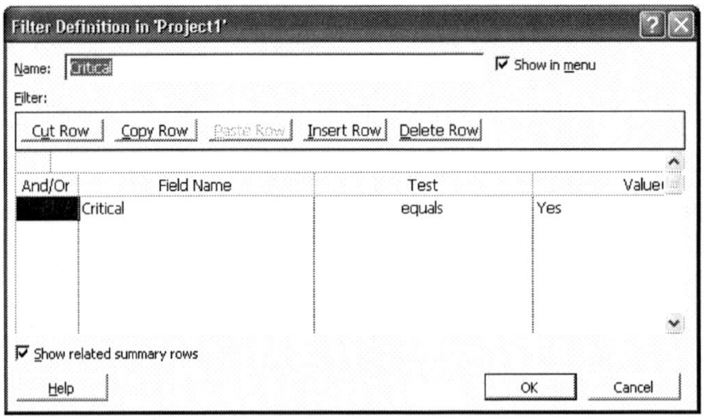

- The filter may then be edited to display or highlight the required tasks.
- Checking the **Show in menu** box will place the **filter** in the menu.
- Checking **Show related summary rows** will display any associated Summary Tasks.
- Select 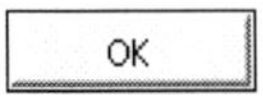 to return to the **More Filters** form.

12.4 Defining Filter Criteria

The filter criteria is determined by four columns of information in the **Filter Definition** form:

- **And/Or**, this function operates when you have two or more lines of data in the **Filter Definition** form to operate on.
- **Field Name** defines the data field you want to operate on.
- **Test** sets the criteria such as "Greater Than" or "Less Than."
- **Value(s)** is/are a date, number or text for the **Test** field to operate on. If more than one value is to be considered, for example when a **Test** is "Between," the two values are separated by a comma ",". The example below is a filter that will only display detail activities that will start between 24 Nov and 1 Dec 02.

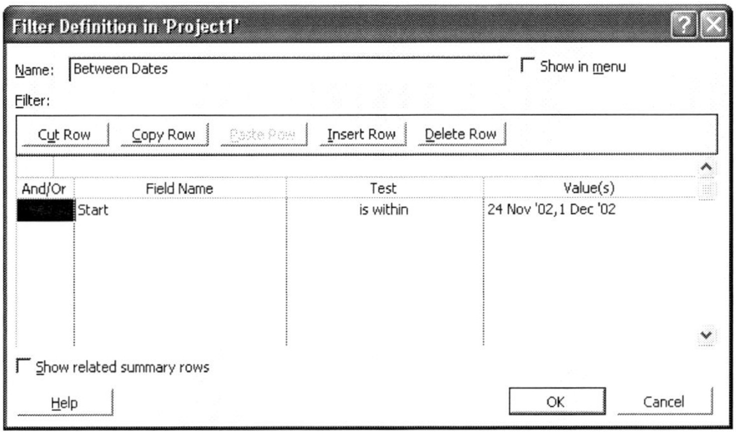

This chapter will not cover all the aspects of filter definitions but will cover the major principles, so you may experiment when you require a filter.

There are a number of predefined filters with a standard Microsoft Project 2003 installation. You should inspect these to gain an understanding of how filters are constructed and applied.

12.4.1 Simple Filters, Operator and Wild Cards

A simple **Filter** contains one line of data and therefore the **And/Or** function is not used. These are the most common filters and meet most common filtering requirements.

These filters are used for such purposes as displaying tasks which:

- Are not started, complete, or in-progress tasks.
- Have a Start or Finish before or after a particular date.
- Contain specific text.
- Are within a range of dates.

There are some operands you should be aware of:

- The **Value(s)** field may have "Yes" or "No" entered. These are used for data items such as Milestones and Critical Tasks. The filter below will select all tasks that are not Milestones:

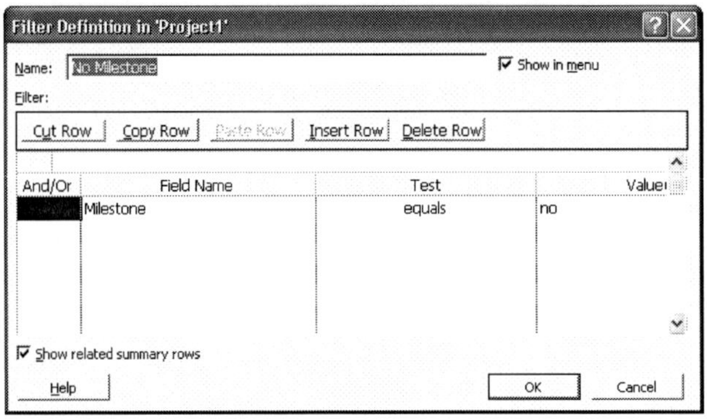

The **Wildcard** functions are similar to the DOS Wildcard functions and are mainly used for filtering text:

- You may replace a single character with a "**?**". Thus, a filter searching for a word containing "b?t" will display words like "bat", "bit" and "but."
- You may replace a group of characters with an *. Thus, a filter searching for a word containing "b*t" will display words like "blot", "blight" and "but."

12.4.2 Calculated Filters

A calculated filter compares one value with another. The **Value(s)** field is selected from the drop-down box. The example below will display those tasks that are scheduled to start later than the baseline:

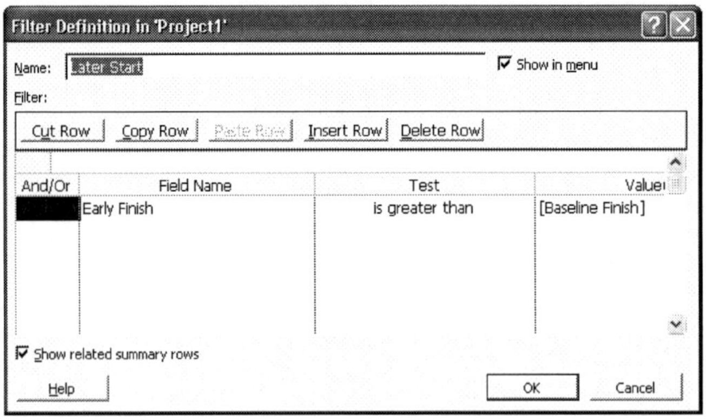

12.4.3 And/Or Filters

The **And/Or** function allows you to search for tasks which:

- Meet more than one criteria by using the **And/Or** option. The filter below displays tasks with Slack less than 5 days **Or** contain the word "Bid" and display the summary tasks:

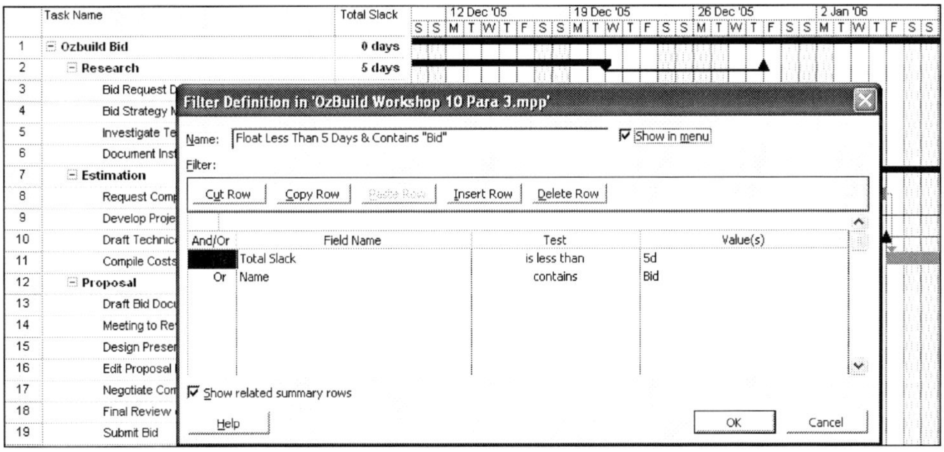

- Meet one criteria or a second by using the **And** function. The filter below displays tasks with Slack less than 5 days **And** contain the word "Bid" and do not display the summary tasks:

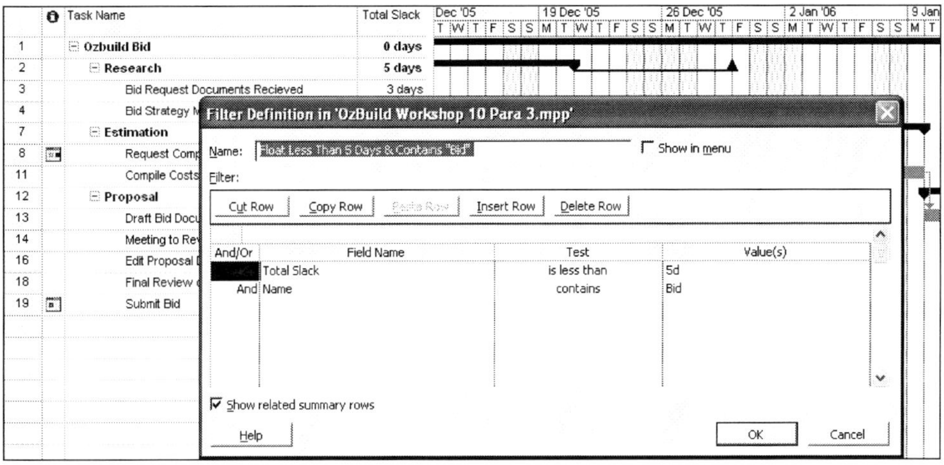

12.5 Multiple And/Or

Multiple **And/Or** statements are possible by placing a line with only an **And/Or** statement. The filter below selects tasks that contain the word "Bid" and start after 1 Jan 06 or are critical:

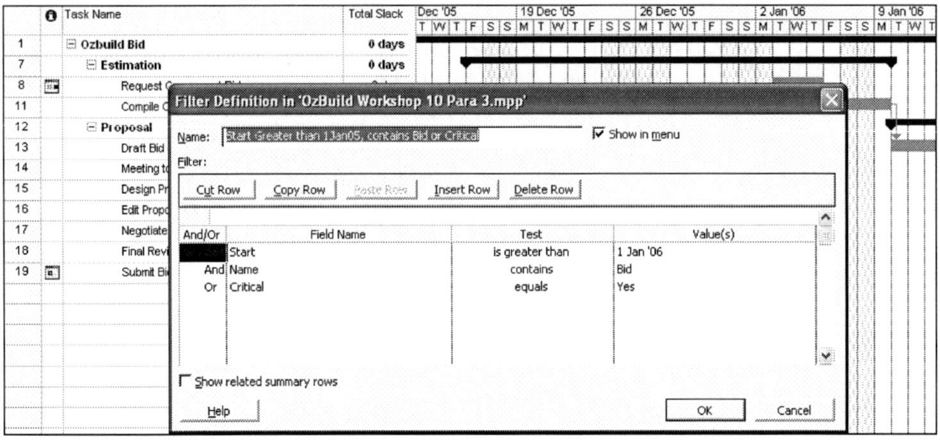

12.5.1 Interactive Filter

These filters allow you to enter the **Value(s)** of the filtered field after applying the filter. The filter is tailored each time it is applied via a user-prompt. The filter below will ask you to enter a word in the task name.

For this function to operate properly, the text in the **Value(s)** field must commence with a double quote **"** and end with a double quote and question mark **"?** :

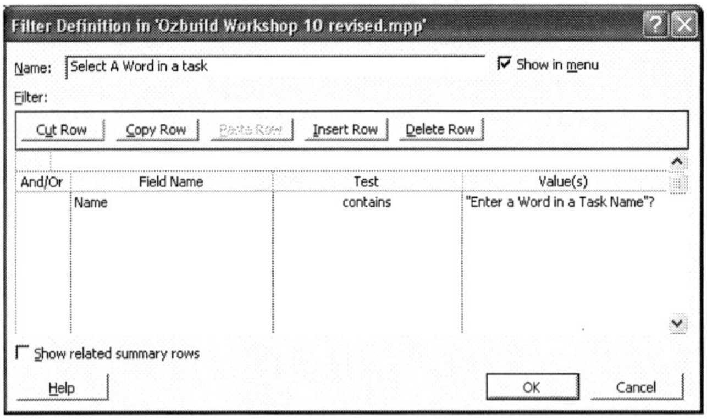

After the filter is applied, you will be presented with the **Interactive Filter** form to enter the required text:

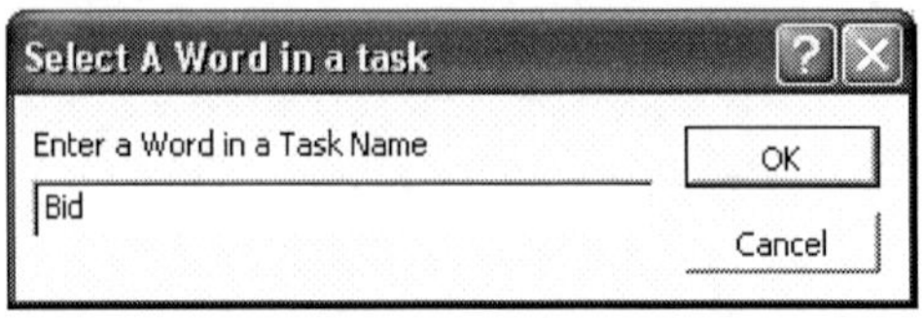

12.6 AutoFilters

AutoFilters in Microsoft Project 2003 are similar to the Excel **AutoFilter** function and allow you to select the filter criteria from drop-down menus in the column headers.

To create an **AutoFilter** based on one parameter:

- Use one of the following methods to turn on the **AutoFilter** function:
  - **Project**, **Filter for: All tasks**, **AutoFilter**, or
  - Click on the **AutoFilter** icon located on the Formatting toolbar.
- The column headers will display the icon in the column header. Click on this icon in one of the columns to display a drop-down box:

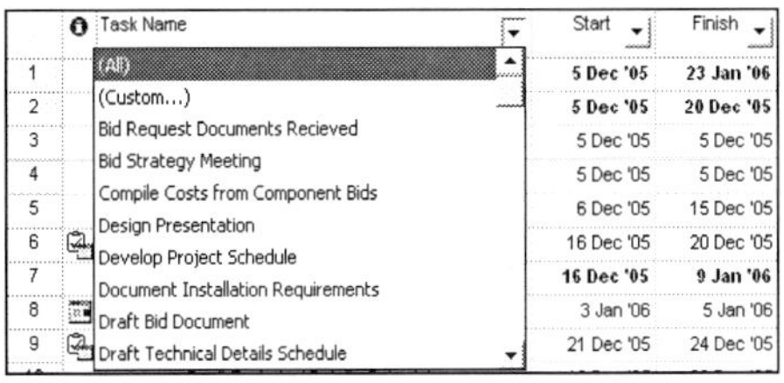

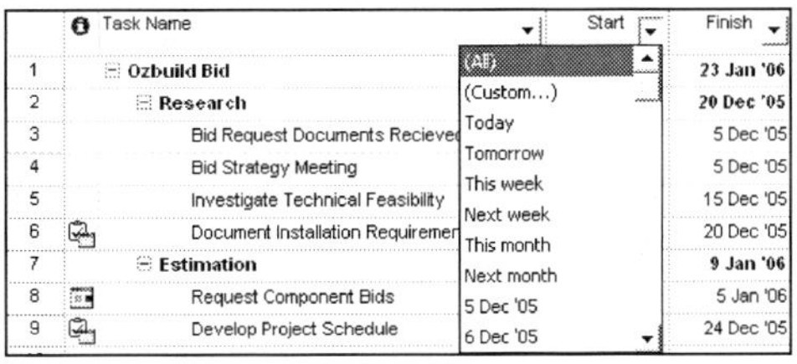

- Select the required criteria from the drop-down box.

You may now select another column and create a filter based on a second parameter to further reduce the number of tasks displayed.

To create a **Custom AutoFilter**, which is based on two parameters:

- Click on the icon in the column headers.
- Select the **(Custom…)** option to open the **Custom AutoFilter** form:

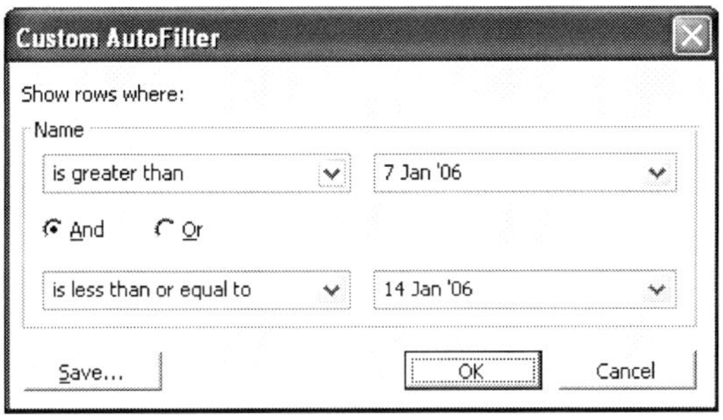

- Select the parameters you want to operate on from the four drop-down boxes and check the **And** or **Or** radio button.
- To save a drop-down filter as a normal filter, click on the Save... button, which opens the **Filter Definition** form.

WORKSHOP 10

Filters

Preamble

Management has asked for reports on tasks to suit their requirements.

Assignment

Open the **OzBuild Bid** project to complete this exercise and ensure the **OzBuild No Resources** is closed.

1. They would like to see all the critical tasks.
   - Apply the **Critical** tasks filter. The Filter command is found under the **Project** menu item.

   You will see only tasks that are on the critical path and their associated summary tasks.

2. They would like to see all the tasks with float less than 5 days:
   - Make a copy of the Critical filter and edit it:
   - Assign a title to the filter: **Float Less Than 5 Days**.
   - Change the condition to display a **Total Slack** of less than 5 days.
   - Show the filter in the menu.

   You should find that one extra task is now shown.

3. They would like to see all the tasks with float less than 5 days or contain the word "Bid."
   - Copy the **Float Less Than 5 Days** filter.
   - Assign a title to the filter: **Float Less Than 5 Days or Contains "Bid"**.
   - Add the condition: **Or** Name (Task Name) contains **Bid**, and
   - Do not show in the menu.
   - Apply the filter.

   You should find that additional tasks are now shown.

4. We now wish to create an **AutoFilter** that displays all the activities containing the word "Meeting."
   - Apply the **All Tasks** Filter.
   - Click on the icon to activate the Auto Filters.
   - Create a filter to select tasks containing the word "Meeting."

5. Now apply the **All Tasks** Filter, remove the **AutoFilter** and save your **OzBuild Bid** project.

ANSWERS TO WORKSHOP 10

After applying the Critical Filter:

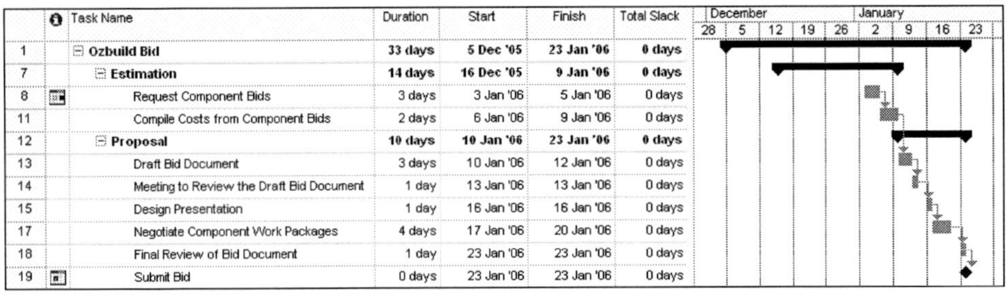

	Task Name	Duration	Start	Finish	Total Slack
1	Ozbuild Bid	33 days	5 Dec '05	23 Jan '06	0 days
7	Estimation	14 days	16 Dec '05	9 Jan '06	0 days
8	Request Component Bids	3 days	3 Jan '06	5 Jan '06	0 days
11	Compile Costs from Component Bids	2 days	6 Jan '06	9 Jan '06	0 days
12	Proposal	10 days	10 Jan '06	23 Jan '06	0 days
13	Draft Bid Document	3 days	10 Jan '06	12 Jan '06	0 days
14	Meeting to Review the Draft Bid Document	1 day	13 Jan '06	13 Jan '06	0 days
15	Design Presentation	1 day	16 Jan '06	16 Jan '06	0 days
17	Negotiate Component Work Packages	4 days	17 Jan '06	20 Jan '06	0 days
18	Final Review of Bid Document	1 day	23 Jan '06	23 Jan '06	0 days
19	Submit Bid	0 days	23 Jan '06	23 Jan '06	0 days

Creating the **Float Less Than 5 Days** Filter:

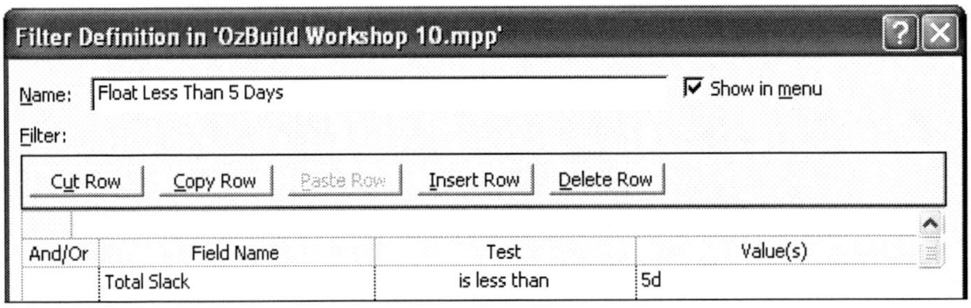

After applying the **Float Less Than 5 Days & Contains "Bid"** Filter:

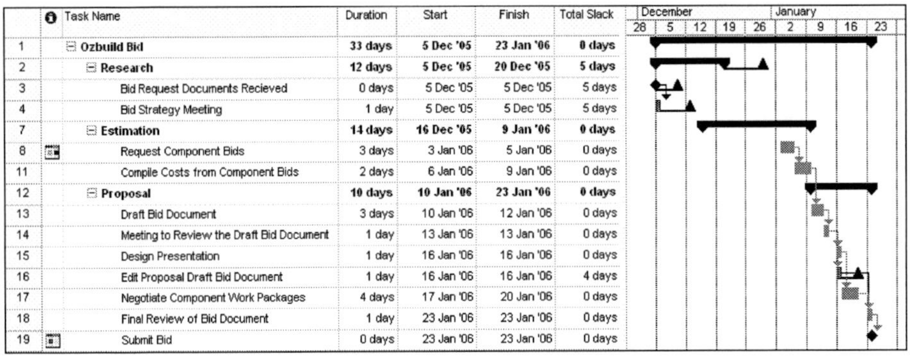

	Task Name	Duration	Start	Finish	Total Slack
1	Ozbuild Bid	33 days	5 Dec '05	23 Jan '06	0 days
2	Research	12 days	5 Dec '05	20 Dec '05	5 days
3	Bid Request Documents Recieved	0 days	5 Dec '05	5 Dec '05	5 days
4	Bid Strategy Meeting	1 day	5 Dec '05	5 Dec '05	5 days
7	Estimation	14 days	16 Dec '05	9 Jan '06	0 days
8	Request Component Bids	3 days	3 Jan '06	5 Jan '06	0 days
11	Compile Costs from Component Bids	2 days	6 Jan '06	9 Jan '06	0 days
12	Proposal	10 days	10 Jan '06	23 Jan '06	0 days
13	Draft Bid Document	3 days	10 Jan '06	12 Jan '06	0 days
14	Meeting to Review the Draft Bid Document	1 day	13 Jan '06	13 Jan '06	0 days
15	Design Presentation	1 day	16 Jan '06	16 Jan '06	0 days
16	Edit Proposal Draft Bid Document	1 day	16 Jan '06	16 Jan '06	4 days
17	Negotiate Component Work Packages	4 days	17 Jan '06	20 Jan '06	0 days
18	Final Review of Bid Document	1 day	23 Jan '06	23 Jan '06	0 days
19	Submit Bid	0 days	23 Jan '06	23 Jan '06	0 days

Creating an AutoFilter that displays Tasks Name containing **Meeting**:

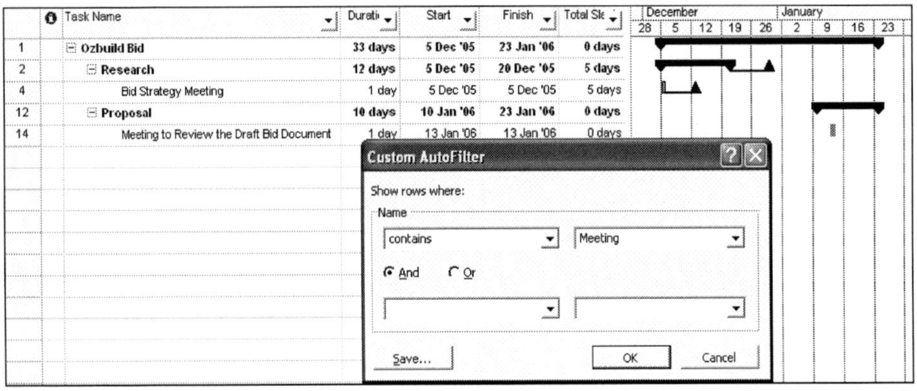

	Task Name	Durati	Start	Finish	Total Sl
1	Ozbuild Bid	33 days	5 Dec '05	23 Jan '06	0 days
2	Research	12 days	5 Dec '05	20 Dec '05	5 days
4	Bid Strategy Meeting	1 day	5 Dec '05	5 Dec '05	5 days
12	Proposal	10 days	10 Jan '06	23 Jan '06	0 days
14	Meeting to Review the Draft Bid Document	1 day	13 Jan '06	13 Jan '06	0 days

13 VIEWS, TABLES AND DETAILS

A **View** is a function where formatting and other parameters such as the **Table**, **Details** and **Bar** formatting are saved. A **View** may be saved and reapplied later on. A filter may be saved as part of a **View**.

The views displayed in the menu and on the **View Bar** with a standard load of Microsoft Project 2003 are listed below. There are more views available from the **More Views** form which will not be covered in this book.

View Name	Notes on the View
• Calendar	This may be applied to the top window only and displays the tasks overlaid on a calendar.
• Gantt	Displays a Table on the left of the screen and Gantt Chart on the right.
• Network Diagram	This view is a PERT-style display and is covered in the **NETWORK DIAGRAM VIEW** chapter.
• Task Usage*	Similar to Resources Usage, displays a Table with Tasks with associated Resources on the left and the Resources usage data apportioned over time on the right.
• Tracking Gantt	Similar to the **Gantt** view.
• Resource Graph*	A split table displaying the Resource name on the left-hand side and resource usage in bar chart format on the right-hand side.
• Resource Sheet*	A single table displaying Resource information such as costs and calendars.
• Resource Usage*	Similar to the Task usage, displays a Table with Resources with associated Tasks on the left and the Resources usage data apportioned over time on the right.

* The Views containing resource information are covered in more detail in the **STATUSING PROJECTS WITH RESOURCES** chapter.

The **Calendar** view is not covered in detail in this book and **Network** views are covered in the **NETWORK DIAGRAM VIEW** chapter.

13.1 Understanding Views

There are two types of Views:

- A **Single View** is normally applied to the top pane only; when the pane is split, the **Details** form is displayed in the bottom pane.
- A **Combination View** is comprised of two **Single Views**, one displayed in the top pane and one in the bottom pane.

All **Single Views** may be applied to the top pane and when the window is split, most may be applied to the bottom pane. When a **Single View** is applied to the bottom pane, it will only display the information attributed to the task that is highlighted in the top pane.

When a **Single View** is applied to the top or bottom pane, the other pane is left with the contents of the previous **Combination View**.

A **View** is based on a **Screen** when it is created. The **Screen** may not be changed after the **View** is created. There are 14 **Screens**:

- Calendar
- Gantt Chart
- Network Diagram
- Relationship Diagram
- Resource Form
- Resource Graph
- Resource Name Form
- Resource Sheet
- Resource Usage
- Task Details Form
- Task Form
- Task Name Form
- Task Sheet
- Task Usage

The Table in **APPENDIX 1 – SCREENS USED TO CREATE VIEWS** lists the Screens and provides further detail about formatting the screens. A few important points to consider are:

- The **Calendar** screen may only be displayed in the top pane.
- The **Gantt Chart**, **Network Diagram** and **Relationship Diagram** screens are best displayed in the top pane and have limited use when displayed in the bottom pane.
- The **Forms** are best displayed in the bottom pane and display **Task** or **Resource** information about a task that is highlighted in the top pane. **Forms** may be further formatted by selecting **Format**, **Details**.
- The **Task Sheet** is similar to the **Gantt Chart** but does not display bars and is best displayed in the top pane.
- The **Resource** and **Task Usage** screens display resource information and are further discussed in the **RESOURCE HISTOGRAMS AND TABLES** chapter.

13.2 Applying a View

Views may be applied by either:

- Selecting **View** from the menu and then selecting the required View from the list above **More Views…**:
- Displaying the **View Bar** and clicking on the icon of the required View. To display or hide the **View Bar**, select **View**, **View Bar**. The **View Bar** will be displayed on the left-hand side of the screen, and will display the same list of Views as those in the menu.

View Bar

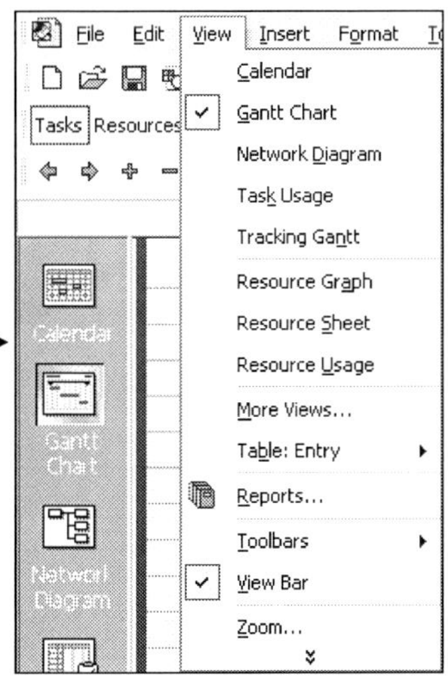

- Selecting **View**, **More Views…** to open the **More Views** form and select the required View from the drop-down list.

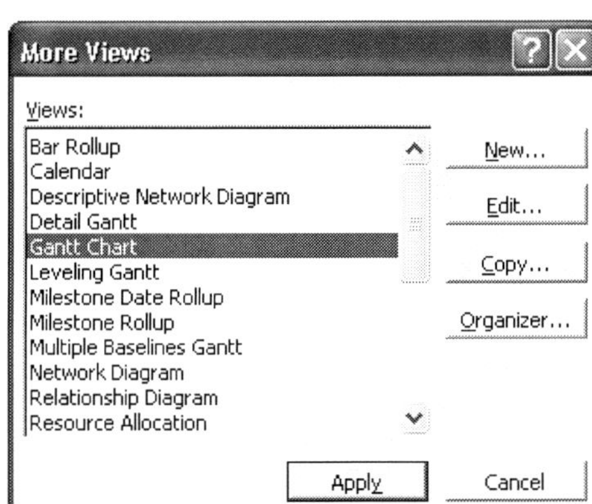

Some Views are displayed in the menu and **View Bar** and these are available from the **More Views** form. To display a View in the menu and the **View Bar**, the View should be edited and the **Show in menu** box checked.

13.3 Creating a New View

A new View may be created by copying and editing an existing View, or by choosing **Copy…** to generate a View.

13.3.1 Creating a New Single View

To create a new Single View:

- Select **View**, **More Views…** to open the **More Views** form,
- Click on the New… button to open the **Define New View** form:

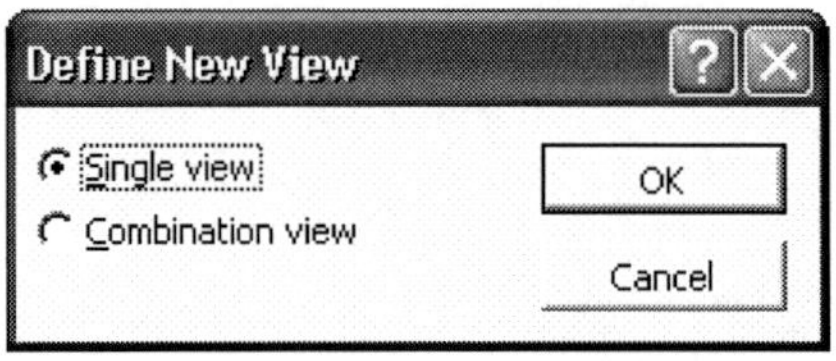

- Click on the **Single view** radio button to open the **View Definition** form:

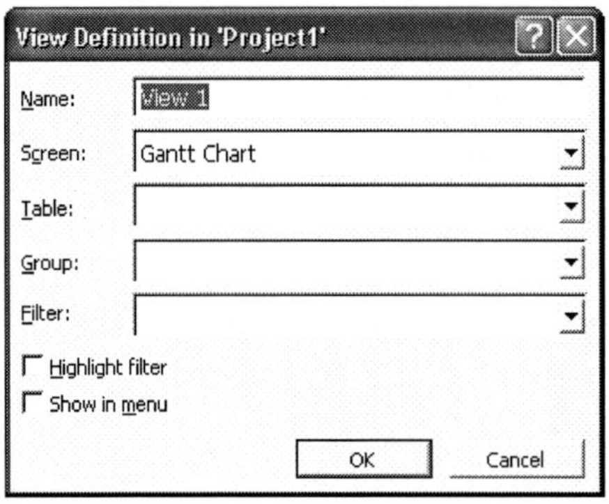

- You may now enter:
  - The new View name in the **Name:** box by overtyping the "**View 1**" name assigned by Microsoft Project 2003,
  - Select the **Screen:** View, see **APPENDIX 1** for more information on the screens. A list of available screens is displayed below:

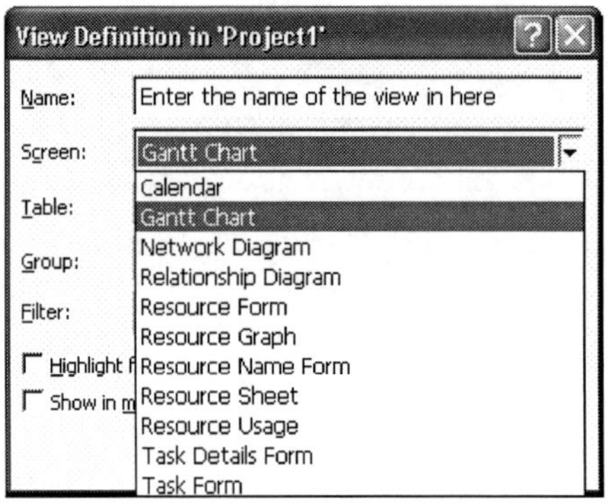

  - Select the **Table:** you want to apply with the view.
  - You may not want to display your tasks using the Outline display mode. The **Group:** function allows you to group your tasks by other data items such as a **Text Column** or **Constraint Type** and is covered in the **GROUPING TASKS** chapter.
  - You may also select a filter to apply with the layout from the **Filter:** drop-down box.
  - Click the **Highlight filter** box to highlight the tasks that meet the filter specification as opposed to applying the filter.
  - Click on the **Show in menu** box to display the View in the **View** menu.
- Click on the OK button to create the new **View**.
- From the **More Views** form click on the Apply button to apply the new **View**.

13.3.2 Creating a Combination View

To create a new Single View:

- Select **View**, **More Views…** to open the **More Views** form,
- Click on the New... button to open the **Define New View** form:

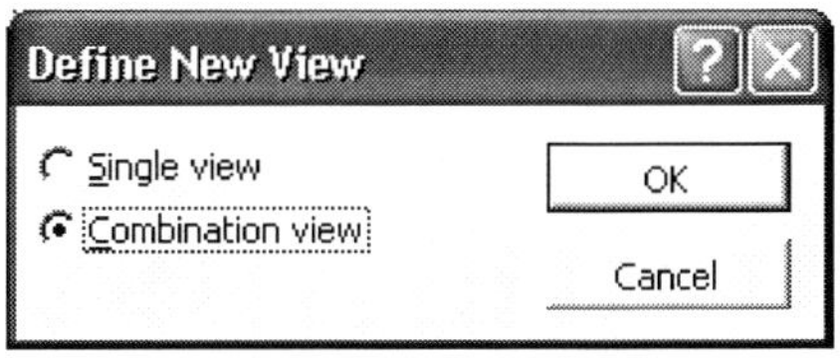

- This will open the **View Definition** form:

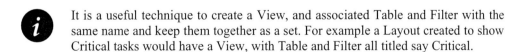

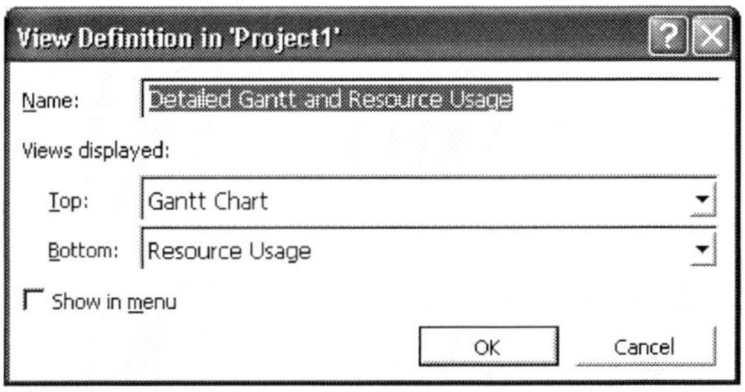

- Enter your View Name and the Views you want to display in the top and bottom pane,
- Click on **Show in menu** to display the View in the menu and on the **View Bar**, and
- Click on the OK button to create the new **View**.

It is a useful technique to create a View, and associated Table and Filter with the same name and keep them together as a set. For example a Layout created to show Critical tasks would have a View, with Table and Filter all titled say Critical.

13.3.3 Copying and Editing a View

To copy an existing View:

- Select **View**, **More Views…**:

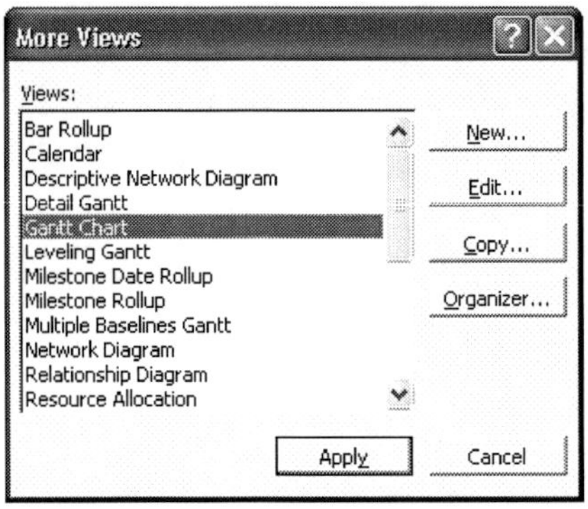

- Select the View you want to copy from the drop-down list and click on the Copy... button to open the **View Definition** form. This form is different for Single and Combination Views. The example below is a **Single View** created from the **Gantt Chart** Screen:

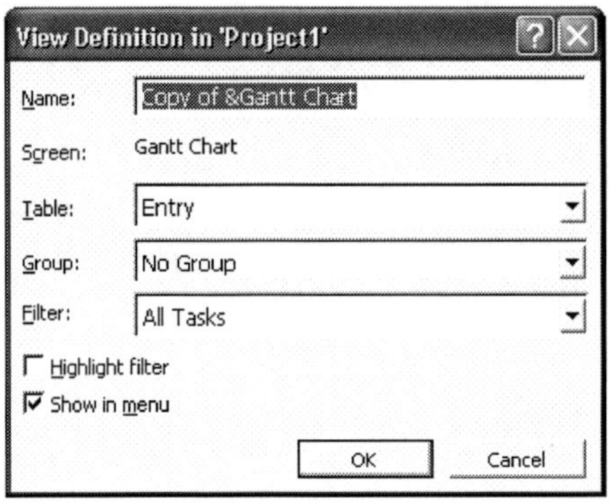

- The **Screen:** may not be changed in an existing or copied view and the screen option is shown as gray. See picture above.
- Change the name and any other parameters and then click on OK to create the **View**.

13.3.4 Copying a View To and From Another Project

Views may be copied from one project to another with the **Organizer** function. This may be accessed by selecting:

- Organizer... from the **More Views** form, or
- **Tools**, **Organizer…**, and selecting the **Views** tab.

13.4 Tables

A table selects and formats the columns displayed with a view. **Tables** are covered in the **FORMATTING THE DISPLAY** chapter. A table is applied to some **Views**. You may change the table applied to a view:

- When the view is active and you assign a different table, or
- By editing the **View**.

There are both Task and Resource tables and these are applied to the relevant view.

When you format a table by adding or removing columns, etc., you are editing the current table. These changes will appear when the table is next applied, even if it is associated with a **View** that is different from the current View. It is therefore recommended that you assign one Table to only one View.

13.4.1 Applying a Table

A **Table** may be applied by:

- Selecting **View**, **Table:** and selecting from the list in the menu, or

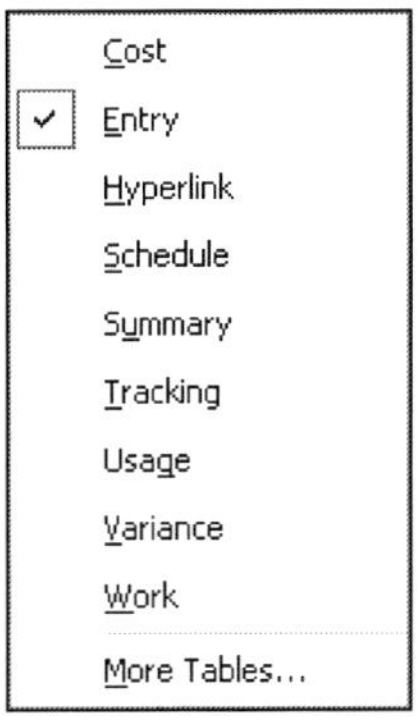

- Selecting **View**, **Table:**, **More Tables…** and selecting a table from the list and clicking on 

, or

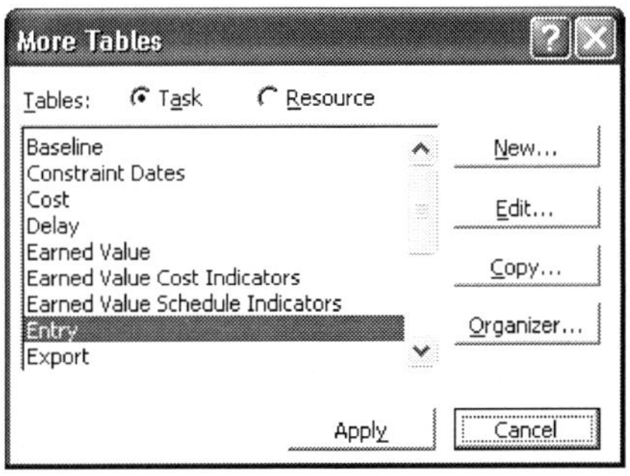

- You may display a menu with the table options by clicking the **Select All** button, the box above the row 1 number, then right-clicking the mouse to display a sub-menu.

13.4.2 Create and Edit a Table

A **Table** may be created or edited by:

- Selecting **View**, **Table:**, **More Tables…** to open the **More Tables** form:

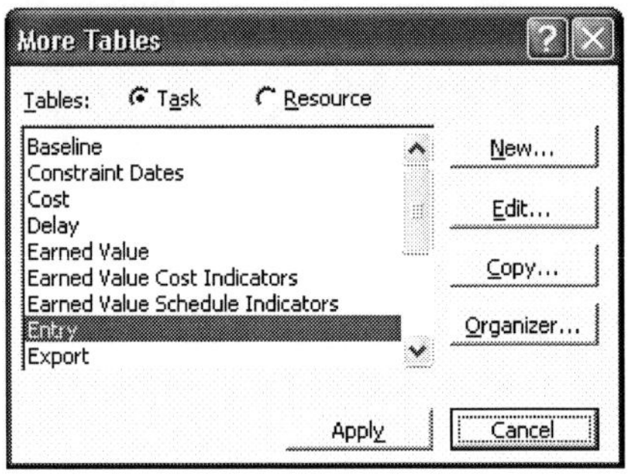

- Clicking on New... to create a new table, Edit... to edit an existing table or Copy... to create a copy of an existing table. All these buttons open the **Table Definition** form shown below:

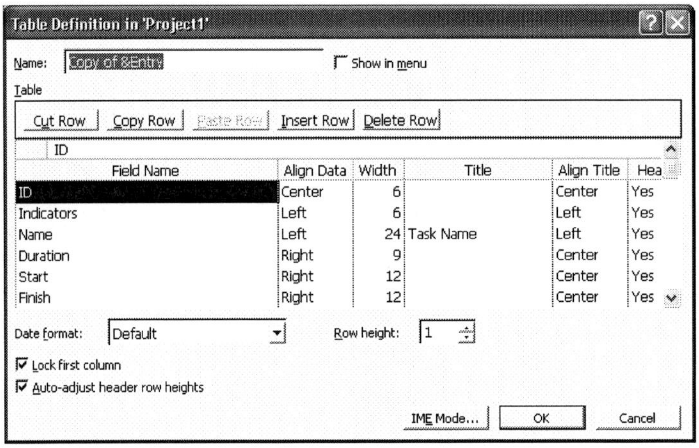

- The functions in this form are similar to those in many other forms. The functions that are unique to this form are listed below:
  - **Date format:** – Changes the format of the dates in this project and table only.
  - **Row height:** – Allows you to specify the row height for this table.
  - **Lock first column** – Ensures the first column is always displayed when scrolling to the right.
  - **Auto-adjust header row heights** – Automatically adjusts the header height when the width of the column is adjusted so the column text wraps.
- Organizer... – Allows a table to be copied from one schedule to another.
- Apply – Applies the table to a view.

13.5 Details Form

Details forms are the third level of formatting that may be assigned in some views. An extensive list of the Details forms is outlined in **APPENDIX 1 – SCREENS USED TO CREATE VIEWS**.

Each view has a number of Details options, which tends to make this aspect of Microsoft Project 2003 difficult for all levels of users.

The **Details** forms may be selected by:

- Selecting **F<u>o</u>rmat**, **<u>D</u>etails**, or
- Right-clicking in the active pane to open the menu.

Example of Details forms:

- Task Form with Resource Costs Details form:

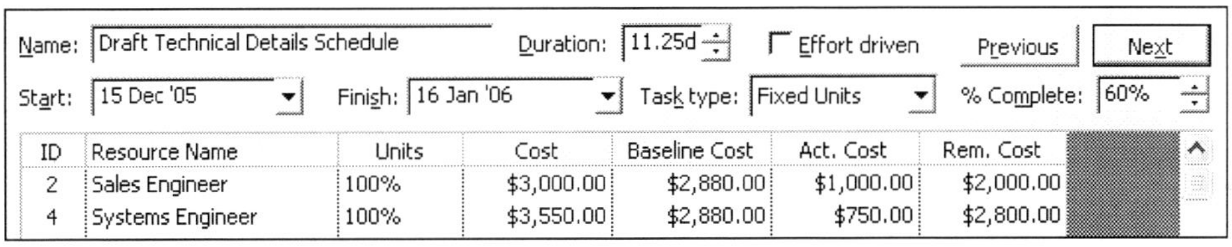

- Task Details with Predecessors and Successors Details form:

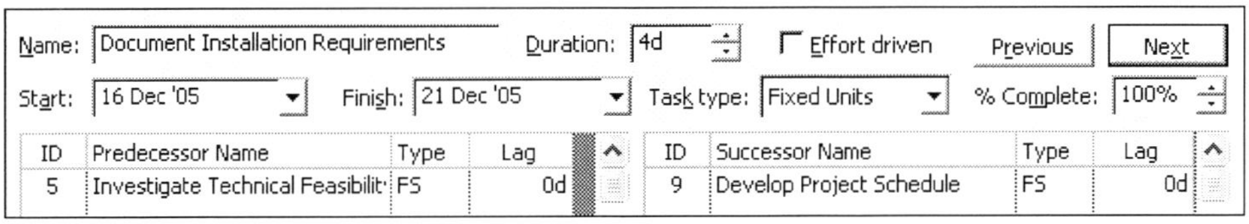

WORKSHOP 11

Organizing your Data

Preamble

Having completed the schedule you may report the information with different views.

Assignment

Display your project in the following formats, noting the different ways you may represent the same data. Open your **OzBuild Bid** project from the previous workshop to complete the following exercise.

1. Display the **Calendar** view and scroll through a few months and see the holidays. This view only displays the Standard calendar.
2. Display the **Network Diagram** view and zoom to 50%, then scroll around the schedule.
3. Display the **Relationship Diagram** view and scroll around the schedule by double-clicking on tasks.
4. Display the **Gantt Chart** view and split the screen.
5. Apply the **Task Details** view in the bottom screen.
6. Apply each one of the following **Details** forms to the **Task Details** view in the bottom screen.

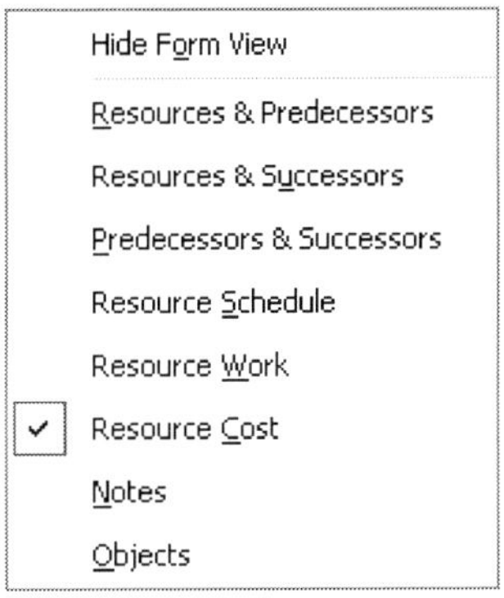

7. Apply the following Tables to the **Gantt Chart** view in the top pane:
   - ➢ Entry, then
   - ➢ Tracking, then
   - ➢ Schedule.
8. Close the split.

14 PRINTING AND REPORTS

You are now at the stage to print the schedule so people may review and comment on it. This chapter will examine one of the many options for printing your project schedule.

There are two tools available to output your schedule to a printer:

- The **Printing** function prints the data displayed in the current View.
- The **Reporting** function prints reports, which are independent of the current View. Microsoft Project 2003 supplies a number of predefined reports that may be tailored to suit your own requirements. Reports will not be covered in detail in this book.

It is recommended that you consider using a product such as Adobe Acrobat to output your schedule in pdf format. You then will be able to email high quality outputs that recipients may print or review on screen without needing a copy of Microsoft Project 2003.

14.1 Printing

When a View is split, only the active View may be printed. The active view has a blue bar down the left-hand side of the screen. Views created from **Forms** (for example, the Task Form) may be not printed so the printing options are shown in gray.

Print settings are applied to the individual Views and the settings are saved with the currently displayed View.

There are three commands used when printing:

- **File**, **Page Setup...**
- **File**, **Print Preview**
- **File**, **Print...** or **Ctrl+P**

Each of these functions will be discussed only for printing the Gantt Chart. Printing all other Views is a similar process.

Some Views will have additional options and others reduced functionality. These other options should be easily mastered once the basics covered in this chapter are understood.

Microsoft Project is sometimes difficult to print a Gantt Chart on one page. In the authors experience adjusting the timescale so the whole project fits into half the screen before selecting **Print Preview** make this process simpler.

Each time you report to the client or management, it is recommended that you save a complete copy of your project and change the name slightly (perhaps by appending a date to the file name or using a version number) or create a subdirectory for this version of the project. This allows you to reproduce these reports at any time in the future and an electronic copy is available for dispute resolution purposes at a later date. You may save the project into the same directory provided you change the file name.

14.2 Print Preview

To preview the printout, use Microsoft Project 2003's **Print Preview** option:

- Select **File**, **Print Preview**, or
- Click the Print Preview icon on the toolbar to view the printout in the window, or
- Click on the [Preview] button from the **Print** form. The **Print** form is covered later in this chapter, or
- Click on the [Print Preview...] button in the **Page** tab of the **Page Setup** form.

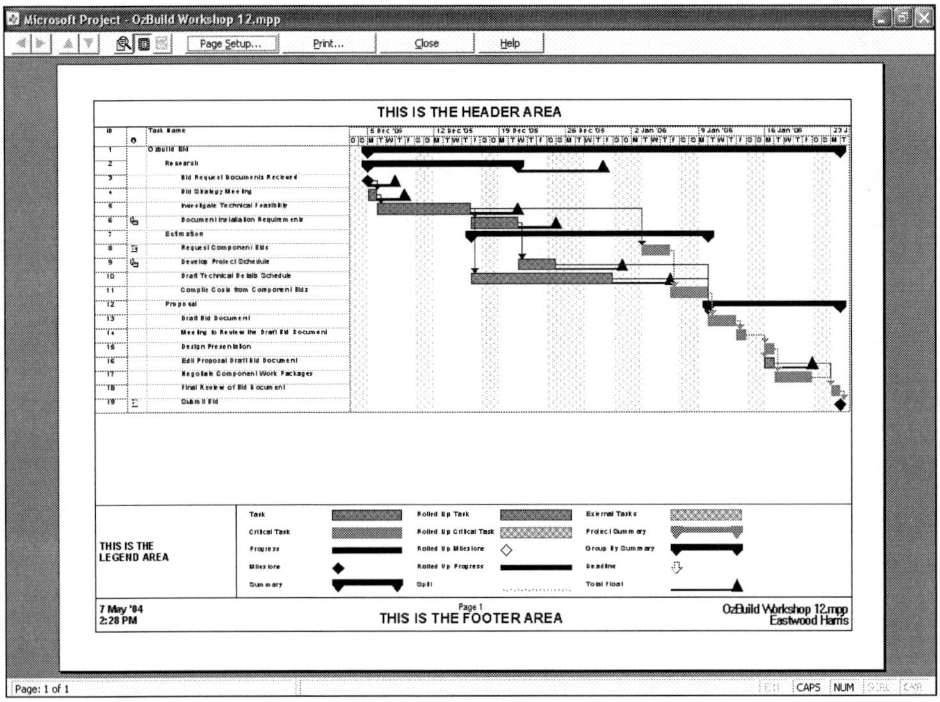

The following paragraphs describe the functions of the icons at the top of the Print Preview screen from left to right:

- The first four icons on the left, ◄ ► ▲ ▼, allow scrolling when a printout has more than one page.
- The magnifying glass zooms in only. You will need to click on the preview screen to zoom out again. You may also click on the preview screen to zoom in.
- The next two icons, and , display one or all pages, respectively.
- The next four are self-explanatory buttons labeled [Page Setup...], [Print...], [Close] and [Help].

14.3 Page Breaks

Pages breaks may be inserted above a task by highlighting the task and selecting **Insert**, **Page Break…**.

Pages breaks may be removed by highlighting the task, or tasks, or all task and selecting **Insert**, **Remove Page Break…**.

14.4 Page Set-up

To open the **Page Setup** form:

- Click the [Page Setup...] button on the **Print Preview** toolbar, or
- Select **File**, **Page Setup…** to display the **Page Setup** form:

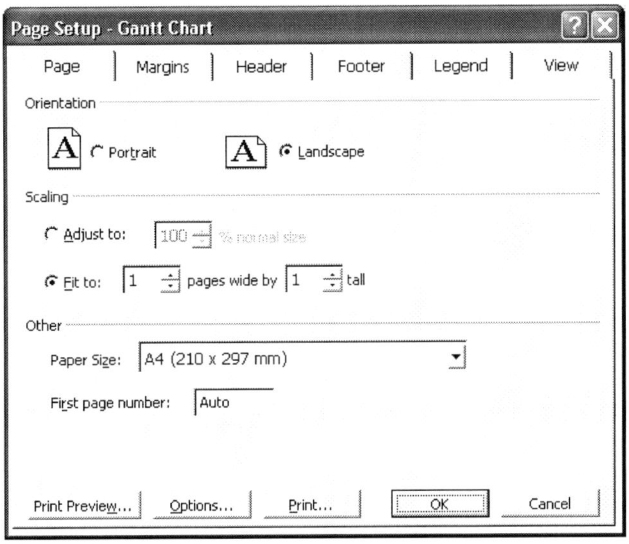

The **Page Setup** form contains the following tabs:

- Page
- Margins
- Header
- Footer
- Legend
- View

Depending on the View being printed, some options in the **Print Setup** form will be unavailable and shown in gray.

The [Options...] button opens the printer setup/options form.

14.4.1 Page Tab

The Microsoft Project 2003 options in the **Page** tab are:

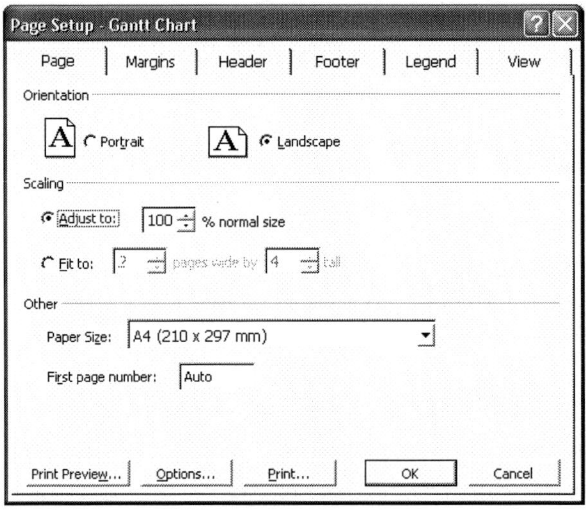

- **Scaling** allows you to adjust the number of pages the printout will fit onto:
  - **Adjust to:** – Allows you to choose the scale of the printout that both the bars and column text are scaled to. Microsoft Project 2003 will calculate the number of pages across and down for the printout.
  - **Fit to:** – Allows you to choose the number of pages across and down and Microsoft Project 2003 will scale the printout to fit.
- The **First page number:** – Allows you to choose the first page number of the printout. This is useful when enclosing the printout as a follow-on to another document.
- Pages are numbered down first and then across:

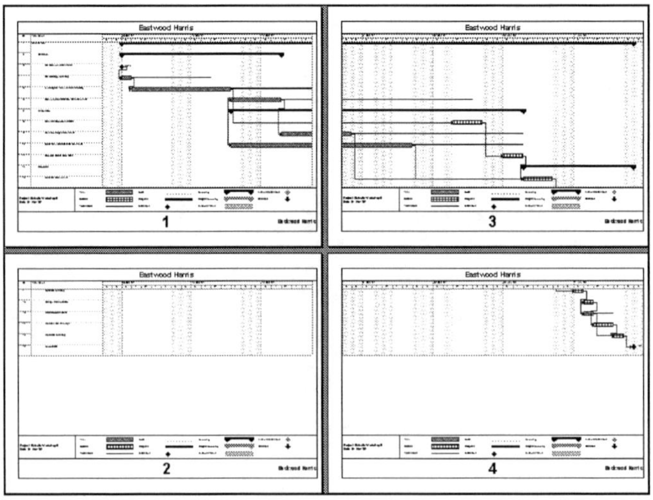

14.4.2 Margins Tab

With this option you may choose the margins around the edge of the printout.

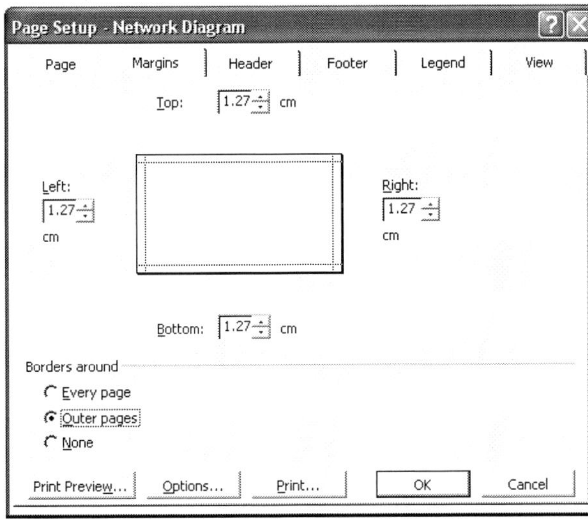

- Type in the margin size around the page. It is best to allow a wider margin for an edge that is to be bound or hole punched. 1" or 2.5cm is usually sufficient. The units of measurement, in. or cm, are adopted from the operating system **Control Panel**, **Regional and Language Option** settings.
- Microsoft Project 2003 will print a border around the Gantt Chart bars and Columns. **Borders around** places a border line around the outside of all text in the Headers and Footers. There are three options:
  - **Every page** – This places a border around every page.
  - **Outer pages** – This capability is only available with a **Network Diagramming** View. It allows you to join all the pages into one large printout with a border only on the outside edge of all the pages once they are joined up.
  - **None** – Does not place a border on any sheets of the printout.

14.4.3 Header and Footer Tabs

Headers appear at the top of the screen above all schedule information and footers are located at the bottom, below the **Legend** if it is displayed. Both the headers and footers are formatted in the same way. We will discuss the setting-up of footers in this chapter.

Click on the **Footer** tab from the **Page Setup** form. This will display the settings of the default footers and headers. You should modify the output to suit your requirements.

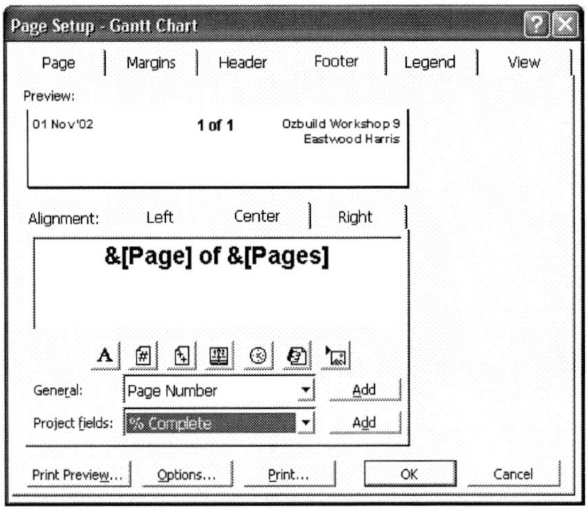

- **Preview:** – The box at the top of the form shows how your Footer will be displayed.
- Text may be placed in all three positions in the Footer and Header – on the Left, Center and Right. Click on the required tab to the right of **Alignment:** to select the alignment position and insert new text or remove existing text in the box below.
- Text may be added by using a number of methods:
  - Freeform text may be typed into the footer box below **Alignment:**
  - You may click one of these icons, **Page Number**, **Total Page Count**, **Current Date**, **Current Time**, **File Name** or **Insert Picture** (for example a corporate logo), to add data in the footer.
  - To insert Microsoft Project field-type information to your footer, click on Add to insert the field into the footer. You may select from the drop-down box to the right of **General:** or **Project fields:**
  - To format the text you must first highlight the text and then click on the **Format Text Font** icon to open the **Font** form.
  - Print Preview... – Returns you to the **Print Preview** form.
  - Options... – Opens the **Printer Properties** form.
  - Print... – Opens the **Print** form.
  - OK – Accepts the changes.
  - Cancel – Cancels any recent changes that have not yet been saved.

14.4.4 Legend Tab

The **Legend Text** is printed to the left of the **Legend**:

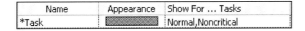

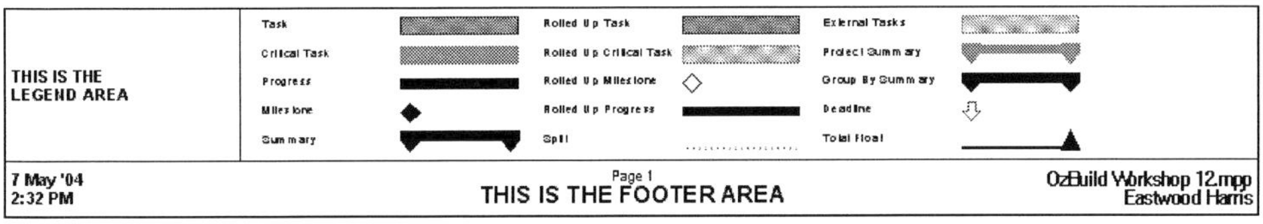

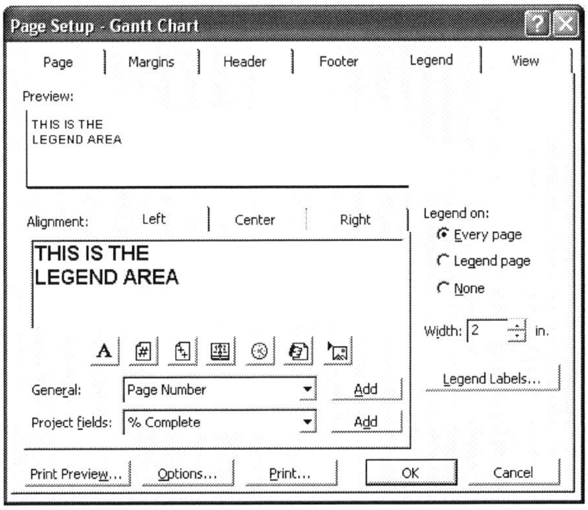

The **Legend Text** is formatted in the same way as the **Header** and **Footer**.

The **Legend** has three additional options not available in the **Header** and **Footer**.

- Click on the radio button below **Legend on:**
  - ➢ **Every page** will print the legend at the bottom of every page.
  - ➢ **Legend page** will print the legend on a separate page with no detailed schedule data, bars, or columns.
  - ➢ **None** will not print a legend.
- **Width:** sets the width of the **Legend Text**.
- Legend Labels... Opens the **Font** form for formatting the font of the text in the legend next to each of the bars and milestones.
- To hide a bar type, type an * in front of the description in the **Bar Styles** form and this bar will not be displayed in the Legend:

Name	Appearance	Show For ... Tasks
*Task		Normal,Noncritical

14.4.5 View Tab

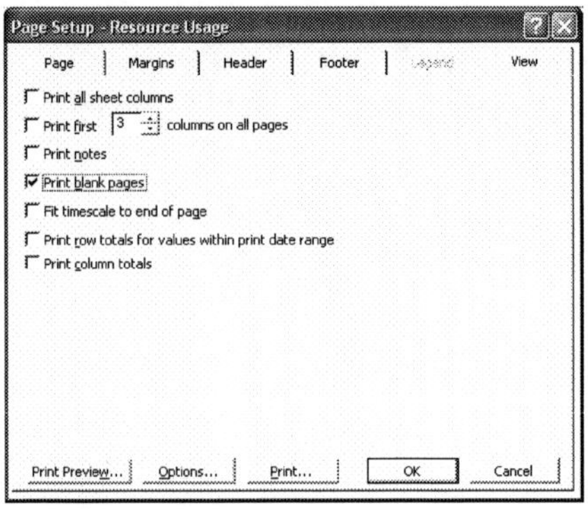

The View tab has five options:

- **Print all sheet columns:** – This option applies to the top-most horizontal pages only when there is more than one page across a printout.
  - ➢ When checked, this option will print all the sheet columns displayed by the current table in the Gantt Chart, even if the columns are hidden by the vertical divider in the normal view.
  - ➢ When unchecked, this option will only print the visible columns.
- **Print first … columns on all pages** – Applies to all pages. The greater of the number of columns between this option and the options above will be printed on the first page.
  - ➢ **When checked**, this will allow repetitiously printing the selected number of columns on all pages.
  - ➢ **When unchecked**, this will not print any columns on the second and subsequent pages across.
- **Print notes** will allow the printing of any **Notes** such as **Task Notes** on a separate page.
- **Print blank pages** will allow the printing of all pages in the Gantt Chart, even when there are no bars.
- **Fit timescale to end of page** will extend the timescale and associated bars so they fit to the end of a page:
  - ➢ Option is unchecked, the bars stop in the middle of a page:

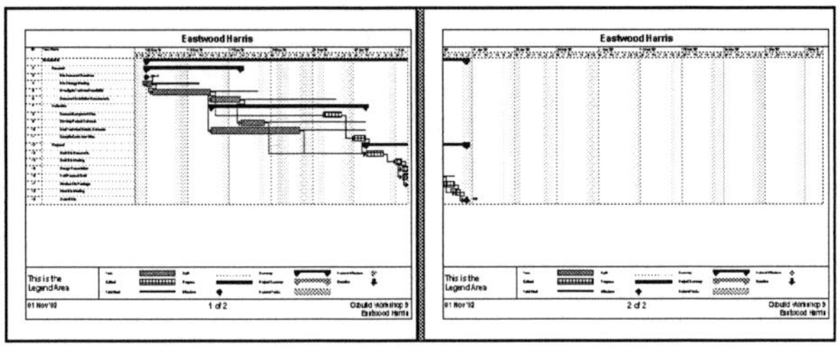

  - ➢ Option is checked, bars and timescale extend to the end of a page:

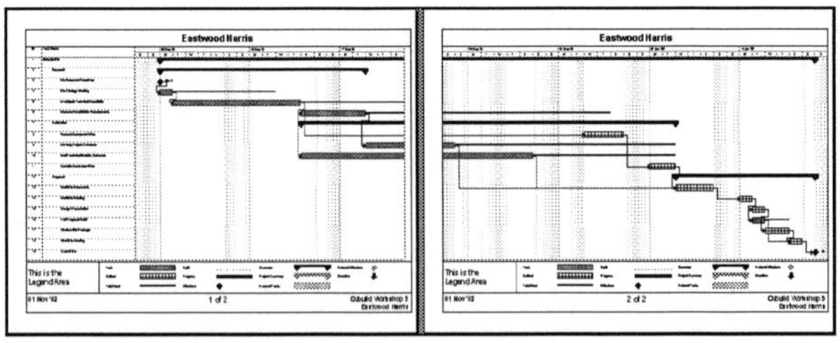

- **Print row totals for values within date range** and **Print column totals** options become available when tabular data is selected for printing such as a Resource Sheet view.

14.5 Print Form

The **Print** form may be opened by:

- Selecting **File**, **Print…**, or
- Executing the keystrokes **Ctrl+P**, while
- Clicking on the  print icon normally found on the Standard toolbar will automatically print out the document .

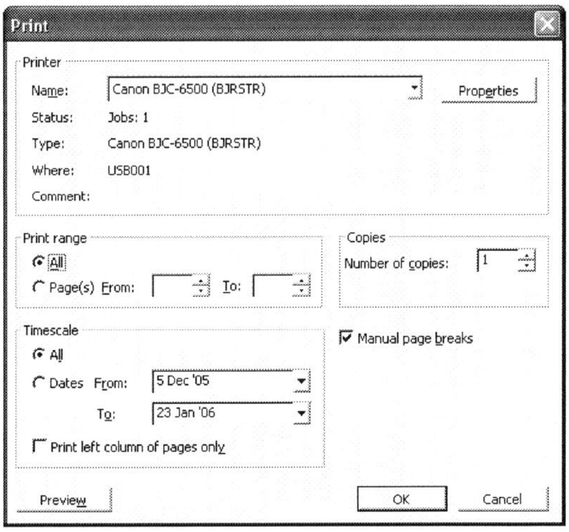

The following options are available:

- The **Printer** options are only available when this form is accessed from the main menu and are shown in gray when accessed from **Print Preview**.
- **Print range** allows you to choose the pages to be printed.
- **Copies** specifies the number of copies to be printed.
- **Timescale** allows specifying a Gantt Chart date range that will be printed. It does not filter out tasks. It is used to reduce the range of the Gantt Chart that will be printed. Unlike earlier versions of Microsoft Project, this setting is saved with a View when the project is saved.
- **Print left column of pages only** will print only the left pages of a printout that are more than one page wide.
- **Manual page breaks:**
  - Manual page breaks are inserted by highlighting the row above where a page break is required. Then, select **Insert**, **Page Break**. A dotted line will indicate the location of the manual page break.
  - To remove a manual page break, highlight the row above where there is a page break and select **Insert**, **Remove Page Break**.
  - The **Manual page breaks** check box in the **Print** form must be checked to acknowledge manual page breaks in the printout.

14.6 Reports

Select **View**, **Reports...** to open the **Reports** form which displays six icons.

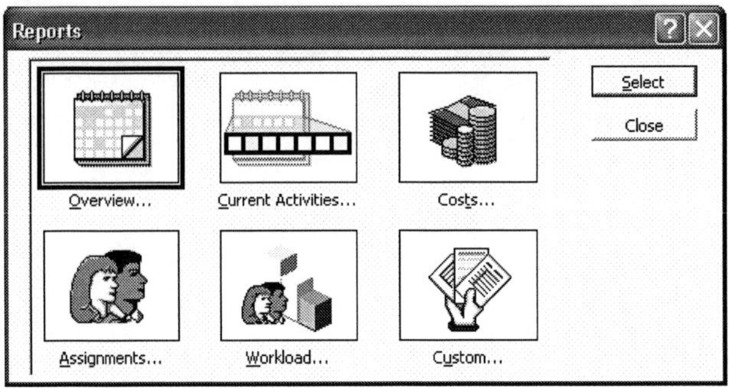

- Select a Report option by:
  - Double-clicking on any one of the icons, or
  - Clicking on an icon and clicking on the Select button.
- This will open up a menu of reports under the chosen heading. The picture below displays the **Current Activities...** menu. The remaining five headings are done similarly and therefore are not explained in detail:

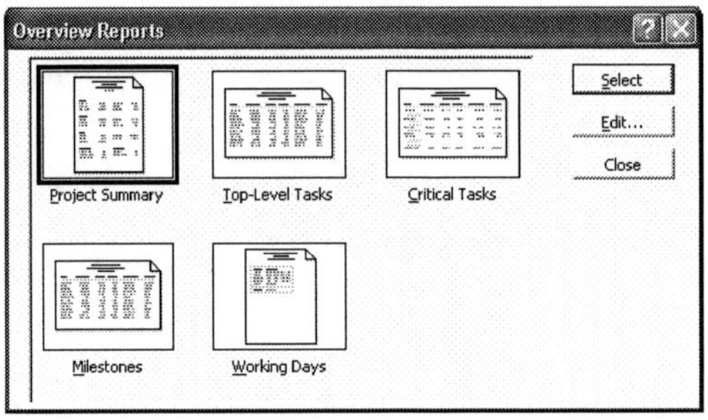

Double-click on any report and it will be sent to **Print Preview.** Here the print settings may be edited prior to printing.

Select **Critical Tasks** and click on the Edit... button to open the **Task Report** form. Some of the report parameters may be edited here:

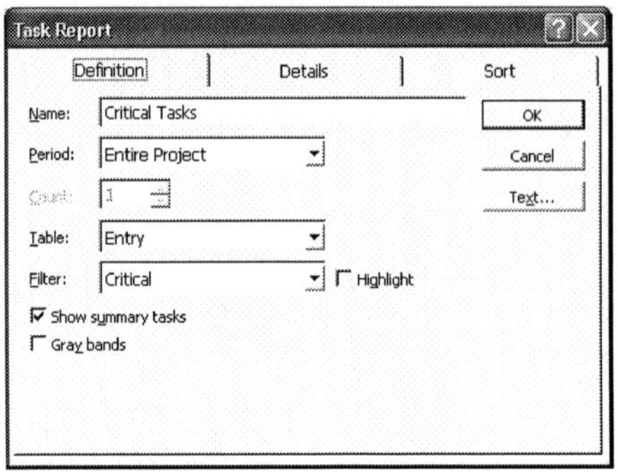

WORKSHOP 12

Reports

Preamble

We want to issue a report for comment by management.

Assignment

Open your **OzBuild Bid** project from the previous workshop to complete the following steps:

1. Apply the **Gantt Chart** view.
2. Select the top pane and apply the Entry Table.
3. Delete the Resources Column.
4. Adjust the columns to the best fit.
5. We wish to fit all the activities on one A4 or Letter size landscape page by:
   - Adjusting the middle tier unit of timescale to **Months** and minor bottom tier to **Weeks** with **Label** of **27,3...** and **Size** of **100%**.
   - Selecting Print Preview and showing all data columns.
   - Setting the page size to fit to 1 page wide by 1 page tall.
6. Hide the Legend by selecting **Legend on: None**.
7. Add the Project Title and Company Name in the footer with a font of Arial Bold 14.
8. Compare your result with the picture on the next page.

ANSWERS TO WORKSHOP 12

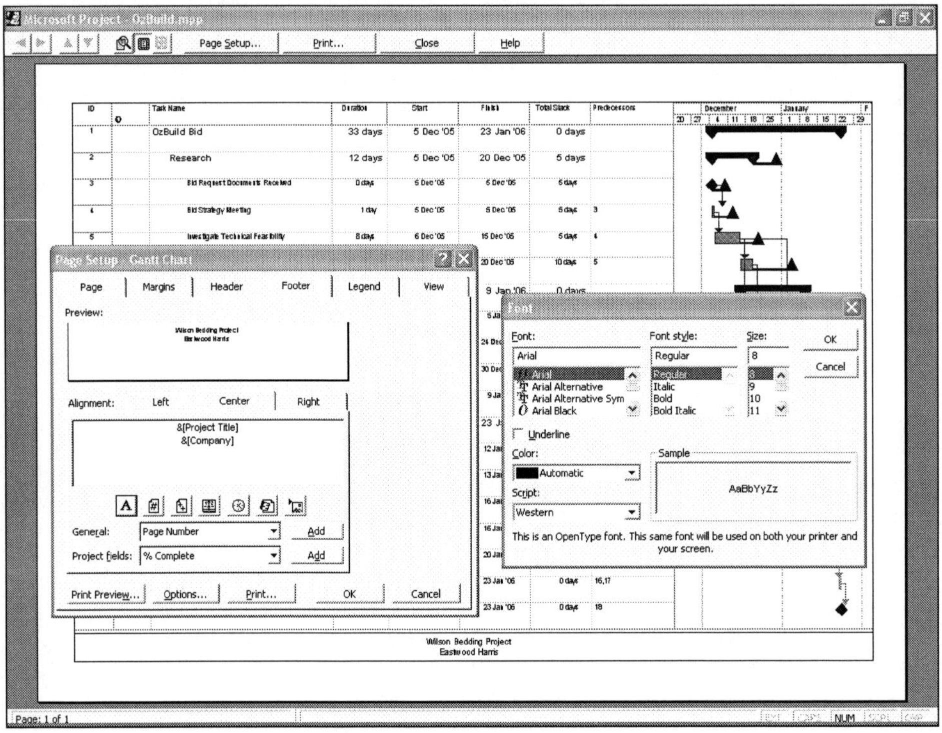

15 TRACKING PROGRESS

The process of tracking progress is used once you have completed the plan, or have completed sufficient iterations to reach an acceptable plan, and the project is progressing. Now the important phase of regular monitoring begins. Monitoring is important to help catch problems as early as possible, and thus minimize the impact of problems on the successful completion of the project.

The main steps for monitoring progress are:

- Set the **Baseline Dates**, also known as **Target Dates**. These are the dates against which progress is compared.
- Record or mark-up progress as of a specific date, often titled the **Data Date**, **Status Date**, **Current Date** and **As-Of-Date**.
- **Update** or **Status** the schedule with **Actual Start** and **Actual Finish** dates where applicable, and adjust the task durations and percent complete.
- Compare and **Report** actual progress against planned progress and revise the schedule, if required.

By the time you get to this phase you should have a schedule that compares your original plan with the current plan, showing where the project is ahead or behind. If you are behind, you should be able to use this schedule to plan appropriate remedial measures to bring the project back on target.

This chapter will cover the following topics:

Topic	Menu Command
• Setting the **Baseline**	Select **Tools**, **Tracking**, **Save Baseline...**
• Recording Progress	Guidelines on how to record progress.
• **Current Date** and **Status Date**	May be edited from the **Project**, **Project Information...** form.
• Updating the Schedule	Select **Tools**, **Tracking**, **Update Project...**
• Move the **Incomplete Work** of an **in-progress** task into the future	Select **Tools**, **Tracking**, **Update Project...**
• Updating the Tasks	Select **Tools**, **Tracking**, **Update Tasks...**

15.1 Setting the Baseline

Setting the Baseline copies the following information into new fields in the existing file:

- **Early Start** and **Finish** dates into fields titled **Baseline Start** and **Baseline Finish**.
- **Original Duration** into the **Baseline Duration**.
- Each resource's **Costs** and **Work** into **Baseline Costs** and **Baseline Work** (the number of hours) when resources are assigned to tasks.

Once the Baseline is set you will be able to compare your progress with your original plan. You will be able to see if you are ahead or behind schedule and by how much.

Some important points you should understand before setting the Baseline:

- The Baseline Dates should be established before you status the schedule for the first time.
- The Baseline Dates and Duration fields are not calculated fields and may be edited in columns and forms. Caution should be used before changing a Baseline Date or Duration, since it is the basis for all project deviation measurements.
- Setting the Baseline Date does not store the logic, float/slack times, or constraints.
- A Baseline is normally applied to all tasks irrespective of their Outline levels.

15.1.1 Setting Baseline Dates

To set the Baseline select **Tools**, **Tracking**, **Save Baseline…** to display the **Save Baseline** form.

- The **Save baseline** option copies the **Early Start** and **Finish** dates into the **Baseline Start** and **Finish** date fields respectively.
- You may highlight some tasks before opening the **Save Baseline** form and clicking on the **Selected tasks** radio button to set the Baseline for the specified tasks only.

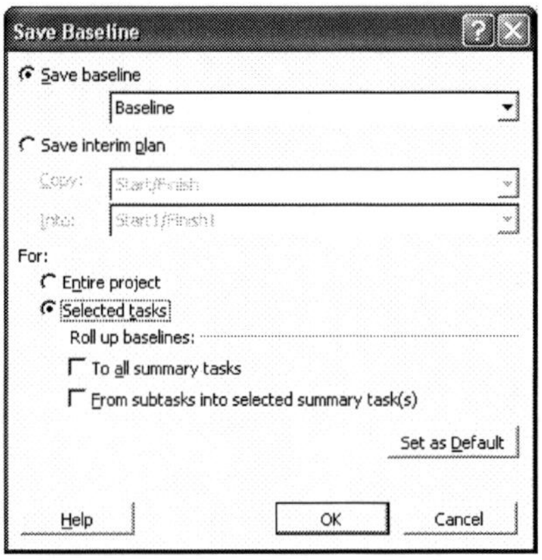

This operation will overwrite any previous Baseline settings.

The **Save Baseline** form shows the default settings, which selects **All Tasks** and the **Early dates** which are titled **Start/Finish**. This is the normal method of setting the Baseline.

After establishing the Baseline, you can input the progress data without fear of losing the original dates. This will also enable you to compare the progress with the original plan.

15.1.2 Setting an Interim Baseline

This was a new feature in Microsoft Project 2002.

It may be necessary to save an interim baseline. This may occur when the scope of a project has changed and a new baseline is required to measure progress against, but at the same time you may also want to keep a copy of the original baseline. This process may also be used to display the effect of scope changes on an original project plan by comparing one baseline with another.

There are two types of data fields to save interim baselines:

- Using one of the 10 additional baselines titled **Baselines 1** to **10**. These Baselines will save start date, finish date, work and cost information.
- Using one of the 10 sets of **Start Date**, **Finish Date** and **Duration** fields. This function will only save the start and finish date information. This is termed an **Interim Plan** by Microsoft Project 2003.

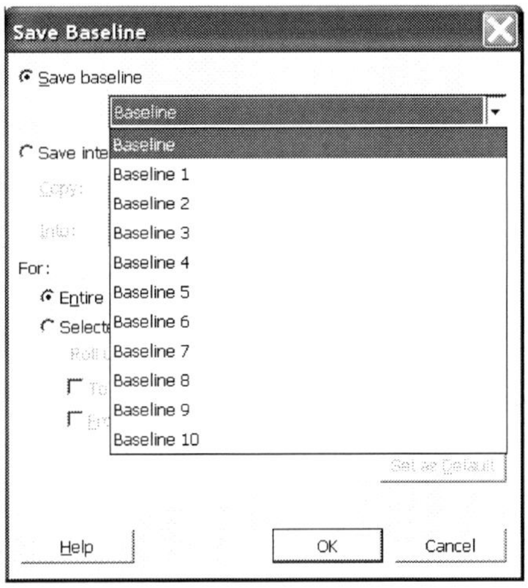

To set the Interim Baseline or Interim Plan, select **Tools**, **Tracking**, **Save Baseline...** to display the **Save Baseline** form, select either:

- **Save baseline** to save the current schedule to one of the 11 Baselines, or
- **Save interim plan** then:
  - Select the dates you want to copy from using the **Copy:** drop-down box, and then
  - Select where you want to copy the dates to using the **Into:** drop-down box.

The **Baseline** data may be reviewed in some Views such as the **Task Details Form** and in columns. You will be able to display the **Baseline 1** to **10** and **Interim Plan** dates and durations in columns and as a bar on the Gantt Chart but not in the forms. Therefore, it is recommended that the current baseline be saved as **Baseline** since the data is more accessible that way. Previous baselines are copied to **Baselines 1** to **10**.

Date fields may be renamed with the **Customize Fields** form. Select **Tools**, **Customize**, **Fields...** to open this form.

15.1.3 Summary Task Interim Baseline Calculation

This is a new function to Microsoft Project 2002. When a Detailed task is moved from one Summary task to another, the Detailed task Baseline dates are moved with the task. It is possible, however, that the original Summary Baseline data will no longer be valid if the Detailed task that has been moved established either the start or the finish date of the Summary task.

When a schedule has costs or work then any movement of a detailed task from one summary task to another summary task should result in a change to the baseline value of the summary tasks. The Costs and Work Baseline calculations are covered in the **STATUSING PROJECT WITH RESOURCES** chapter, paragraph **20.9 Summary Task Interim Baseline Calculation.**

In the picture below, the task Detailed 2.3 was moved from below Detailed 2.2 to below Detailed 1.3 and the baseline of tasks Parent 1 and Parent 2, the lower bar on each task, are no longer an accurate summary of the detailed tasks. When the task was moved in the example below, the option **Autolink inserted or moved tasks** found under **Tools**, **Options...**, **Schedule** tab is unchecked.

	Task Name	Duration	Start	Finish
1	**Parent 1**	**12 days**	**17 Oct '05**	**1 Nov '05**
2	Detail 1.1	2 days	17 Oct '05	18 Oct '05
3	Detail 1.2	2 days	19 Oct '05	20 Oct '05
4	Detail 1.3	2 days	21 Oct '05	24 Oct '05
5	Detail 2.3	2 days	31 Oct '05	1 Nov '05
6	**Parent 2**	**4 days**	**25 Oct '05**	**28 Oct '05**
7	Detail 2.1	2 days	25 Oct '05	26 Oct '05
8	Detail 2.2	2 days	27 Oct '05	28 Oct '05

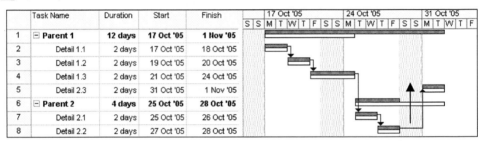

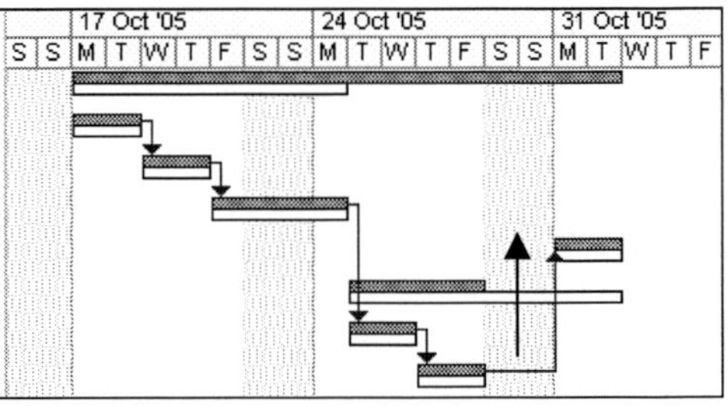

A new function to Microsoft Project 2002 may recalculate the Summary Task Baseline data (but not Interim Plan dates) when a Detailed task is moved by using the **Roll up baselines** option in the **Save Baseline** form:

- **Entire project** – Will recalculate the entire project baseline and recalculate all Summary tasks. This is effectively resetting the baseline.
- **Selected tasks: - Roll up Baselines:** – This has two options which operate on selected tasks:
  - **To all summary tasks** – This option resets the baseline of selected tasks to the current schedule. This does not reset the summary tasks baseline to reflect the new baseline of detailed tasks.
  - **From subtasks to selected summary task(s)** – This option will allow a summary task baseline to be updated to reflect the original baseline of moved tasks.

In the example below a new task Detail 1.4 was added below the Parent 1 task.

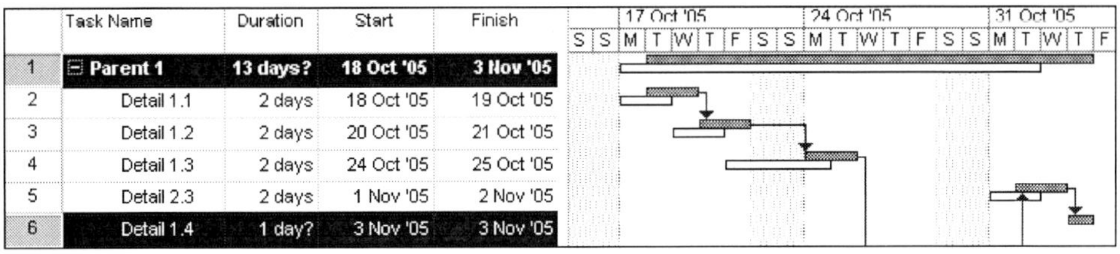

Parent 1 and Detail 1.4 were highlighted and the option **From subtasks to selected summary task(s)** applied. The baseline of Parent 1 now accurately reflects the detailed tasks which it summarizes. See below:

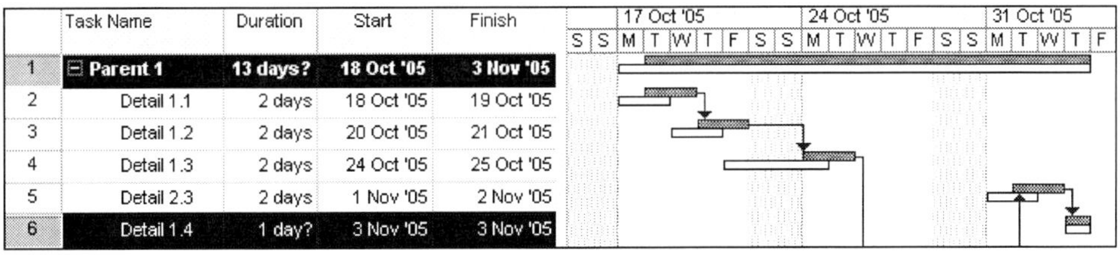

It is recommended to set a revised baseline when a new detailed activity has been added. The new activity and the associated Parent tasks should be highlighted and the option **From subtasks to selected summary task(s)** applied.

15.1.4 Clearing and Resetting the Baseline

You may clear some or all of the Baseline fields using the **Clear Baseline** form. This is opened by selecting **Tools**, **Tracking**, **Clear Baseline…**:

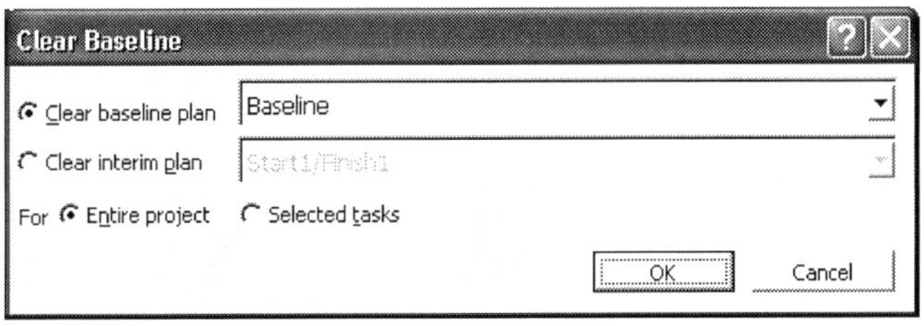

15.1.5 Displaying the Baseline Data

The Baseline dates data may be displayed by:

- Displaying the Baseline columns where the data may be edited, or
- Displaying the dates in a form such as the Resources Details Form. This will only display the **Baseline** data and not for **Baseline 1** to **10**, or
- Show a baseline bar on the Bar Chart.

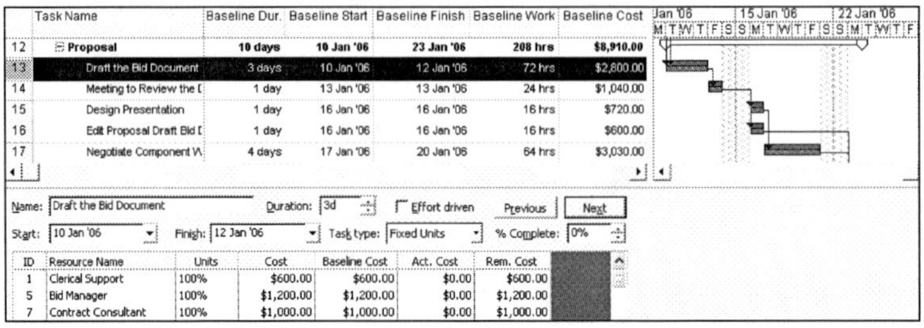

The Baseline Start Dates, Finish Dates and Durations may be edited and are not linked by logic. Therefore, a change to a Baseline Duration will not affect either the Baseline Start Date nor the Baseline Finish date.

15.2 Practical Methods of Recording Progress

Normally a project is statused once a week, bi-weekly, or monthly. Very short projects could be statused daily or even by the shift or hour. As a guide, a project would typically be statused between 12 and 20 times. Progress is recorded on or near the **Data Date** and the scheduler updates the schedule upon the receipt of the information.

The following information is typically recorded for each task when statusing a project:

- The task start date and time if required,
- The number of days/hour the **Task** has to go or when the task is expected to finish,
- The percentage complete, and
- If complete, the task finish date and time, if required.

A marked-up copy recording the progress of the current schedule is often produced prior to updating the data with Microsoft Project 2003. Ideally, the mark-up should be prepared by a physical inspection of the work or by a person who intimately knows the work, although that is not always possible. It is good practice to keep this marked-up record for your own reference at later date. Ensure that you note the date of the mark-up (i.e. the data date) and, if relevant, the time.

Often a Statusing Report or mark-up sheet, such as the one below, is distributed to the people responsible for marking up the project progress. The marked-up sheets are returned to the scheduler for data entry into the software system.

The View, such as the one below, may have a filter applied. In this case, only tasks in-progress or due to start in the next few weeks are displayed. A manual page break could be placed at each responsible person's band, and when the schedule is printed each person would have a personal listing of their tasks that are either in progress or due to commence. This is particularly useful for large projects.

	Task Name	Early Start	Act. Start	Early Finish	Act. Finish	% Comp.	Act. Dur.	Rem. Dur.
	⊟ Responsibility: Angela Low	**02 Jan '02**	**NA**	**08 Jan '02**	**NA**	**0%**	**0 days**	**5 days**
8	Request Component Bids	02 Jan '02	NA	04 Jan '02	NA	0%	0 days	3 days
11	Compile Costs from Bids	07 Jan '02	NA	08 Jan '02	NA	0%	0 days	2 days
	⊟ Responsibility: Carol Peterson	**03 Dec '01**	**NA**	**19 Jan '02**	**NA**	**0%**	**0 days**	**32 days**
3	Bid Document Received	03 Dec '01	NA	03 Dec '01	NA	0%	0 days	0 days
14	Draft Bid Meeting	14 Jan '02	NA	14 Jan '02	NA	0%	0 days	1 day
19	Submit Bid	19 Jan '02	NA	19 Jan '02	NA	0%	0 days	0 days
	⊟ Responsibility: David Williams	**03 Dec '01**	**NA**	**18 Jan '02**	**NA**	**0%**	**0 days**	**32 days**
4	Bid Strategy Meeting	03 Dec '01	NA	03 Dec '01	NA	0%	0 days	1 day
13	Draft Bid Documents	09 Jan '02	NA	11 Jan '02	NA	0%	0 days	3 days
15	Design Presentation	15 Jan '02	NA	15 Jan '02	NA	0%	0 days	1 day
18	Final Bid Meeting	18 Jan '02	NA	18 Jan '02	NA	0%	0 days	1 day
	⊟ Responsibility: Melinda Young	**15 Jan '02**	**NA**	**17 Jan '02**	**NA**	**0%**	**0 days**	**3 days**
16	Edit Proposal Draft	15 Jan '02	NA	15 Jan '02	NA	0%	0 days	1 day
17	Finalise Bid Package	16 Jan '02	NA	17 Jan '02	NA	0%	0 days	2 days
	⊟ Responsibility: Scott Morrison	**04 Dec '01**	**NA**	**28 Dec '01**	**NA**	**0%**	**0 days**	**17 days**
5	Investigate Technical Feasibility	04 Dec '01	NA	13 Dec '01	NA	0%	0 days	8 days
6	Document Installation Requirements	14 Dec '01	NA	18 Dec '01	NA	0%	0 days	4 days
9	Develop Project Schedule	19 Dec '01	NA	22 Dec '01	NA	0%	0 days	4 days
10	Draft Technical Details Schedule	14 Dec '01	NA	28 Dec '01	NA	0%	0 days	9 days

Other electronic methods, discussed next, may be employed to collect the data, but irrespective of the method used, the same data needs to be collected.

The previous View has been created by:

- Entering the responsible person in Text 1,
- Renaming Text 1 as Responsibility using **Tools, Customize, Fields...** and
- Grouping by Text 1, which has been renamed **Responsibility**.

There are several methods of collecting the project status:

- By sending a sheet to each responsible person to mark up and return to the scheduler.
- By cutting and pasting the data from Microsoft Project 2003 into another document, such as Excel, and emailing the data to them as an attachment.
- By giving the responsible party direct access to the schedule software to update it. This approach is not recommended, however, unless the project is broken into subprojects. By using the subproject method, only one person updates each part of the schedule.
- Microsoft Central, a companion Microsoft Project product that allows collaborative scheduling, could be implemented. This topic is beyond the scope of this book.

Some projects involve a number of people. In such cases, it is important that procedures be written to ensure that the status information is collected:

- In a timely manner,
- Consistently,
- In a complete manner, and
- In a usable format.

i It is important for a scheduler to be aware that some people have great difficulty in comprehending a schedule. When there are a number of people with different skill levels in an organization, it is necessary to provide more than one method of updating the data. You even may find that you have to sit down with some people to obtain the correct data, yet others are willing and comfortable to email you the information.

15.3 Understanding the Concepts

There are some terms and concepts used in Microsoft Project 2003 that must be understood before we update a project schedule:

15.3.1 Task Lifecycle

There are three stages of a task's lifecycle:

- **Not Started** – The **Early Start** and **Early Finish** dates are calculated from the logic, **Constraints** and **Task Duration**.
- **In-Progress** – The task has an **Actual Start** but is not complete.
- **Complete** – The task is in the past, the **Actual Start** and **Actual Finish** dates have been entered into Microsoft Project 2003, and they override the logic and constraints.

15.3.2 Actual Start Date Assignment of an In-Progress Task

This section will explain how Microsoft Project 2003 assigns the **Early Start** of an **In-Progress** task.

- A task **Duration** is the duration from the **Early Start** or **Actual Start** to the **Early Finish** or **Actual Finish** and is calculated over the **Task Calendar**.
- When an **Actual Start** is entered into the **Actual Start** field, this date overrides the **Early Start date**. The predecessor logic and start date constraints are ignored.
- An un-started task **Actual Start date** is set to equal the **Early Start** date when:
  - A **% Complete** between 1% and 99% is entered,, or
  - An **Actual Duration** is entered, or
  - An **Actual Finish** is entered, or
  - When a % Complete of 100% is entered, then an Actual Start date equal to the Early Start and an Actual Finish date equal to the Early Finish date are both set.

15.3.3 Calculation of Actual and Remaining Durations of an In-Progress Task

- The **Actual Duration** is the worked duration of a task and **Remaining Duration** is the unworked duration of a task.
- **Duration** = **Actual Duration** + **Remaining Duration**. Before a task is commenced the **Actual Duration** is zero and the **Remaining Duration** equals the **Duration** assigned to the task.
- There is an in-built proportional link between **Duration**, **Actual Duration**, **Remaining Duration** and **% Complete**. It is not possible to unlink these fields (as in other scheduling software) and therefore not possible to enter the **Remaining Duration** independently of the **% Complete**.
  - Change the **Duration**, the **Actual Duration** remains constant and the **% Complete**, and the **Remaining Duration** changes proportionally.
  - Change the **% Complete**, the **Duration** remains constant and the **Actual Duration** and the **Remaining Duration** change proportionally.
  - Change the **Actual Duration**, the **Duration** remains constant and the **% Complete** and the **Remaining Duration** changes proportionally.
  - Change the **Remaining Duration**, the **Actual Duration** remains constant and the **% Complete** and the **Duration** changes proportionally.

- When a **% Complete** is assigned, then the **Actual Start** date is set to be equal to the **Early Start** date and durations are recalculated as per the above rules.
- When the **% Complete** is set to 100 or the **Remaining Duration** is set to zero, the **Actual Finish** date is set to the **Early Finish date** and the **Actual Duration is** set to the **Duration**.
- An **Actual Finish** date overrides an **Early Finish** date, and finish date constraints are ignored.

The example below shows three tasks, the first unstarted, the second in-progress and the third complete. You should observe:

- The relationship between the **Duration**, **Actual Duration**, **Remaining Duration** and **% Complete** in each of the tasks, and
- How the **Actual Start** and **Actual Finish** are set.

	Start	Actual Start	Early Finish	Actual Finish	% Complete	Duration	Actual Duration	Remaining Duration
1	18 Oct '05	NA	31 Oct '05	NA	0%	10 days	0 days	10 days
2	18 Oct '05	18 Oct '05	31 Oct '05	NA	25%	10 days	2.5 days	7.5 days
3	18 Oct '05	18 Oct '05	31 Oct '05	31 Oct '05	100%	10 days	10 days	0 days

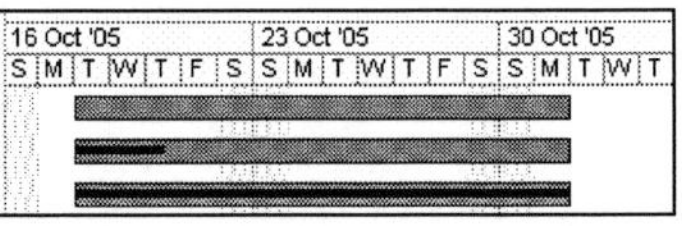

Unlike other scheduling software, it is not possible in Microsoft Project 2003 to unlink the **Remaining Duration** and **% Complete**. This will prove frustrating to some schedulers as you will not be able use the **% Complete** column to represent the amount of work complete when the task is not progressing at a linear rate.

15.3.4 Calculating the Early Finish Date of an In-Progress Task

Retained Logic and Progress Override are not terms used by Microsoft Project 2003 documentation. These terms are used by other scheduling software and are used here to help clarify how Microsoft Project 2003 performs its calculations. In the example below, there are two tasks with a Finish-to-Start relationship:

	Actual Start	Actual Finish	% Complete
1	NA	NA	0%
2	NA	NA	0%

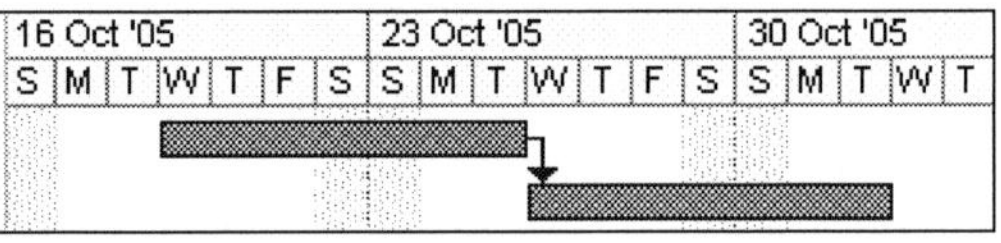

There are two options for calculating the finish date of the successor when the successor task starts before the predecessor task is finished:

- **Retained Logic**. In the example below, the logic relationship is maintained between the predecessor and successor for the unworked portion of the Task, the Remaining Duration, and continues after the predecessor has finished.

	Actual Start	Actual Finish	% Complete
1	19 Oct '05	NA	50%
2	21 Oct '05	NA	10%

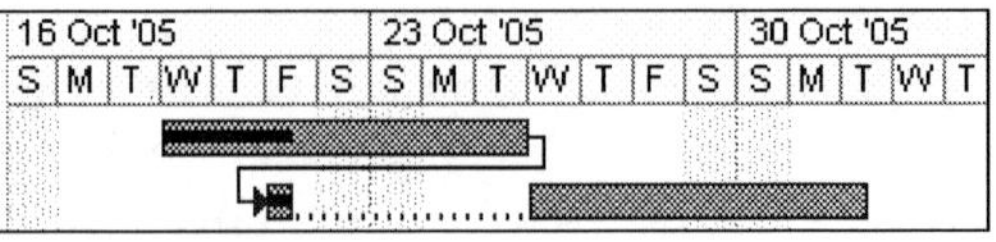

This option will operate when the following conditions are met:

- The Project Option **Split in-progress tasks** is checked, and
- There is an **Actual Start**, and
- A **% Complete** between 1% and 99% is assigned to the successor task.

- **Progress Override**. In the example below, the Finish-to-Start relationship between the predecessor and successor is disregarded, and the unworked portion of the Task, the Remaining Duration, continues before the predecessor has finished:

	Act. Start	Act. Finish	% Comp.
1	19 Oct '05	NA	50%
2	21 Oct '05	NA	10%

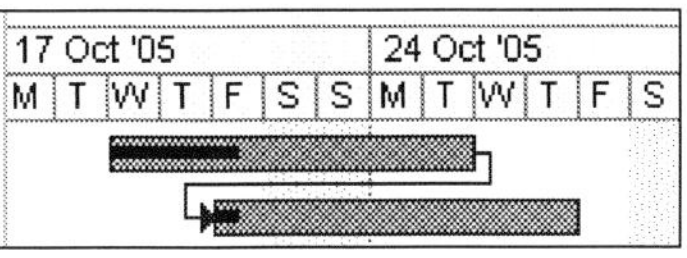

Progress Override option will operate when:

- The Project Option **Split in-progress tasks IS NOT** checked,
  or
- The task has an **Actual Start** and 0% Complete and Project Option **Split in-progress tasks IS** checked.

This function may result in some problems in reporting a schedule when the **Split in-progress tasks** option is used in combination with some Tasks set to 0 % Complete and other tasks set between 1% and 99% Complete. The two examples below are from the same schedule, both with the **Split in-progress tasks** option checked, one with 0% and one with 1%. You will notice the task assigned 0% has an earlier Finish Date than the task assigned 1% complete, which has split.

	Act. Start	Act. Finish	% Comp.
1	19 Oct '05	NA	50%
2	19 Oct '05	NA	0%

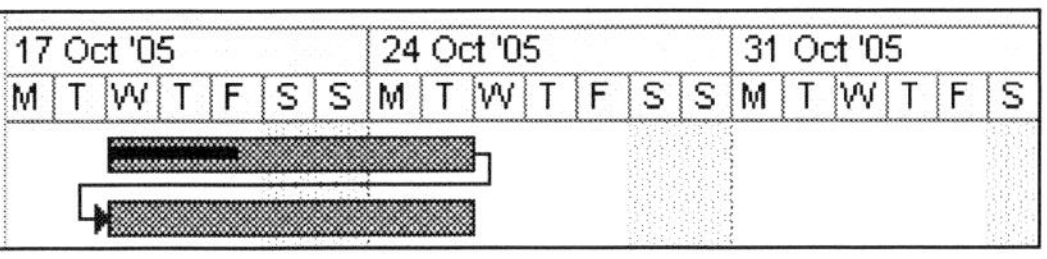

	Act. Start	Act. Finish	% Comp.
1	19 Oct '05	NA	50%
2	19 Oct '05	NA	1%

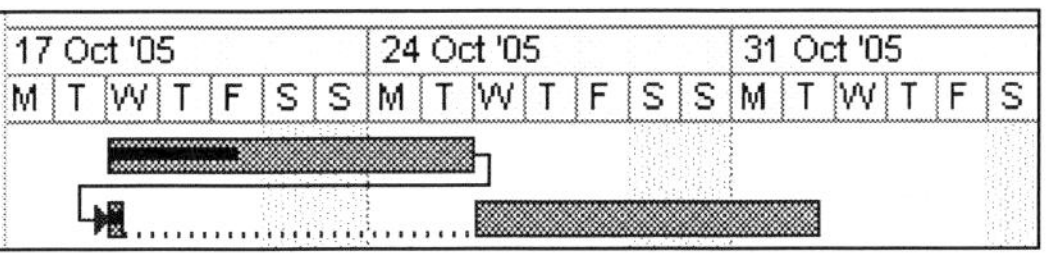

You therefore need to pay careful attention to any warning messages Microsoft Project 2003 presents.

15.3.5 Summary Bars Progress Calculation

Summary bars are not normally statused by entering the **Actual Start** date, **Actual Finish** date, or **% Complete** against them. This is possible in Microsoft Project and is covered later in this chapter in the section titled **Marking Up Summary Tasks.**

This status information is usually entered against the detailed tasks, and the summary tasks inherit the status data from the detailed tasks.

- An **Actual Start** is assigned against a Summary Task when any Child Task has an **Actual Start**.
- A Summary Task's **% Complete** is calculated from the total of all the child tasks' **Actual Durations** divided by the total of all the child tasks' **Durations**.
- A Summary Task's **Actual** and **Remaining Durations** are calculated from the **Duration** and **% Complete**.
- An **Actual Finish** is assigned against a Summary task when all Detail tasks have an **Actual Finish**.

15.3.6 Understanding the Current Date, Status Date and Update Project Date

Microsoft Project 2003 has two project data date fields that may be displayed as vertical lines on the schedule, and these dates may be edited from the **Project**, **Project Information...** form:

- **Current Date** – This date is set to the computer's date each time a project file is opened. It is used for calculating **Earned Value** data when a **Status Date** has not been set.
- **Status Date** – This field is blank by default. When this date is set, it will not change when the project is saved and reopened at a later date. When set, this date overrides the **Current Date** for calculating **Earned Value** data.

It is recommended that the **Status Date** be set and displayed as a vertical line on a progressed schedule, and that the **Current Date** not be displayed, as the **Current Date** represents the date today and does not normally represent any scheduling significance.

The **Update Project Date** may also influence how Microsoft Project 2003 calculates the end date of some activities. This date may not be displayed as a vertical line on the screen but may be used in conjunction with the **Reschedule Uncompleted Work To Start After** function covered later in this chapter.

NEITHER the **Current Date** nor the **Status Date** is used to calculate the **Early Finish** of an **In-Progress** task of a schedule using **F9** or **Automatic Scheduling**, unless one of the following two functions is used:

- The **Tools, Tracking, Update Project..., Reschedule incomplete to start after: Current Date**, or
- The **Status Date** may be used to move the start of incomplete and the finish of completed parts of a task back or forward to the **Status Date** with the new to Microsoft Project 2002 function **Status Date Calculation Options**. This option will be discussed at the end of this chapter. It also has restrictions which could make the function difficult to use on a project.

Ideally, scheduling software has one **Data Date** and its function is to:

- Separate the completed parts of tasks from incomplete parts of tasks,
- Calculate or record all costs and hours to date before the data date, and to forecast costs and hours to go after the data date.
- Calculate the **Finish Date** of an in-progress task from the **Data Date** plus the **Remaining Duration** over the **Task Calendar**.

Therefore, it is relatively simple in Microsoft Project 2003 to be in a situation where you have complete or in-progress tasks with start dates later than the data date, and/or incomplete or un-started tasks with a finish date earlier than the data date. This is an unrealistic situation, which is more difficult to achieve in other scheduling software packages. Care should be taken to avoid this situation.

15.4 Updating the Schedule

The next stage is to update the schedule by entering the mark-up information against each task.

When dealing with large schedules it is normal to create a look-ahead schedule by creating a filter to display incomplete and un-started tasks commencing in the near future only.

Microsoft Project 2003 provides several methods of statusing the schedule:

- **Update Project** – This is an automated process that assumes all tasks have progressed as planned. After updating a project, you may adjust the dates and percent complete to the recorded progress.
- **Update Tasks** – This function is used to status tasks one at a time.
- Update tasks using the **Task** or **Task Details** form.
- Update tasks by displaying the appropriate tracking columns by:
  - Selecting the **Tracking** table, or
  - Creating your own table, or
  - Inserting the required columns in an existing table.

Each of these four methods is discussed in the following sections. Then we will discuss the following two functions which are designed to assist you in moving tasks to their logical places in relation to the **Status Date**:

- **Move Incomplete Work into the Future**, and
- **Status Date Calculation Options**.

15.4.1 Using Update Project

Microsoft Project 2003 has a facility titled **Update Progress** for updating a project as if it had progressed according to plan. This function sets **Actual Start** and **Actual Finish** dates, **% Complete** and **Renaming Durations** in proportion to a user-assigned date.

Select **Tools**, **Tracking**, **Update Project…** to open the **Update Project** form:

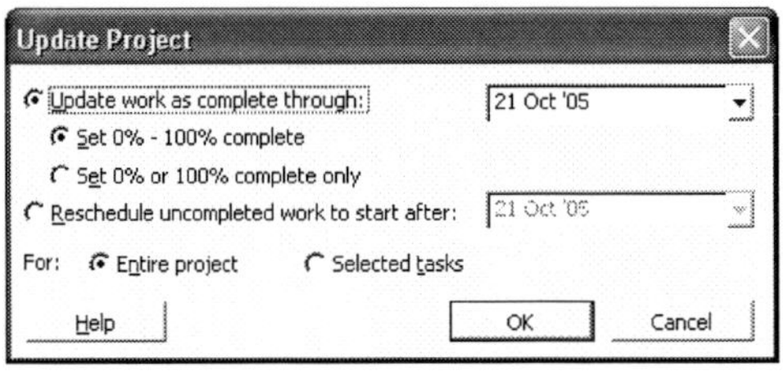

There are two options under **Update work as complete through:** which apply to in-progress tasks only.

- **Set 0% – 100 % Complete**:
  - ➢ This option sets the **Actual Start** to the **Early Start**, and
  - ➢ Sets the **% Complete** and **Actual Duration** in proportion to the amount of time worked for any in-progress tasks.
  - ➢ Sets the **Status Date** to the date you have nominated as the update date, the vertical line in the picture below:

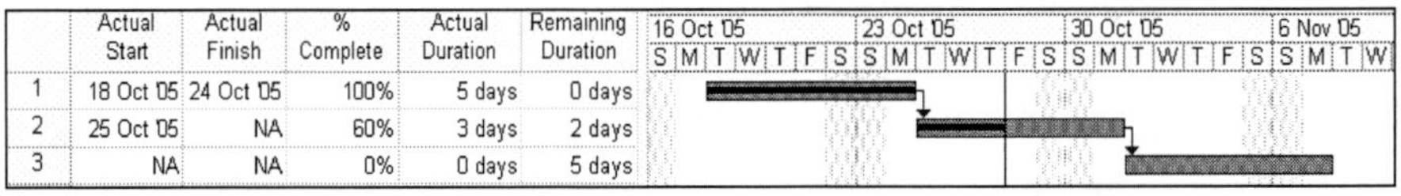

	Actual Start	Actual Finish	% Complete	Actual Duration	Remaining Duration
1	18 Oct '05	24 Oct '05	100%	5 days	0 days
2	25 Oct '05	NA	60%	3 days	2 days
3	NA	NA	0%	0 days	5 days

- **Set 0% or 100 % Complete only**. This option sets:
  - ➢ The **Actual Start** to the **Early Start** but leaves the **% Complete** and **Actual Duration** at zero.
  - ➢ The **% Complete** is set to 100% only when the task is complete.
  - ➢ Does not set the **Status Date** (or reset if it has be set in a previous update) to the date you have nominated as the update date. Therefore is no vertical line representing the **Status Date** in the picture below:

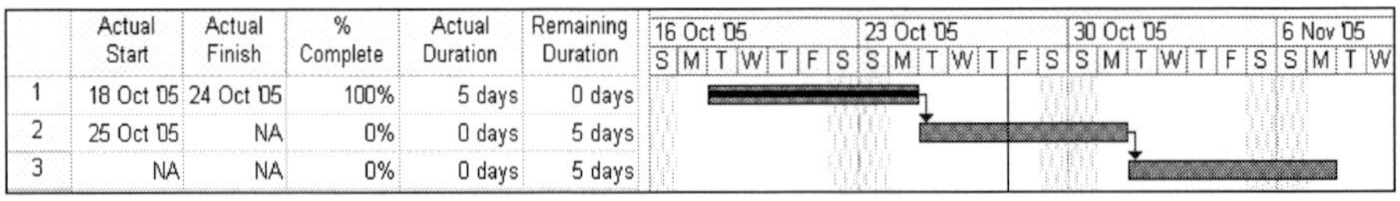

	Actual Start	Actual Finish	% Complete	Actual Duration	Remaining Duration
1	18 Oct '05	24 Oct '05	100%	5 days	0 days
2	25 Oct '05	NA	0%	0 days	5 days
3	NA	NA	0%	0 days	5 days

You may use the **Selected tasks** option when you highlight the tasks to be progressed before opening the **Update Project** form. Unselected tasks will not be progressed, but these tasks may be scheduled to occur after the **Status Date** by using the **Reschedule uncompleted work to start after:** function. This is covered later in this section.

You may not reverse progress with this option.

15.4.2 Update Tasks

Microsoft Project 2003 has a function that may be used for updating tasks one at a time or may be used for updating selected tasks to be updated with the same information. For example, there may be several tasks with the same Actual Start or % Complete.

- Select one or more tasks that you want to update with the same information such as the same Actual Start date.
- Select **Tools**, **Tracking**, **Update Tasks…** to open the **Update Tasks** form:

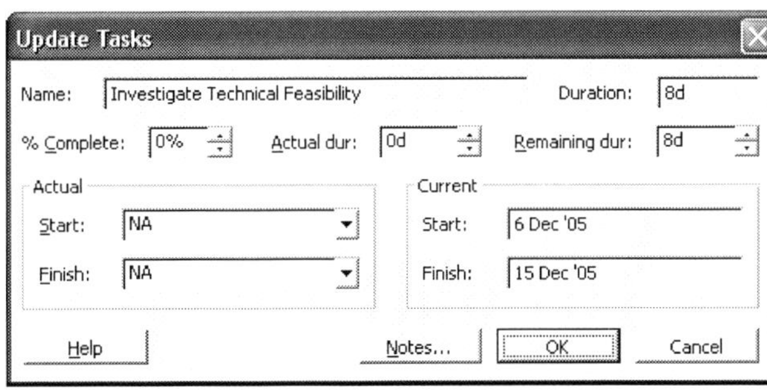

- Enter the required data and click on the 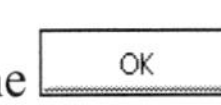 button to close the form.

Using the **Update Tasks** form is a cumbersome method for updating a large number of individual tasks since the form has to be closed after each task has been updated, the cursor moved to the next task and the form reopened. If you have a large number of tasks to be updated, it is quicker to use columns or a **Detailed** form in the lower pane.

15.4.3 Updating Tasks Using the Task Information, Task or Task Details Form

You may use either the **Task Information**, **Task** form or the **Task Details** form to update progress:

- Open the **Task** form by double-clicking on a task:

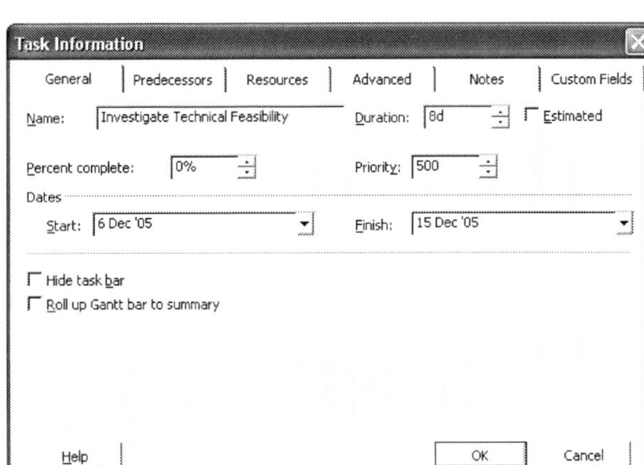

- To open the **Task Details** form, open the dual-pane view by either:
  - ➢ Selecting **Window**, **Split**,
  - ➢ Dragging the divider line with the mouse, or
  - ➢ Selecting a **View** with a dual pane or right-click.
  - ➢ Then, click in the bottom pane to make it active. There will be a blue bar down the left-hand side when the bottom pane is active.

  Now select **View**, M**ore Views…**, **Task Details Form** if an acceptable view is not available:

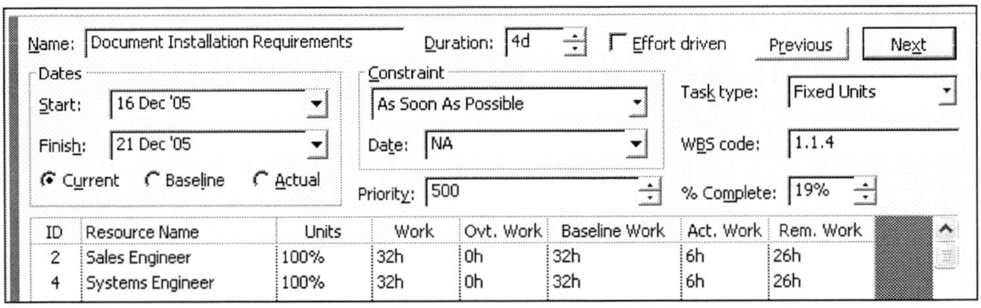

- Updating a task:
  - ➢ Enter a **% Complete**. An **Actual Start** date will be set by Microsoft Project 2003 and may be edited from the **Start:** drop-down box.
  - ➢ The **Duration** may be edited in order to calculate a new finish date. You will note that the % Complete will change when the duration is edited.
  - ➢ When a date is typed into the **Finish:** drop-down box, a **Finish Constraint** will be set without warning. These may be edited from the **Task Information** form's **Advanced** tab.

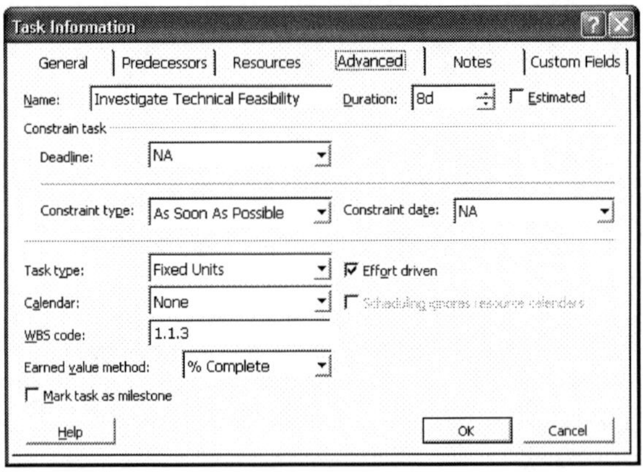

Using the **Task Information** form function is a cumbersome method for updating a large number of tasks, since the form has to be closed after each task has been updated, the cursor moved to the next task and the form reopened. If you have a large number of tasks to be updated, it is quicker to use columns or a **Detailed** form in the lower pane.

15.4.4 Updating Tasks Using Columns

An efficient method of updating tasks is by displaying the data in columns. This may be achieved by:

- Applying the **Tracking Table**, or
- Creating your own table, or
- Inserting the required columns.

To apply the tracking table, select **View**, **Table:**, **Tracking** table. This table may not exist if you are working on a schedule that has had this table deleted or if you have a non-standard load of Microsoft Project 2003 in which this table has been deleted from the Global.mpt. If you do not use the tracking table, be sure to add the following columns: **Actual Start**, **Actual Finish**, **% Complete**, **Actual Duration**, and **Remaining Duration**.

	Task Name	Act. Start	Act. Finish	% Comp.	Act. Dur.	Rem. Dur.
1	**Ozbuild Bid**	**5 Dec '05**	**NA**	**19%**	**6.13 days**	**26.87 days**
2	**Research**	**5 Dec '05**	**NA**	**60%**	**7.2 days**	**4.8 days**
3	Bid Request Documents Recieved	5 Dec '05	5 Dec '05	100%	0 days	0 days
4	Bid Strategy Meeting	5 Dec '05	5 Dec '05	100%	1 day	0 days
5	Investigate Technical Feasibility	6 Dec '05	NA	85%	6.8 days	1.2 days
6	Document Installation Requirements	NA	NA	0%	0 days	4 days

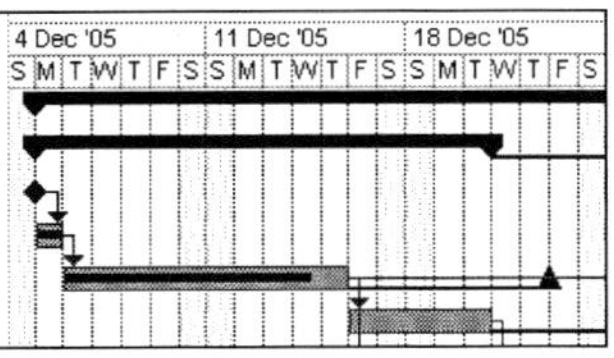

The status data may now be entered in the columns.

15.4.5 Marking Up Summary Tasks

It is not normal to mark up Summary Tasks, and Actual Dates may not be entered against Summary tasks. % Complete may be entered against a Summary Task and all the subordinate tasks will be auto-statused to match the summary % Complete.

When a **% Complete** is entered against a Summary task, all subordinate tasks, both Summary and Detailed, inherit a value depending upon the setting of the **Updating task status updates resource status** option found in **Tools**, **Options…**, **Calculation** tab.

To explain this concept further, see the following example. Both schedules pictured below had 60% entered against Task 2, the Research Phase of the OzBuild Bid task. All other Actual dates and % Complete were calculated by Microsoft Project 2003.

- With the **Updating task status updates resource status CHECKED** the activities are statused as if they were completed according to plan:

	Task Name	% Complete
1	**Ozbuild Bid**	**19%**
2	**Research**	**60%**
3	Bid Request Documents Recieved	100%
4	Bid Strategy Meeting	100%
5	Investigate Technical Feasibility	85%
6	Document Installation Requirements	0%

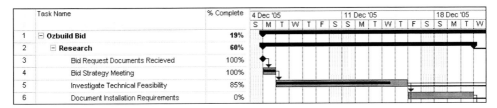

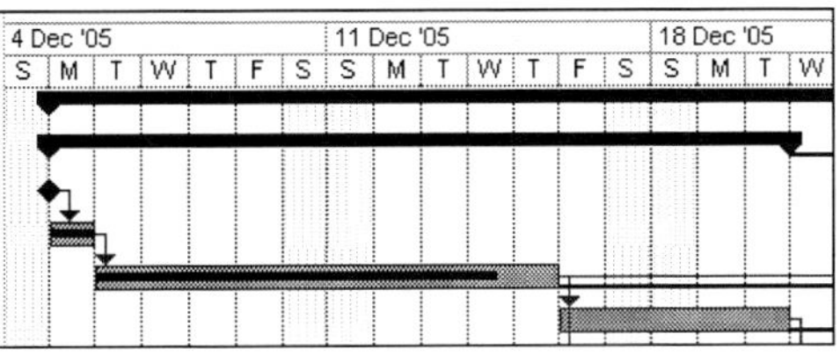

- With the **Updating task status updates resource status UNCHECKED** all activities are assigned to 60% complete, which is not the logical progression of the project:

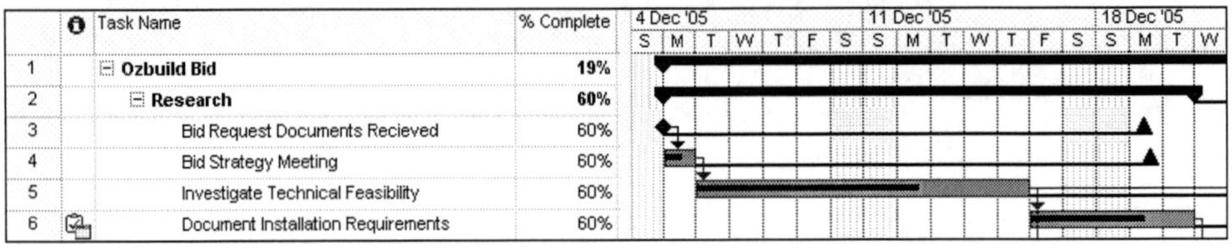

15.4.6 Reschedule Uncompleted Work To Start After

There is a feature available in the **Update Project** form titled **Reschedule uncompleted work to start after:** which will schedule the **Incomplete Work** of an **In-Progress** task to start on a specific date in the future.

- If you want to apply this operation to some of the tasks, then these should be selected first.
- Select **Tools**, **Options…**, **Schedule** tab and ensure the **Split in-progress tasks option** is checked. If this option is not checked, then this function will not operate.

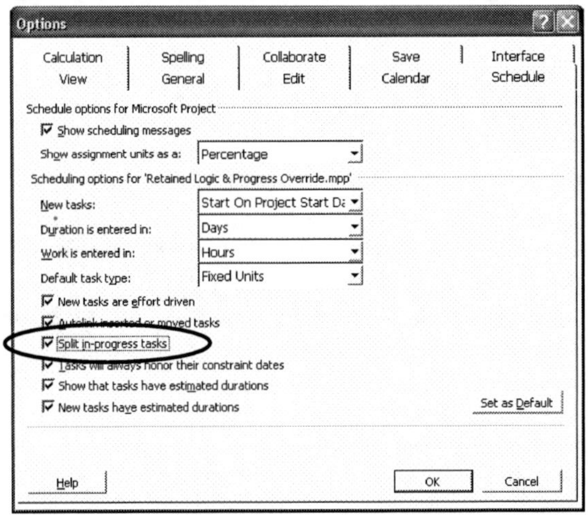

- Select **Tools**, **Tracking**, **Update Project…** to open the **Update Project** form.

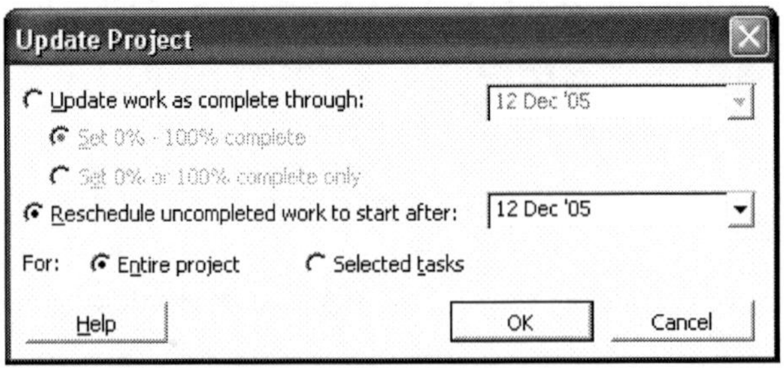

- Click on the **Reschedule uncompleted work to start after:** radio button.
- In the drop-down box to the right, specify the date after which incomplete work should commence. Then click on the OK button.

A task may be split without the split being displayed in the Gantt Chart by unchecking the **Show bar splits** in the **Layout** form. In the examples below, the **Incomplete Work** has been moved after 26 October:

- The task has been split in the picture below and the splits displayed:

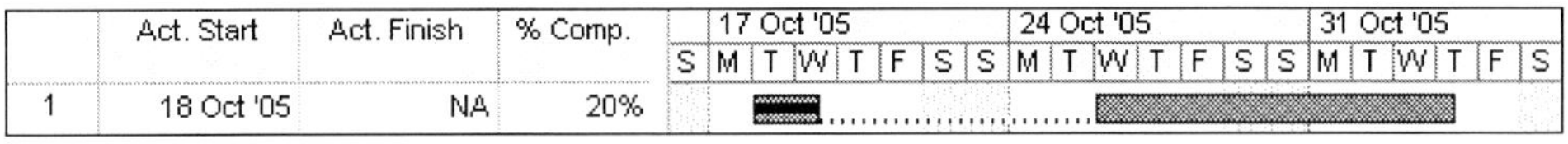

Act. Start
Act. Finish
% Comp.
17 Oct '05
24 Oct '05
31 Oct '05
S M T W T F S S M T W T F S S M T W T F S
1
18 Oct '05
NA
20%

- The task has been split in the picture below and the splits **NOT** displayed:

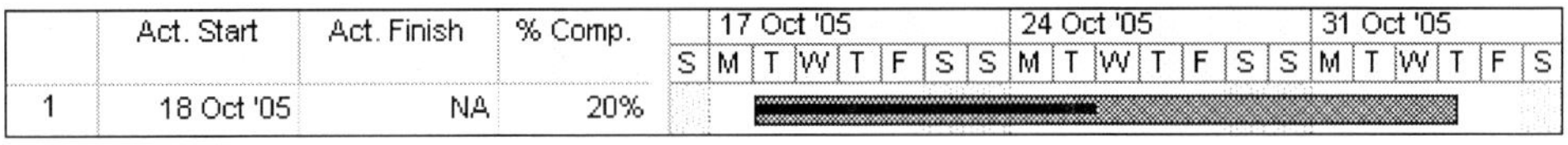

Act. Start
Act. Finish
% Comp.
17 Oct '05
24 Oct '05
31 Oct '05
S M T W T F S S M T W T F S S M T W T F S
1
18 Oct '05
NA
20%

Un-started tasks will have a **Start No Earlier Constraint** set to the **Update Project** date. This constraint will have to be removed from tasks if it is required to move the **Update Date** back in time and task splits will have to be manually dragged back in time.

When a project is updated using the **Update Project** function, the **Status Date DOES NOT** change the **Project Update Date**. You may end up with two dates that represent the **Data Date**—the **Update Project** date and the **Status Date**, which may both reflect different dates. The picture below shows the Status Date in the black vertical line at 18 Dec 02 and the **Update Project** date set at 11 Nov 05, which has split the last activity.

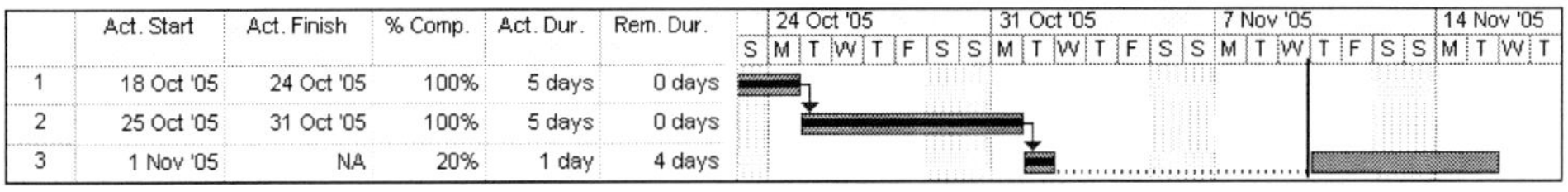

Act. Start
Act. Finish
% Comp.
Act. Dur.
Rem. Dur.
24 Oct '05
31 Oct '05
7 Nov '05
14 Nov '05
S M T W T F S S M T W T F S S M T W T F S S M T W T
1 18 Oct '05 24 Oct '05 100% 5 days 0 days
2 25 Oct '05 31 Oct '05 100% 5 days 0 days
3 1 Nov '05 NA 20% 1 day 4 days

You may use the **Update Project** function to reschedule tasks that are behind schedule and reschedule them after the **Update Project** date. This date will effectively become the **Data Date**. It is not possible to show the **Update Project** date as a vertical line on the bar chart. You may set and display the **Status Date** at the same date to achieve a position where incomplete work is all scheduled after a line displayed on the Gantt Chart.

> *i* It is recommended when you use the **Update Project** function that you should open the **Project Information** form and set the **Status Date** to the same date as the **Update Project** date. This enables you to display a **Data Date** in the bar chart in the correct place.

15.4.7 Status Date Calculation Options - New Tasks

New functions were introduced in Microsoft Project 2002 intended to assist schedulers to place the new tasks as they are added in a logical position with respect to the **Status Date**. If the **Status Date** has not been set the **Current Date** is used. In the example below the vertical black line is the **Status Date**:

- Task 1 has complete work in the future which is not logical,
- Task 2 would normally be considered in the logical position with respect to the **Status Date**, and
- Task 3 has incomplete work in the past which is also not logical.

	Duration	Start	% Complete
1	6 days	1 Nov '05	50%
2	6 days	25 Oct '05	50%
3	6 days	18 Oct '05	50%

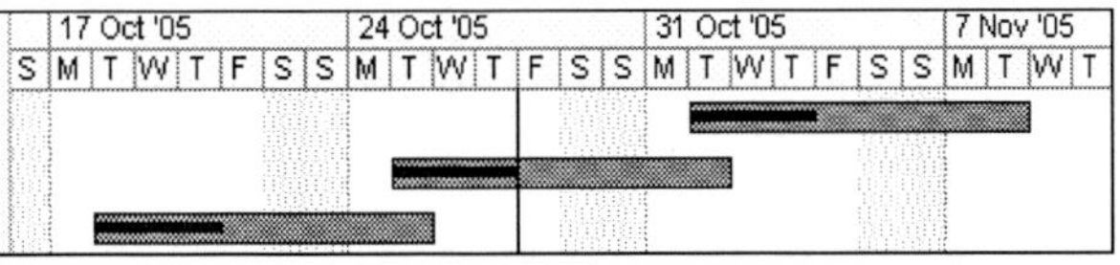

Select **Tools**, **Options**, **Calculation** tab and the new options are found under the **Calculation options for 'Project 1'**:

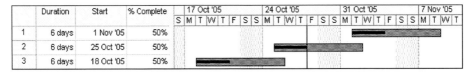

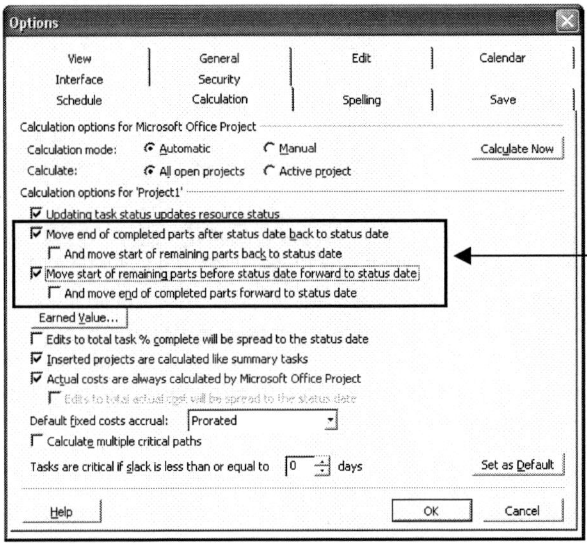

New Calculation Options in this box with 2002.

For these options to operate:

- The **Split in-progress tasks** option in the **Schedule** tab must be checked, and
- The required option on the **Calculation** tab under **Calculation** tab must be checked before the task is added, and
- The **Updating task status updates resource status** option on the **Calculation** tab must be checked.

The documentation found in the Microsoft Project help files is not precise and does not make it clear to the user that these options may NOT be turned on and off to recalculate all tasks. The options only work on new tasks when they are added to a schedule or when a Task is updated by changing the % Complete.

The following four options are available:

- Check only **Move end of completed task parts after status date back to status date** and this is the effect on task 1 in the picture on the previous page:

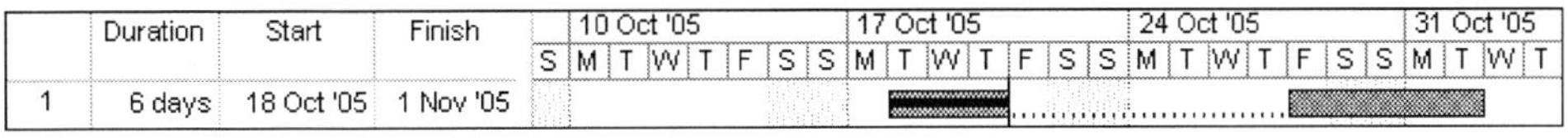

- The **And move start of remaining parts back to status date** option is only available after the option above is checked:

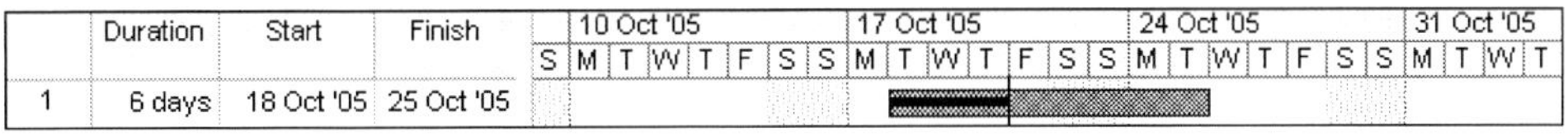

- The **Move start of remaining parts before status date forward to status date** may be checked at any time, the picture below shows the effect on task 3 in the picture on the previous page:

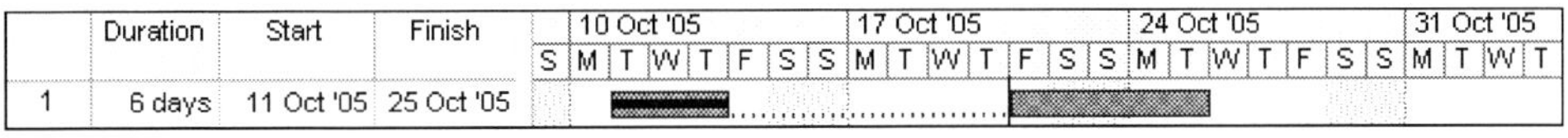

- The **And move end of completed parts forward to status date** may be checked after the option above is checked:

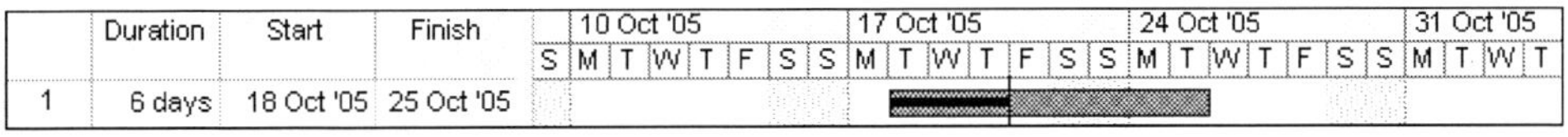

Duration
Start
Finish
10 Oct '05
17 Oct '05
24 Oct '05
31 Oct '05
S M T W T F S S M T W T F S S M T W T F S S M T W T
1
6 days
18 Oct '05
25 Oct '05

This function will ignore constraints even when the Schedule Option **Tasks will always honor their constraint date** has been set.

Experimentation by the author indicates that the activities will adopt the rules that are set in the options form when the task is added or when a task Percentage Complete is changed. Therefore this function may not be applied to existing schedules, but only to new tasks if the options are set before the tasks are added or when a task Percentage is updated.

This function in its current form has some restrictions that schedulers may find unacceptable:

- Existing schedules may not be opened and the function applied.
- When the **Move start of remaining parts before status date forward to status date** is used, it overrides any **Actual Start** date that you have entered prior to entering a % Complete.

This option should be used with caution and users should ensure they fully understand how this function operates by statusing a simple schedule multiple times.

15.4.8 Status Date Calculation Options - When Statusing a Schedule

The Status Date Calculation Options also operates when a schedule is statused by changing the Percent Completes of Tasks. The example below shows a task with the Status Date, the dark vertical line, is set to the next period:

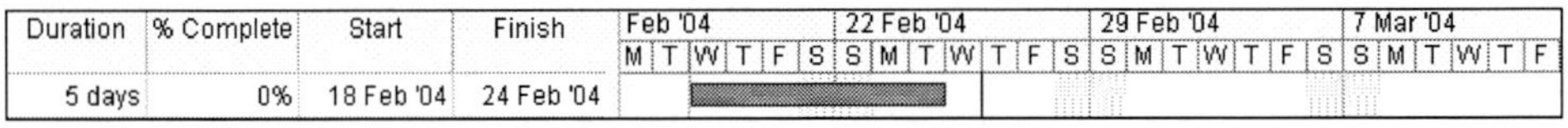

When all the **Status Date Calculation Option** options are checked and the percent complete is changed from 0% to 50% the task is aligned to the new Status Date:

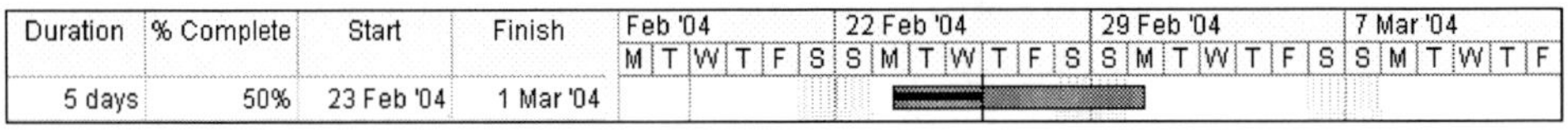

Duration
% Complete
Start
Finish
Feb '04
22 Feb '04
29 Feb '04
7 Mar '04
M T W T F S S M T W T F S S M T W T F S S M T W T F
5 days
50%
23 Feb '04
1 Mar '04

Now the status date has been moved to the next week and the tasks have not moved in respect to the Status date and will not move when the project is recalculated:

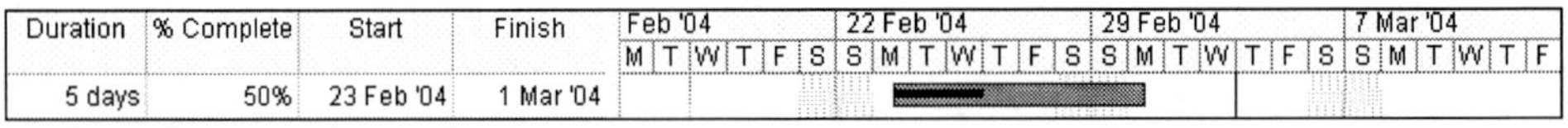

Duration
% Complete
Start
Finish
Feb '04
22 Feb '04
29 Feb '04
7 Mar '04
M T W T F S S M T W T F S S M T W T F S S M T W T F
5 days
50%
23 Feb '04
1 Mar '04

After percent complete changed to 60% the task splits and the unworked portion of the task is set to be completed after the new Status Date:

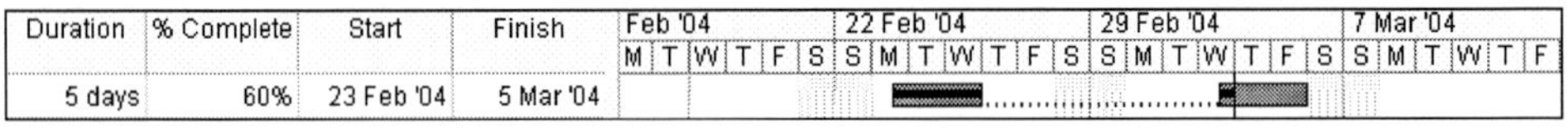

Duration
% Complete
Start
Finish
Feb '04
22 Feb '04
29 Feb '04
7 Mar '04
M T W T F S S M T W T F S S M T W T F S S M T W T F
5 days
60%
23 Feb '04
5 Mar '04

15.5 Comparing Progress with Baseline

There will normally be changes to the schedule dates and more often than not these are delays. The full extent of the change is not apparent without having a Baseline bar to compare with the statused schedule.

- To display the **Baseline Bar** in the **Bar Chart** you may use either the **Format, Bar Styles…** function covered in the **FORMATTING THE DISPLAY** chapter, or
- You may use the **Gantt Chart Wizard** as follows:
  - ➢ Select **Format, Gantt Chart Wizard…** to open the Gantt Chart Wizard.
  - ➢ Select the **Baseline** option. This will display both the current schedule and the baseline.

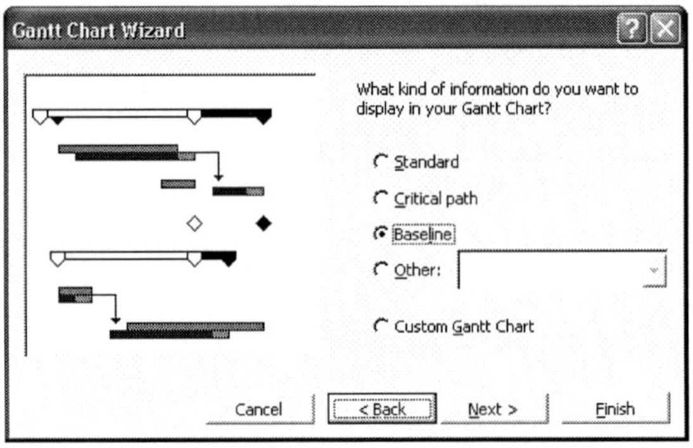

➢ Hit the Next > button and follow the remainder of the instructions to complete the formatting. You will bc givcn options for applying tcxt and rclationships in thc Gantt Chart.

This wizard will overwrite any customized formatting you have made using the **Format**, **Bar Styles...** option.

If you want to see the Start and Finish Date variances, they are available by displaying the **Start Variance** and **Finish Variance** columns. These variance columns use the **Baseline** data and variance columns and are not available for **Baseline 1** to **10**.

15.6 Corrective Action

There are two courses of action available with date slippage:

- The first is to accept the slippage. This is rarely acceptable, but it is the easiest answer.
- The second is to examine the schedule and evaluate how you could improve the end date.

Solutions to return the project to its original completion date must be cleared with the person responsible for the project, since they can have the most impact on the work.

Suggested solutions to bring the project back on track include:

- Reducing the durations of tasks on, or almost on, the critical path. When tasks have applied resources, this may include increasing the number of resources working on the tasks. Changing longer tasks is often more achievable than changing the length of short duration tasks.
- Changing calendars, say from a five-day to a six-day calendar, so that tasks are being worked on for more days per week.
- Reducing the project scope and deleting tasks.
- Changing task relationships so tasks takes place concurrently. This may be achieved by introducing negative lags to Finish to Start relationships which maintains a Closed Network.
- Introducing Start-to-Start relationships which has the potential of creating an open network. Should maintaining the critical path be important then this option should be avoided.
- Changing the logic or sequencing of tasks to reduce the overall length of the critical path. This may take a lot of work and reviews.

WORKSHOP 13

Baseline Comparison

Preamble

At the end of the first week you have to update the schedule and report progress and slippage.

Assignment

Open your **OzBuild Bid** project file and complete the following steps:

1. Save the Baseline for all the tasks on your project using the **Tools**, **Tracking** command.
2. Display the Baseline Bars using the **Format**, **GanttChartWizard**. Do not display dates or resources and display links.
3. Apply the **Gantt Chart** view and the **Tracking** table.
4. To make the display clearer increase the Timescale as required and display dotted Gridlines for the Bottom Tier.
5. Use **Tools**, **Tracking**, **Update Project…** to update progress to 12 Dec 05. Ignore the Planning Wizard as no Tasks with constraints are before the 12 Dec 05.
6. Check the **Status date** in the **Project Information** form; it should now be 12 Dec 05.
7. Display the Status Date as a solid black line, use the **Format**, **Gridlines…** command.
8. Update the following tasks and note the change in the end date of all the project tasks:
   - **Bid Strategy Meeting** finish date set to 08 Dec 05.
   - Change the **Actual Start** date of **Investigate Technical Feasibility** to 09 Dec 05.
   - Set the **Actual Duration** of **Investigate Technical Feasibility** to 3 Days.
   - Observe the **Remaining Duration**, it should be 5 Days, and
   - The **% Complete** should be 38%.

8. Observe the activities that have been delayed and save your **OzBuild Bid** project.

ANSWERS TO WORKSHOP 13

After using **Tools**, **Tracking**, **Update Project…** to update progress to 12 Dec 05:

	Task Name	Act. Start	Act. Finish	% Comp.	Act. Dur.	Rem. Dur.
1	**Ozbuild Bid**	**5 Dec '05**	**HA**	**14%**	**4.71 days**	**28.29 days**
2	**Research**	**5 Dec '05**	**HA**	**46%**	**5.54 days**	**6.46 days**
3	Bid Request Documents Recieved	5 Dec '05	5 Dec '05	100%	0 days	0 days
4	Bid Strategy Meeting	5 Dec '05	5 Dec '05	100%	1 day	0 days
5	Investigate Technical Feasibility	6 Dec '05	NA	63%	5 days	3 days
6	Document Installation Requirements	NA	NA	0%	0 days	4 days
7	**Estimation**	**HA**	**HA**	**0%**	**0 days**	**14 days**
8	Request Component Bids	NA	NA	0%	0 days	3 days
9	Develop Project Schedule	NA	NA	0%	0 days	4 days
10	Draft Technical Details Schedule	NA	NA	0%	0 days	9 days
11	Compile Costs from Component Bids	NA	NA	0%	0 days	2 days
12	**Proposal**	**HA**	**HA**	**0%**	**0 days**	**10 days**
13	Draft Bid Document	NA	NA	0%	0 days	3 days
14	Meeting to Review the Draft Bid Document	NA	NA	0%	0 days	1 day
15	Design Presentation	NA	NA	0%	0 days	1 day
16	Edit Proposal Draft Bid Document	NA	NA	0%	0 days	1 day
17	Negotiate Component Work Packages	NA	NA	0%	0 days	4 days
18	Final Review of Bid Document	NA	NA	0%	0 days	1 day
19	Submit Bid	NA	NA	0%	0 days	0 days

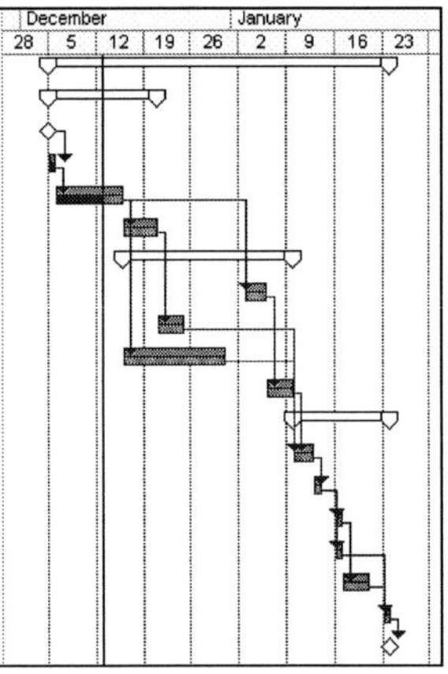

After Updating the tasks:

	Task Name	Act. Start	Act. Finish	% Comp.	Act. Dur.	Rem. Dur.
1	**Ozbuild Bid**	**5 Dec '05**	**HA**	**16%**	**5.13 days**	**27.87 days**
2	**Research**	**5 Dec '05**	**HA**	**44%**	**6.56 days**	**8.44 days**
3	Bid Request Documents Recieved	5 Dec '05	5 Dec '05	100%	0 days	0 days
4	Bid Strategy Meeting	5 Dec '05	8 Dec '05	100%	4 days	0 days
5	Investigate Technical Feasibility	9 Dec '05	NA	38%	3 days	5 days
6	Document Installation Requirements	NA	NA	0%	0 days	4 days
7	**Estimation**	**HA**	**HA**	**0%**	**0 days**	**11 days**
8	Request Component Bids	NA	NA	0%	0 days	3 days
9	Develop Project Schedule	NA	NA	0%	0 days	4 days
10	Draft Technical Details Schedule	NA	NA	0%	0 days	9 days
11	Compile Costs from Component Bids	NA	NA	0%	0 days	2 days
12	**Proposal**	**HA**	**HA**	**0%**	**0 days**	**10 days**
13	Draft Bid Document	NA	NA	0%	0 days	3 days
14	Meeting to Review the Draft Bid Document	NA	NA	0%	0 days	1 day
15	Design Presentation	NA	NA	0%	0 days	1 day
16	Edit Proposal Draft Bid Document	NA	NA	0%	0 days	1 day
17	Negotiate Component Work Packages	NA	NA	0%	0 days	4 days
18	Final Review of Bid Document	NA	NA	0%	0 days	1 day
19	Submit Bid	NA	NA	0%	0 days	0 days

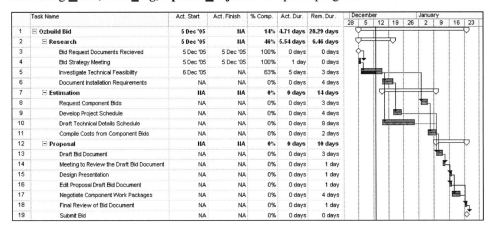

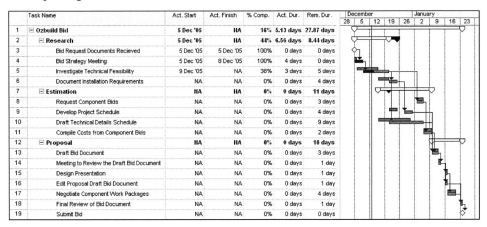

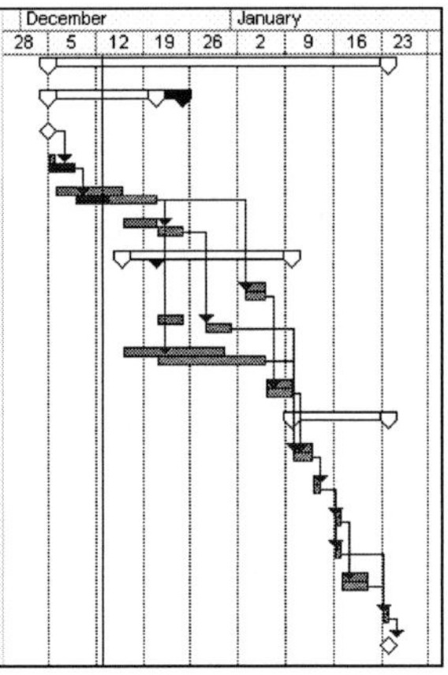

16 GROUPING TASKS, OUTLINE CODES AND WBS

Outlining was discussed earlier as a method of organizing detailed tasks under summary tasks. There are alternative features available in Microsoft Project 2003 for organizing, grouping and displaying task information:

- Grouping
- Custom Outline Codes
- Outline Codes
- User-Defined WBS (Work Breakdown Structure)

These functions are addressed in this book but are not examined in detail.

16.1 Understanding a Project Breakdown Structure

A Project Breakdown Structure represents a hierarchical breakdown of a project into logical functional elements. Some organizations have highly organized and disciplined structures with "rules" for creating and coding the elements of the structure. Some clients also impose a WBS code on a contractor for reporting and/or claiming payments. The following are examples of such structures:

- WBS **Work Breakdown Structure**, breaking down the project into the elements of work required to complete a project.
- OBS **Organization Breakdown Structure**, showing the hierarchical management structure of a project.
- CBS **Contract Breakdown Structure**, showing the breakdown of contracts.
- SBS **System Breakdown Structure**, showing the elements of a complex system.

Microsoft Project 2003 has functionality enabling you to represent your Project Breakdown structures using the **Grouping**, **Outlining** and the **WBS** functions.

We will discuss the following functions available in Microsoft Project 2003 to represent these structures in your schedule.

Topic	Menu Command
• Grouping	Select **Project**, **Group by:**
• Custom Outline Codes	Select **Tools**, **Customize**, **Fields...** and select the **Custom Outline Codes** tab to create a Custom Outline Code, Assign the codes by displaying a column and Select **Project**, **Group by:**, **More Groups...** option to create a grouping to display the tasks under the Custom Outline Code.
• Outline Codes	Displaying the **Outline Code** column or the **Outline Code** with the **Task Name** by selecting **Tools**, **Options...**, **View** tab and checking the **Show outline number**.
• WBS	Select **Project**, **WBS**, **Define Code...**

16.2 Grouping

Grouping allows you to group tasks under data items such as Customized Outline Codes, Text columns, Durations, Constraints, etc. It is particularly useful when you have schedules with a large number of tasks and you want to work with a related group of tasks that are spread throughout a project. The picture below displays a simple project where the relationship between each Task is difficult to check by inspection of the Gantt Chart organized by Phase:

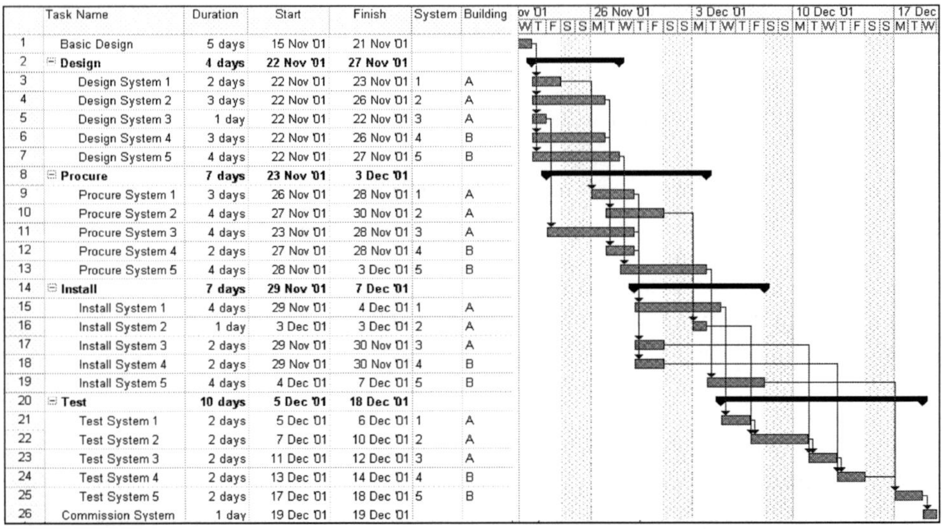

	Task Name	Duration	Start	Finish	System	Building
1	Basic Design	5 days	15 Nov '01	21 Nov '01		
2	**Design**	**4 days**	**22 Nov '01**	**27 Nov '01**		
3	Design System 1	2 days	22 Nov '01	23 Nov '01	1	A
4	Design System 2	3 days	22 Nov '01	26 Nov '01	2	A
5	Design System 3	1 day	22 Nov '01	22 Nov '01	3	A
6	Design System 4	3 days	22 Nov '01	26 Nov '01	4	B
7	Design System 5	4 days	22 Nov '01	27 Nov '01	5	B
8	**Procure**	**7 days**	**23 Nov '01**	**3 Dec '01**		
9	Procure System 1	3 days	26 Nov '01	28 Nov '01	1	A
10	Procure System 2	4 days	27 Nov '01	30 Nov '01	2	A
11	Procure System 3	4 days	23 Nov '01	28 Nov '01	3	A
12	Procure System 4	2 days	27 Nov '01	28 Nov '01	4	B
13	Procure System 5	4 days	28 Nov '01	3 Dec '01	5	B
14	**Install**	**7 days**	**29 Nov '01**	**7 Dec '01**		
15	Install System 1	4 days	29 Nov '01	4 Dec '01	1	A
16	Install System 2	1 day	3 Dec '01	3 Dec '01	2	A
17	Install System 3	2 days	29 Nov '01	30 Nov '01	3	A
18	Install System 4	2 days	29 Nov '01	30 Nov '01	4	B
19	Install System 5	4 days	4 Dec '01	7 Dec '01	5	B
20	**Test**	**10 days**	**5 Dec '01**	**18 Dec '01**		
21	Test System 1	2 days	5 Dec '01	6 Dec '01	1	A
22	Test System 2	2 days	7 Dec '01	10 Dec '01	2	A
23	Test System 3	2 days	11 Dec '01	12 Dec '01	3	A
24	Test System 4	2 days	13 Dec '01	14 Dec '01	4	B
25	Test System 5	2 days	17 Dec '01	18 Dec '01	5	B
26	Commission System	1 day	19 Dec '01	19 Dec '01		

With the Grouping function it is possible to Group on a text field to reorganize the data. In this example, the schedule has been reorganized by the Text 1 and 2 fields, which have been renamed using the **Tools**, **Customize**, **Field...** to System & Building. You may now clearly see the logic between the Items:

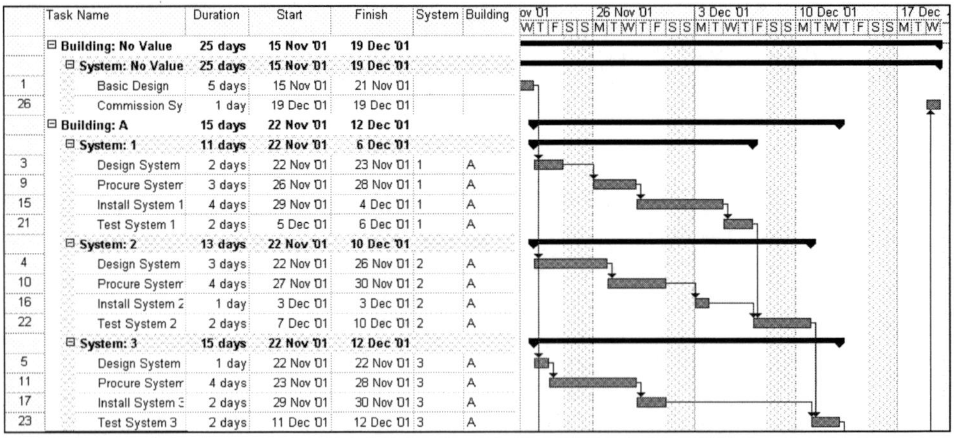

	Task Name	Duration	Start	Finish	System	Building
	⊟ Building: No Value	**25 days**	**15 Nov '01**	**19 Dec '01**		
	⊟ System: No Value	**25 days**	**15 Nov '01**	**19 Dec '01**		
1	Basic Design	5 days	15 Nov '01	21 Nov '01		
26	Commission Sy	1 day	19 Dec '01	19 Dec '01		
	⊟ Building: A	**15 days**	**22 Nov '01**	**12 Dec '01**		
	⊟ System: 1	**11 days**	**22 Nov '01**	**6 Dec '01**		
3	Design System	2 days	22 Nov '01	23 Nov '01	1	A
9	Procure Systen	3 days	26 Nov '01	28 Nov '01	1	A
15	Install System 1	4 days	29 Nov '01	4 Dec '01	1	A
21	Test System 1	2 days	5 Dec '01	6 Dec '01	1	A
	⊟ System: 2	**13 days**	**22 Nov '01**	**10 Dec '01**		
4	Design System	3 days	22 Nov '01	26 Nov '01	2	A
10	Procure Systen	4 days	27 Nov '01	30 Nov '01	2	A
16	Install System 2	1 day	3 Dec '01	3 Dec '01	2	A
22	Test System 2	2 days	7 Dec '01	10 Dec '01	2	A
	⊟ System: 3	**15 days**	**22 Nov '01**	**12 Dec '01**		
5	Design System	1 day	22 Nov '01	22 Nov '01	3	A
11	Procure Systen	4 days	23 Nov '01	28 Nov '01	3	A
17	Install System 3	2 days	29 Nov '01	30 Nov '01	3	A
23	Test System 3	2 days	11 Dec '01	12 Dec '01	3	A

When you group by a Text Field, the first few characters of the field determine the sort order. If you want the items to be ordered differently than the fields' text values, then you should place a number or letter at the start of the description to make the order of the groups suit your sorting requirements, or create a **Custom Outline Codes** which will take a little more effort but provide a more satisfactory result.

16.2.1 Customize Group

The **Customize Group** option allows you to create a temporary Group where the settings are not saved. Select **Project**, **Group by:** and there will be a predefined list of Groups for you to chose from:

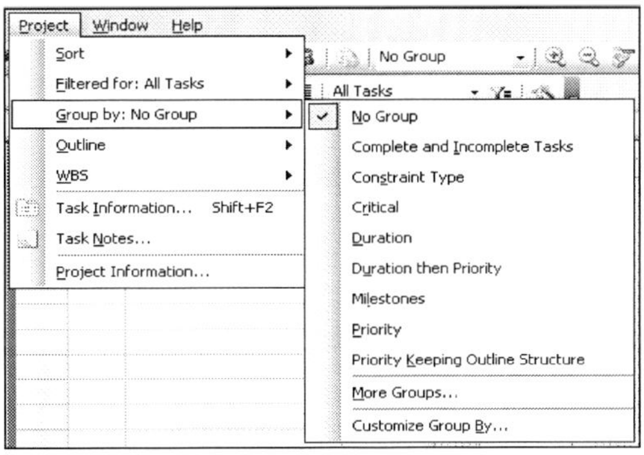

Click on **Customize Group By...** to open the **Customize Group By** form:

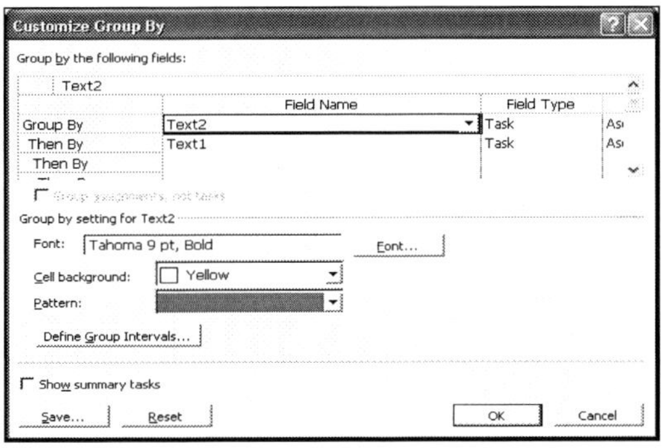

A hierarchical multiple grouping is available by selecting how you require the Tasks to be Grouped. The form above produces the result below:

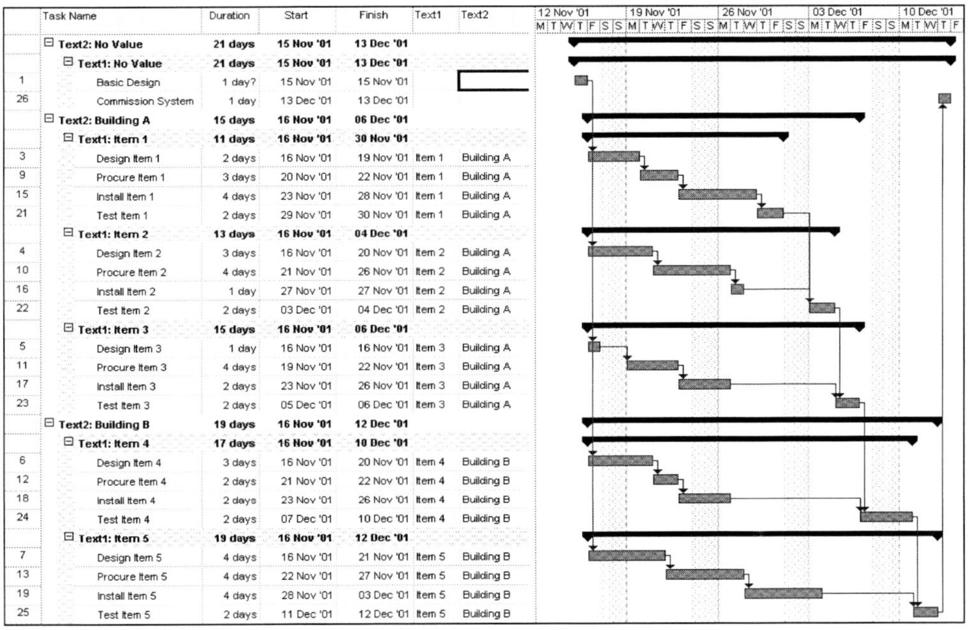

	Task Name	Duration	Start	Finish	Text1	Text2
	Text2: No Value	21 days	15 Nov '01	13 Dec '01		
	Text1: No Value	21 days	15 Nov '01	13 Dec '01		
1	Basic Design	1 day?	15 Nov '01	15 Nov '01		
26	Commission System	1 day	13 Dec '01	13 Dec '01		
	Text2: Building A	15 days	16 Nov '01	06 Dec '01		
	Text1: Item 1	11 days	16 Nov '01	30 Nov '01		
3	Design Item 1	2 days	16 Nov '01	19 Nov '01	Item 1	Building A
9	Procure Item 1	3 days	20 Nov '01	22 Nov '01	Item 1	Building A
15	Install Item 1	4 days	23 Nov '01	28 Nov '01	Item 1	Building A
21	Test Item 1	2 days	29 Nov '01	30 Nov '01	Item 1	Building A
	Text1: Item 2	13 days	16 Nov '01	04 Dec '01		
4	Design Item 2	3 days	16 Nov '01	20 Nov '01	Item 2	Building A
10	Procure Item 2	4 days	21 Nov '01	26 Nov '01	Item 2	Building A
16	Install Item 2	1 day	27 Nov '01	27 Nov '01	Item 2	Building A
22	Test Item 2	2 days	03 Dec '01	04 Dec '01	Item 2	Building A
	Text1: Item 3	15 days	16 Nov '01	06 Dec '01		
5	Design Item 3	1 day	16 Nov '01	16 Nov '01	Item 3	Building A
11	Procure Item 3	4 days	19 Nov '01	22 Nov '01	Item 3	Building A
17	Install Item 3	2 days	23 Nov '01	26 Nov '01	Item 3	Building A
23	Test Item 3	2 days	05 Dec '01	06 Dec '01	Item 3	Building A
	Text2: Building B	19 days	16 Nov '01	12 Dec '01		
	Text1: Item 4	17 days	16 Nov '01	10 Dec '01		
6	Design Item 4	3 days	16 Nov '01	20 Nov '01	Item 4	Building B
12	Procure Item 4	2 days	21 Nov '01	22 Nov '01	Item 4	Building B
18	Install Item 4	2 days	23 Nov '01	26 Nov '01	Item 4	Building B
24	Test Item 4	2 days	07 Dec '01	10 Dec '01	Item 4	Building B
	Text1: Item 5	19 days	16 Nov '01	12 Dec '01		
7	Design Item 5	4 days	16 Nov '01	21 Nov '01	Item 5	Building B
13	Procure Item 5	4 days	22 Nov '01	27 Nov '01	Item 5	Building B
19	Install Item 5	4 days	28 Nov '01	03 Dec '01	Item 5	Building B
25	Test Item 5	2 days	11 Dec '01	12 Dec '01	Item 5	Building B

16.2.2 Using a Predefined Group

You may assign a predefined Group by:

- Selecting **Project**, **Group by:**

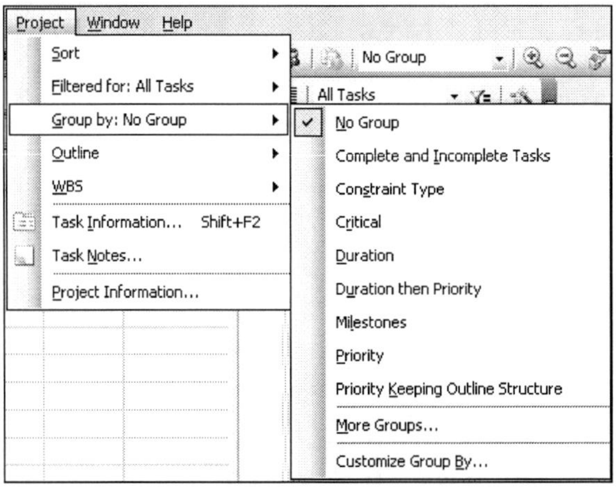

- Then, you may either:
  - Select a grouping from the list, or
  - Select **More Groups…** to open the **More Groups** form and then select one from the list after clicking on the **Task** or **Resource** radio button.

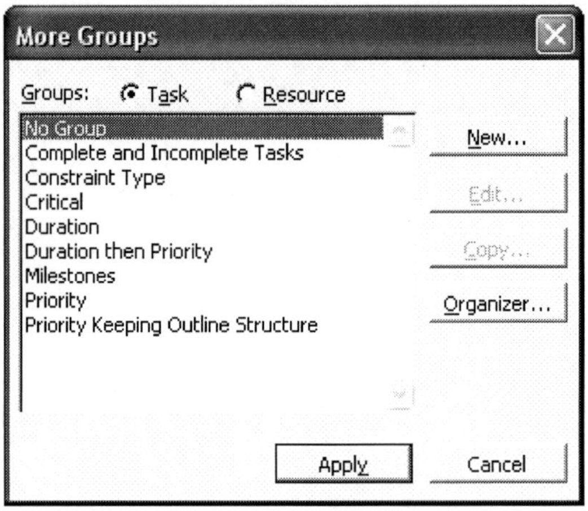

16.2.3 Creating a New Group

You may create a new Group by carrying out the following sequence:

- Selecting **Project**, **Group by:**, **More Groups…** to open the **More Groups** form,
- Click on the New... button to open the **Group Definition** form
- You may now create a "Grouping" which you may reapply at a later date or copy to another project using **Organizer**.

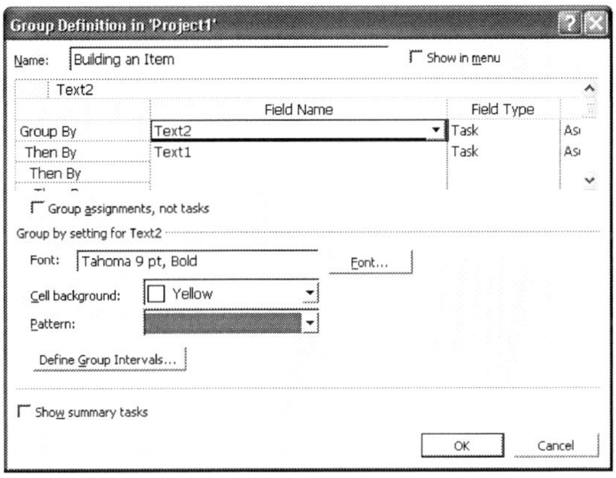

Grouping is similar to the Primavera P3 and SureTrak organize function. It is possible to mimic this Primavera function using the text columns as Activity Code dictionaries. Projects converted from Primavera P3 or SureTrak using the MPX format often translate Primavera's Activity Codes to Microsoft Project 2003's Text 1 through Text 10 fields. After conversion, the project may be **Grouped** by Text 1–10. **Custom Outline Codes** may produce a better result as bands may be ordered with this function.

16.2.4 Grouping Resources in the Resource Sheet

Resources may be created in the **Resource Sheet**. Then the resources may be grouped by a number of attributes. The standard options are shown below:

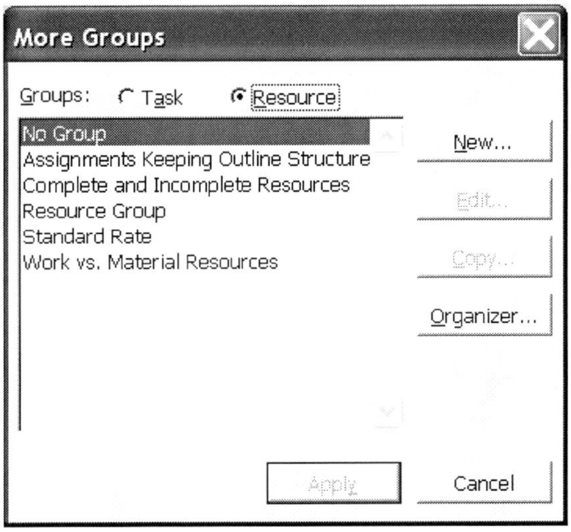

Resources are covered in more detail in the **RESOURCES** chapters.

A project hierarchical organizational structure may be created using this function and resources displayed under their department for example.

16.3 Custom Outline Codes

Custom Outline Codes is a function new to Microsoft Project 2000 that allows the creation of up to 10 hierarchical Project Breakdown Structures. The process to use this function has three steps:

- Define the new Custom Outline Code structure, then
- Assign the codes to the tasks, then
- Create a Group to organize the tasks under the new Custom Outline Code structure.

16.3.1 Define a Custom Outline Code Structure

Select **Tools**, **Customize**, **Fields...** and select the **Custom Outline Codes** tab to open the form:

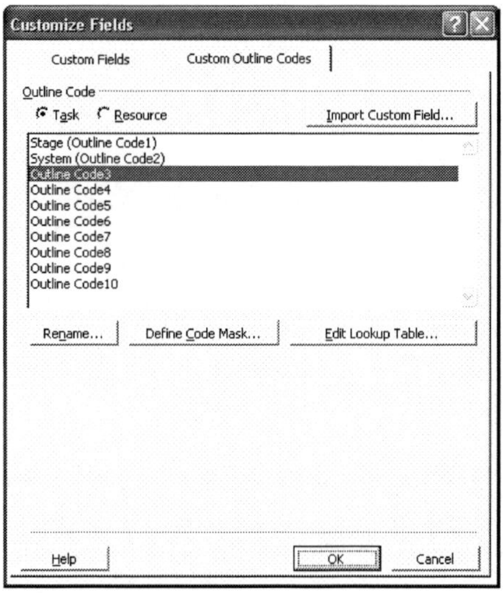

- An Outline Code may be created for either Task or Resource data by clicking on the appropriate radio button under the title **Outline Code**.
- The [Import Custom Field...] function allows you to copy a code structure from another project in a similar method to Organizer.
- The [Rename...] button opens a form to edit the name of the Outline Code. Two codes in the picture have been renamed Stage and System.
- The [Define Code Mask...] button opens the **Outline Code Definition** form where the code structure is defined.
  - ➢ As each **Level** is created it is assigned a number.
  - ➢ The **Sequence** defines the type of text that may be entered for the code: Numbers, Upper Case, Lower Case or Characters (text).
  - ➢ The **Length** specifies how many characters the Code Level may have: any, or a number between 1 and 10.
  - ➢ The **Separator** defines the character that separates each level in the structure.

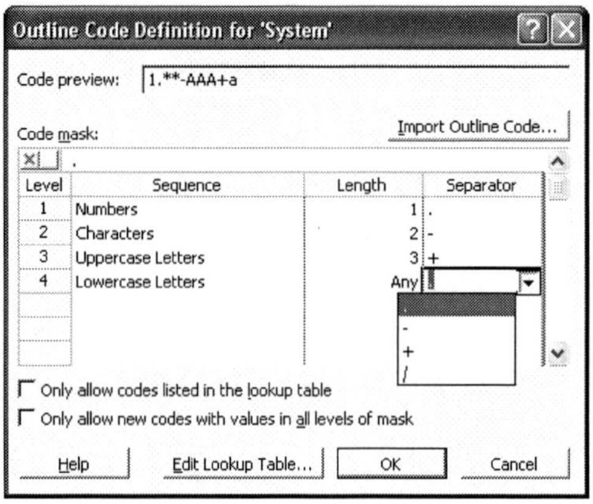

- The Edit Lookup Table... opens a form to enter the codes and descriptions for the Outline Code.
- The picture shows three units at level 1 and their subsystems at level 2.
- The icons along the top of the form have a similar function to their use in outlining and may be used to indent and outdent codes and copy and paste code groups.

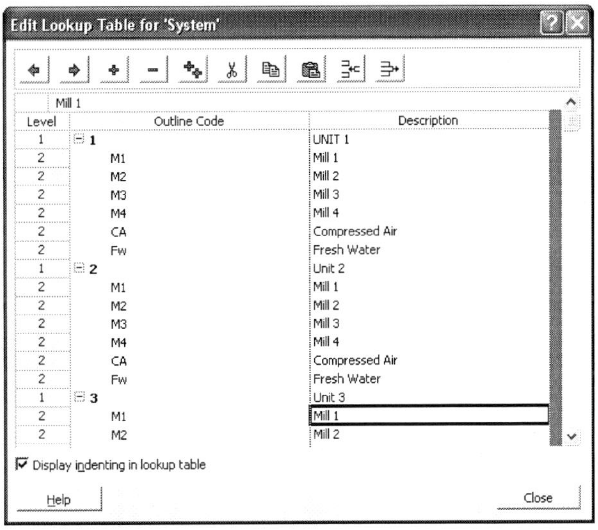

16.3.2 Assign Custom Codes to Tasks

The codes are assigned by displaying the appropriate column:

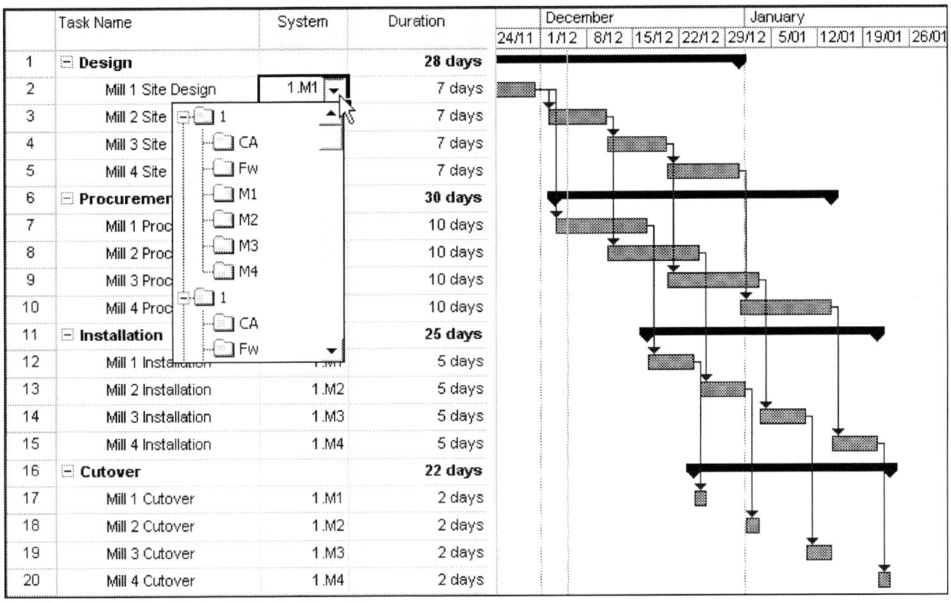

16.3.3 Organize Tasks Under a Custom Outline Code Structure

Select **Project**, **Group by:**, **More Groups…** option to create a grouping to display the tasks under the Custom Outline Code. The same process is used as outlined in the previous section to create a grouping:

	Task Name	System	Duration
	⊟ **UNIT 1**	**1**	**45 days**
	⊟ **Mill 1**	**1.M1**	**25 days**
2	Mill 1 Site Design	1.M1	7 days
7	Mill 1 Procurement	1.M1	10 days
12	Mill 1 Installation	1.M1	5 days
17	Mill 1 Cutover	1.M1	2 days
	⊟ **Mill 2**	**1.M2**	**24 days**
3	Mill 2 Site Design	1.M2	7 days
8	Mill 2 Procurement	1.M2	10 days
13	Mill 2 Installation	1.M2	5 days
18	Mill 2 Cutover	1.M2	2 days
	⊟ **Mill 3**	**1.M3**	**24 days**
4	Mill 3 Site Design	1.M3	7 days
9	Mill 3 Procurement	1.M3	10 days
14	Mill 3 Installation	1.M3	5 days
19	Mill 3 Cutover	1.M3	2 days
	⊟ **Mill 4**	**1.M4**	**24 days**
5	Mill 4 Site Design	1.M4	7 days
10	Mill 4 Procurement	1.M4	10 days
15	Mill 4 Installation	1.M4	5 days
20	Mill 4 Cutover	1.M4	2 days

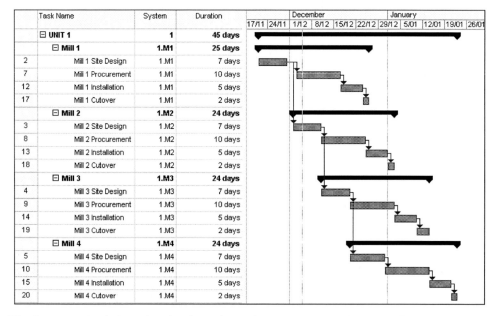

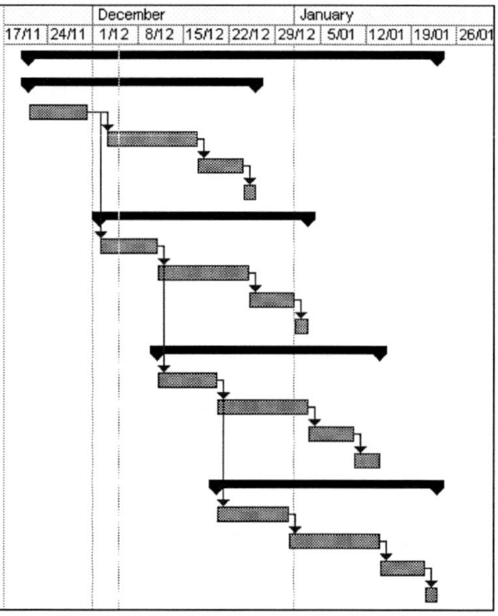

The Summary Tasks are virtual tasks and may have resources and costs assigned to them.

16.4 Outline Codes

The Microsoft Project 2003 **Outlining** function may be used to code the Project Breakdown Structure. It is possible to display the Outline Code against each task in the schedule by:

- Displaying the **Outline Code** column, or
- Displaying the **Outline Code** with the **Task Name** by selecting **Tools**, **Options…**, **View** tab and checking the **Show outline number**.

An **Outline Level** displays how many levels down in the Outline Code structure the Task lies.

	Outline Number	Outline Level	Task Name
1	**1**	**1**	⊟ **1 OzBuild Bid**
2	**1.1**	**2**	⊟ **1.1 Research**
3	1.1.1	3	1.1.1 Bid Request Documents Received
4	1.1.2	3	1.1.2 Bid Strategy Meeting
5	1.1.3	3	1.1.3 Investigate Technical Feasibility
6	1.1.4	3	1.1.4 Document Installation Requirements
7	**1.2**	**2**	⊟ **1.2 Estimation**
8	1.2.1	3	1.2.1 Request Component Bids
9	1.2.2	3	1.2.2 Develop Project Schedule
10	1.2.3	3	1.2.3 Draft Technical Details Schedule
11	1.2.4	3	1.2.4 Compile Costs from Component Bids
12	**1.3**	**2**	⊟ **1.3 Proposal**
13	1.3.1	3	1.3.1 Draft Bid Document
14	1.3.2	3	1.3.2 Meeting to Review the Draft Bid Docu
15	1.3.3	3	1.3.3 Design Presentation
16	1.3.4	3	1.3.4 Edit Proposal Draft Bid Document
17	1.3.5	3	1.3.5 Negotiate Component Work Packages
18	1.3.6	3	1.3.6 Final Review of Bid Document
19	1.3.7	3	1.3.7 Submit Bid

16.5 User Defined WBS Function

There are occasions when information must be presented under a predefined WBS Code structure. This structure could be a company standard, project-specific need, or defined by a client. This function is a method of tailoring the display of the **Outline Code** in a column titled **WBS**; but it does not provide an additional set of codes to organize your project tasks.

In a new project, the default WBS code is identical to the Outline Code. The example below displays both the WBS and Outline codes of the OzBuild schedule before doing any tailoring of the WBS code:

	WBS	Outline Number	Outline Level	Task Name
1	**1**	**1**	**1**	**1 OzBuild Bid**
2	**1.1**	**1.1**	**2**	**1.1 Research**
3	1.1.1	1.1.1	3	1.1.1 Bid Request Documents Received
4	1.1.2	1.1.2	3	1.1.2 Bid Strategy Meeting
5	1.1.3	1.1.3	3	1.1.3 Investigate Technical Feasibility
6	1.1.4	1.1.4	3	1.1.4 Document Installation Requirements
7	**1.2**	**1.2**	**2**	**1.2 Estimation**
8	1.2.1	1.2.1	3	1.2.1 Request Component Bids
9	1.2.2	1.2.2	3	1.2.2 Develop Project Schedule
10	1.2.3	1.2.3	3	1.2.3 Draft Technical Details Schedule
11	1.2.4	1.2.4	3	1.2.4 Compile Costs from Component Bids

The WBS Codes structure may be tailored to suit your own requirements. Below is an example of a WBS code which has been tailored using the WBS code function. The Outline sequential numbers may be replaced by user-defined sequences of numbers or letters.

	WBS	Task Name
1	**OzBuild1**	**Ozbuild Bib**
2	**OzBuild1.AA**	**Research**
3	OzBuild1.AA-aaa	Bid Document Received
4	OzBuild1.AA-aab	Bid Strategy Meeting
5	OzBuild1.AA-aac	Investigate Technical Feasibility
6	OzBuild1.AA-aad	Document Installation Requirements
7	**OzBuild1.AB**	**Estimation**
8	OzBuild1.AB-aaa	Request Component Bids
9	OzBuild1.AB-aab	Develop Project Schedule
10	OzBuild1.AB-aac	Draft Technical Details Schedule
11	OzBuild1.AB-aad	Compile Costs from Bids
12	**OzBuild1.AC**	**Proposal**
13	OzBuild1.AC-aaa	Draft Bid Documents
14	OzBuild1.AC-aab	Draft Bid Meeting
15	OzBuild1.AC-aac	Design Presentation
16	OzBuild1.AC-aad	Edit Proposal Draft
17	OzBuild1.AC-aae	Finalise Bid Package
18	OzBuild1.AC-aaf	Final Bid Meeting
19	OzBuild1.AC-aag	Submit Bid

To create your own user-defined WBS codes, select **Project**, **WBS**, **Define Code…** to open the **WBS Code Definition** form:

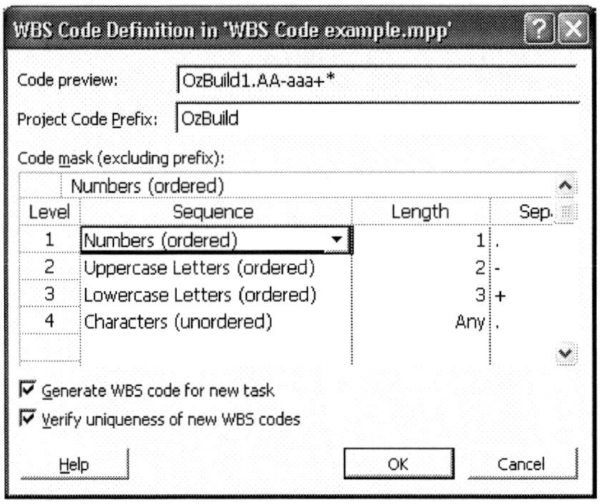

- The **Project Code Prefix:** is where you enter a text string that will precede all **WBS** codes. See the example on the previous page.
- Each code level is defined in the **Sequence** column. There are four code types. Each is displayed above:
  - Numbers (ordered)
  - Uppercase Letters (ordered)
  - Lowercase Letters (ordered)
  - Characters (unordered)

  The **Numbers**, **Uppercase Letters** and **Lowercase Letters** are all numbered automatically.

  The **Characters** option will initially define all WBS codes at this level as an * and these may be overwritten with text. Microsoft Project 2003 will not renumber them.

- A **Length** of 1 to 10 or "Any" may be selected as the length of the code.
- Check **Generate WBS code for new task** and each new task will be automatically coded when created. When left unchecked, no WBS code is assigned to new tasks.
- Check **Verifies uniqueness of new WBS codes** to prevent the creation of duplicate WBS codes.

By selecting **All** or Selected Tasks, you may renumber the WBS codes. Select **Project**, **WBS**, **Renumber…** from the menu.

WORKSHOP 14

Reorganization of the Schedule

Preamble

We want to issue another report for comment by management by grouping the activities without float and showing the WBS columns.

Assignment

1. Grouping – to group tasks without float:
   - Apply the **Entry** table and ensure the **Total Slack** column is displayed between the **Task Name** and **Duration**.
   - Create a new Group titled **Total Float** and group the tasks by **Total Slack**.
   - Check the **Show in menu** option.
   - Apply this grouping.
   - All the tasks with zero days float are grouped at the top under the heading **Total Slack: 0 days**.
2. Outline Codes – we wish to display the Outline Codes and Level:
   - Remove the above grouping by selecting **Project**, **Group by: No Group**.
   - Copy the **Schedule** Table and rename it by **Outline Codes**.
   - Insert the **Outline Level** column between the Task ID and Task Name, then
   - Align the content of the inserted column to the left and apply the table.
   - Select **Tools**, **Options...**, **View** tab and **Show outline number**.
3. WBS – Work Breakdown Structure – we wish to display the WBS Predecessors & Successors
   - Hide the Outline Level column.
   - Display the **WBS**, **WBS Predecessors** and **WBS Successors** columns to the left of the descriptions. Align the column data to the left as shown in the following page. Note the WBS Predecessor and Successor columns may not be edited.

ANSWERS TO WORKSHOP 14

Grouped by Total Float:

	ℹ	Task Name	Total Slack	Duration	Start	Finish
		⊟ **Total Slack: 0 days**	**0 days**	**33 days**	**5 Dec '05**	**23 Jan '06**
3	✓	Bid Request Documents Recieved	0 days	0 days	5 Dec '05	5 Dec '05
4	✓	Bid Strategy Meeting	0 days	4 days	5 Dec '05	8 Dec '05
8		Request Component Bids	0 days	3 days	3 Jan '06	5 Jan '06
11		Compile Costs from Component Bids	0 days	2 days	6 Jan '06	9 Jan '06
13		Draft Bid Document	0 days	3 days	10 Jan '06	12 Jan '06
14		Meeting to Review the Draft Bid Document	0 days	1 day	13 Jan '06	13 Jan '06
15		Design Presentation	0 days	1 day	16 Jan '06	16 Jan '06
17		Negotiate Component Work Packages	0 days	4 days	17 Jan '06	20 Jan '06
18		Final Review of Bid Document	0 days	1 day	23 Jan '06	23 Jan '06
19		Submit Bid	0 days	0 days	23 Jan '06	23 Jan '06
		⊟ **Total Slack: 2 days**	**2 days**	**17 days**	**9 Dec '05**	**5 Jan '06**
5		Investigate Technical Feasibility	2 days	8 days	9 Dec '05	20 Dec '05
10		Draft Technical Details Schedule	2 days	9 days	21 Dec '05	5 Jan '06
		⊟ **Total Slack: 4 days**	**4 days**	**1 day**	**16 Jan '06**	**16 Jan '06**
16		Edit Proposal Draft Bid Document	4 days	1 day	16 Jan '06	16 Jan '06
		⊟ **Total Slack: 6 days**	**6 days**	**6 days**	**21 Dec '05**	**31 Dec '05**
6		Document Installation Requirements	6 days	4 days	21 Dec '05	24 Dec '05
9		Develop Project Schedule	6 days	4 days	28 Dec '05	31 Dec '05

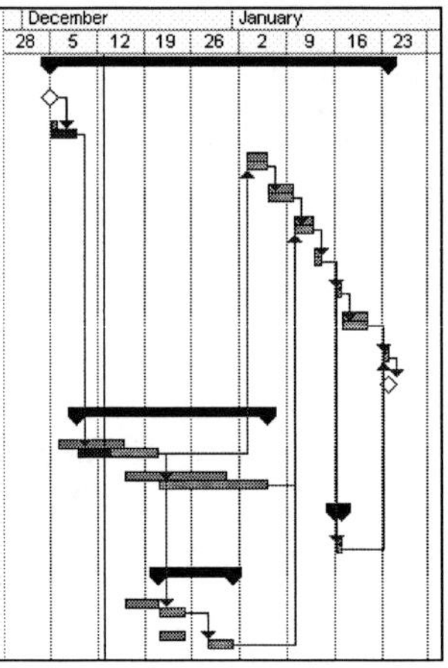

Showing Outline Level in the Task Name:

	Outline Level	Task Name
1	**1**	⊟ **1 Ozbuild Bid**
2	**2**	⊟ **1.1 Research**
3	3	1.1.1 Bid Request Documents Recieved
4	3	1.1.2 Bid Strategy Meeting
5	3	1.1.3 Investigate Technical Feasibility
6	3	1.1.4 Document Installation Requirements
7	**2**	⊟ **1.2 Estimation**
8	3	1.2.1 Request Component Bids
9	3	1.2.2 Develop Project Schedule
10	3	1.2.3 Draft Technical Details Schedule

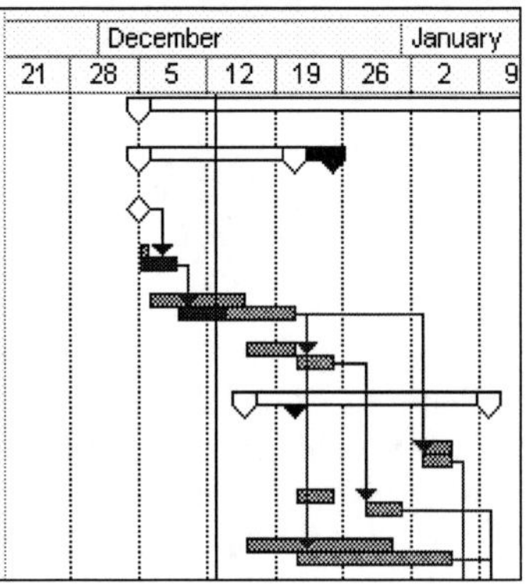

Showing WBS, WBS Predecessors and WBS Successors:

	WBS	WBS Predecessors	WBS Successors	Task Name
1	**1**			⊟ **1 Ozbuild Bid**
2	**1.1**			⊟ **1.1 Research**
3	1.1.1		1.1.2	1.1.1 Bid Request Documents Recieved
4	1.1.2	1.1.1	1.1.3	1.1.2 Bid Strategy Meeting
5	1.1.3	1.1.2	1.1.4,1.2.1,1.2.3	1.1.3 Investigate Technical Feasibility
6	1.1.4	1.1.3	1.2.2	1.1.4 Document Installation Requirements
7	**1.2**			⊟ **1.2 Estimation**
8	1.2.1	1.1.3	1.2.4	1.2.1 Request Component Bids
9	1.2.2	1.1.4	1.3.1	1.2.2 Develop Project Schedule
10	1.2.3	1.1.3	1.3.1	1.2.3 Draft Technical Details Schedule

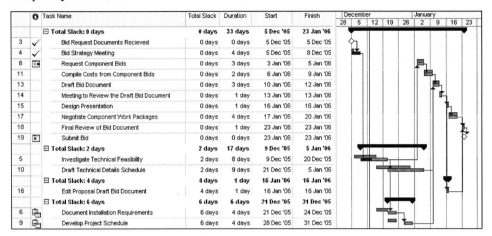

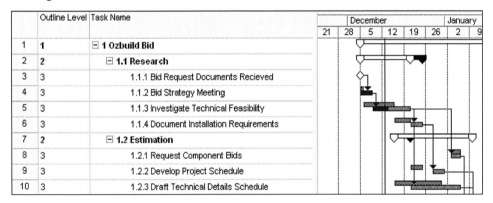

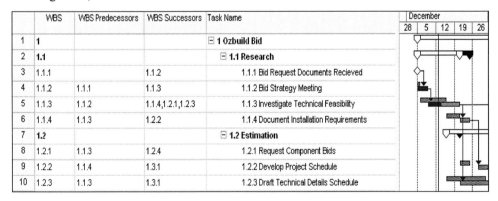

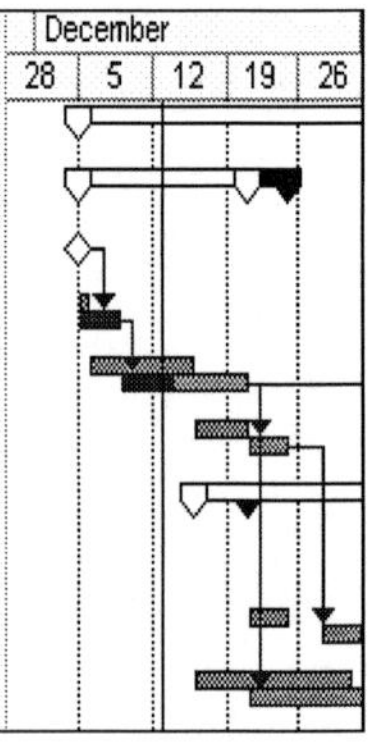

17 OPTIONS

17.1 Options

The **Options** forms allow you to decide how Microsoft Project 2003 calculates and displays information. Most of the options are self-explanatory. Under **Tools**, **Options…**, there are ten tabs:

Interface	Security		
Schedule	Calculation	Spelling	Save
View	General	Edit	Calendar

There are many options which are not essential for scheduling. Unfortunately, it is often difficult for new users to determine which are important and which are not, which can lead to confusion. This chapter will explain most of the functions identified on each Options form tab and indicate the important options.

- **View** Controls the display of data.
- **General** Controls the display of software productivity tools and help, and default resource settings.
- **Edit** Mainly controls edit and display functions.
- **Calendar** These options **SHOULD** be understood especially if multiple **Task** calendars are used.
- **Schedule** These options affect how your schedules calculate and **SHOULD** be understood.
- **Calculation** These options **SHOULD** be understood if you are progressing your schedule.
- **Spelling** Controls the spelling check options.
- **Security** This is a new feature in Microsoft Project 2003 allowing file attributes to be removed on save and security options for the running of Macros.
- **Save** Specifies where and how your files are saved and do not affect how your project calculates.
- **Interface** This new form has options for the Indicator and Options button settings and the Project Guide, which are intended to assist the user but do not provide further scheduling options.

Some options apply to the current project only and some to all projects:

- Those that apply to the current project have a subheading **Calculation options for 'Project 1'**.
- Those that apply to all projects have a subheading of **Calculation options for Microsoft Office Project**.

17.2 View

The options in this form control the display of data and not the results of calculation.

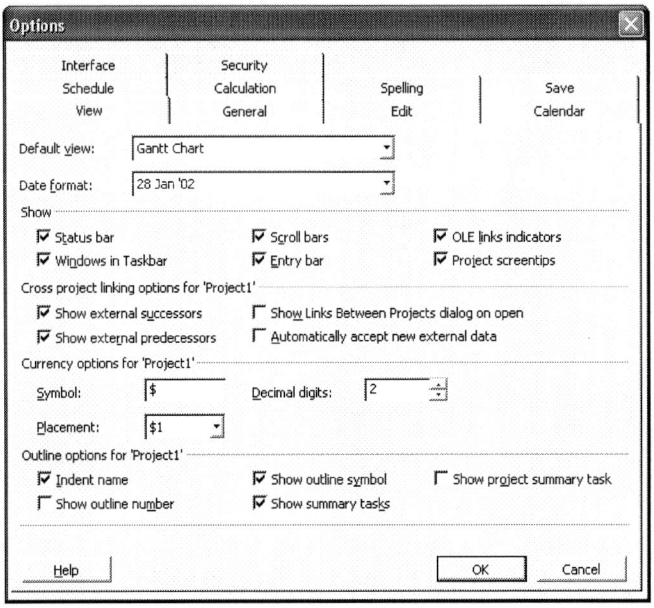

- **Default view:** Selecting the View applied when a new project is created.
- **Date format:** A selection of date formats are available from the drop-down box for the display of data in columns. This format is applied to all projects and all views.
  - The date format will be displayed according to a combination of your system default settings and the Microsoft Project 2003 Options settings. You may adjust your date format under the system Control Panel, Regional and Language Options. Changing the Control Panel setting will change the options available in the Microsoft Project 2003 **Options** form.
  - The dates on the bars are formatted in the **Format**, **Layout…** option.
  - Individual Tables may also have their dates formatted differently to the default format.

On international projects there can be confusion between the numerical US date style (mmddyy) and the numerical European date style (ddmmyy). For example, in the United States 020705 is read as 07 Feb 05 and in many other countries as 02 Jul 05. You should consider adopting the ddmmmyy style, **30 Jan 00** or mmmddyy style, **Jan 30 00**.

- **Show** allows each of the display options to be hidden or displayed.
- **Cross project linking options for 'Project 1'** are options for displaying predecessor and successor task information from other projects. Inter-project relationships are not covered in this book.
- **Currency options for 'Project 1'** allows you to specify the:
  - Currency sign by inserting a sign into the box next to **Symbol:**,
  - How it is placed(before or after the value) as selected from the **Placement:** box, and
  - The number of decimal places displayed from the **Decimal digits:** box.
- **Outline options for 'Project 1':**
  - **Indent name** Indents the title for each Outline level to the right.
  - **Show outline number** Shows the Outline Number 1, 1.1, 1.1.1, etc.
  - **Show outline symbol** – Displays the - and + sign by the Task Name.
  - **Show summary tasks** Uncheck this to hide the summary tasks.
  - **Show project summary task** Displays the Project Title as Task – **OzBuild Bid** and a Summary bar that spans the duration of the project.

	Task Name
1	⊟ **1 OzBuild Bid**
2	⊟ **1.1 Research**
3	1.1.1 Bid Request Documents Received
4	1.1.2 Bid Strategy Meeting

17.3 General

These options control the display of software productivity tools, display of help, and default resource settings when resources are calculated.

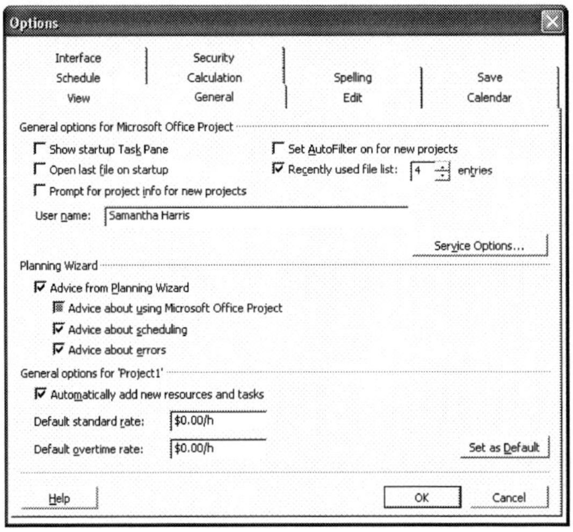

- **General options for Microsoft Project** – These options are self-explanatory and apply to all projects.
- **Planning Wizard** – These options specify what advice the Planning Wizard will offer you and apply to all projects.
- **General options for 'Project 1'**
  - ➢ **Automatically add new resources and tasks** – This option confirms if a resource exists in the resource pool. If it does not, a form will ask you to confirm if you want the resource to be added.
  - ➢ **Default standard rate:** – The default standard hourly rate assigned to new resources.
  - ➢ **Default overtime rate**: – The default overtime rate assigned to new resources.
- **Set as Default** – Makes these rates the default rates for each new project.

17.4 Edit

This form mainly controls the edit and display functions.

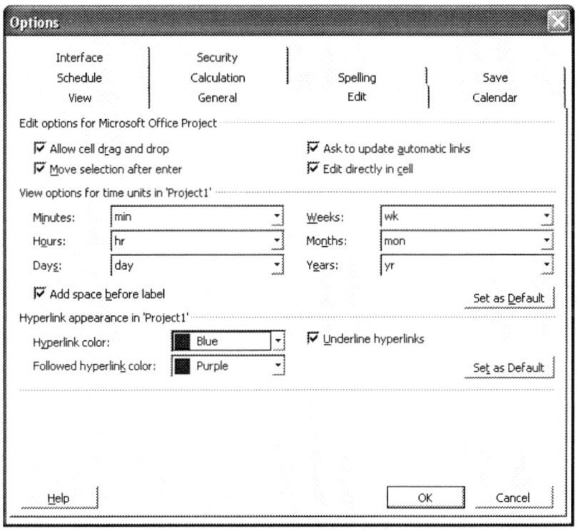

- **Edit options for Microsoft Project** when checked:
  - **Allow cell drag and drop** – Allows you to select a cell and drag the contents to another location.
  - **Move selection after enter** – After completing an entry in a cell, pressing the **Enter Key** will result in the cursor moving down a line to the cell below.
  - **Ask to update automatic links** – On opening a file which has links to other programs such as Excel, you will be asked if you want to refresh the links to the most recent data. To create a link, copy the selected data from another program. Then, in Microsoft Project 2003, select **Edit**, **Paste Special…**, select the **Paste Link** option, and the data will be placed in the bar chart area.
  - **Edit directly in cell** – Allows you to edit the data directly in the cell. Otherwise data has to be edited in the **Edit Bar** near the top of the screen.
- **View options for time units in 'Project 1'**
  - **Minutes:**, **Hours:**, **Days:**, **Weeks:**, **Months:**, **Years:** – From the drop-down boxes, select your preferred designators for these units. It is **RECOMMENDED** changing "days" to "d" and "hr" to "h" to make the duration columns narrower.
  - **Add space before label** – Places a space between the value and the label; this makes the data column wider. It is **RECOMMENDED** to uncheck this to make date columns narrower.
- **Hyperlink appearance in 'Project 1'**
  - **Hyperlink color:** and **Followed hyperlink color:** – Select from the drop-down box the color you require for these data items.
  - **Underline hyperlinks** – Select if you want these data items to be underlined.
- Three phonetics options are available with some Asian operating systems that have IME (Input Method Editor) loaded: Katakana Half , Katakana, or Hiragana.
- **Add space before label** when checked allows a gap to be placed between the amount of days and the word 'day' in the Duration column.

17.5 Calendar

The title of these options does not clearly indicate their functionality. Options for **Hours per day:**, **Hours per week:** and **Days per month: SHOULD** be understood, especially if multiple Task calendars are used.

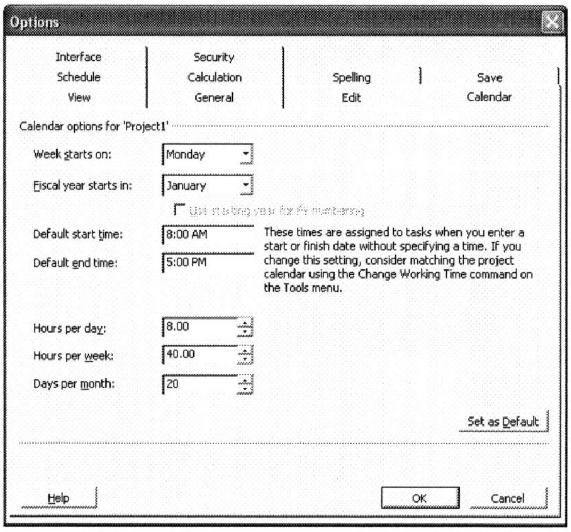

- **Calendar options for 'Project 1'**
  - **Week starts on:** – Sets the first day of the week. This is the day of the week that is displayed in the timescale. This setting affects the display of the calendar in the Change Working Time form.
  - **Fiscal year starts in:** – A month other than January may be selected as the first month of the fiscal year for companies that want to schedule projects using their financial years.
  - **Use starting year for FY numbering** – This option becomes available when a month other than January is selected as a **Fiscal year starts on:**. The year in the timescale may be assigned as the year that the fiscal starts or the year that the fiscal year finishes.

- **Default times**:
  - **Default start time:** – This is the time of day that tasks are scheduled to start when a start date constraint (without a time) is applied. It is also the default time for an Actual Start when an Actual Start (without a time) date is entered.
  - **Default end time:** – This is the time of day that tasks are scheduled to finish when a finish date constraint (without a time) is applied. It is also the default time for an Actual finish when an Actual finish date (without a time) is entered.

It is best to match both of these times with your project calendar normal start and finish times. When the **Default start time:** is set later than the calendar normal start time, a task set with a constraint without a time will appear to finish one day later than scheduled. Therefore a one-day task will span two days and a two-day task, three days.

- The following options need to be understood as they affect how summary durations are displayed. **NOTE: THESE OPTIONS ARE IMPORTANT.** Microsoft Project 2003 effectively calculates in hours. Task durations may be displayed in days, weeks and months. These summarized durations are calculated based on the parameters set in **Options, Calendar**. These options work fine when all project calendars are based on the same number of work hours per day. When tasks are scheduled with calendars that do not conform to the **Options, Calendar** settings (e.g. when the **Options, Calendar** settings are set for 8 hours per day and there are tasks scheduled on a 24 hour/day calendar), the results often create confusion for new users. The examples below use the **Options, Calendar** settings from the previous page:
  - **Hours per day:** – This setting is used to convert the calendar hours to days. For example, a task entered as a 3-day duration and assigned a 24 hour per day calendar will be displayed as 1-day elapsed duration. (This is 1/3 of the "normal" duration since each day is an 8-hour equivalent on a 24-hour calendar.) See Task 1, below:
  - **Hours per week:** – This setting is used to convert the calendar hours to weeks. For example, a task entered as 2 weeks (10 working days) and assigned a 24 hour per day calendar is displayed with a 3-day and 6-hour duration in the bar chart. This is 1/3 of the "normal" duration. See Task 2, below:
  - **Days per month:** – This option is used to convert the displayed task duration from days to months. For example, a 0.5-month task on a 24 hour per day calendar is displayed as 3-day and 6-hour duration in the bar chart. See Task 3, below:

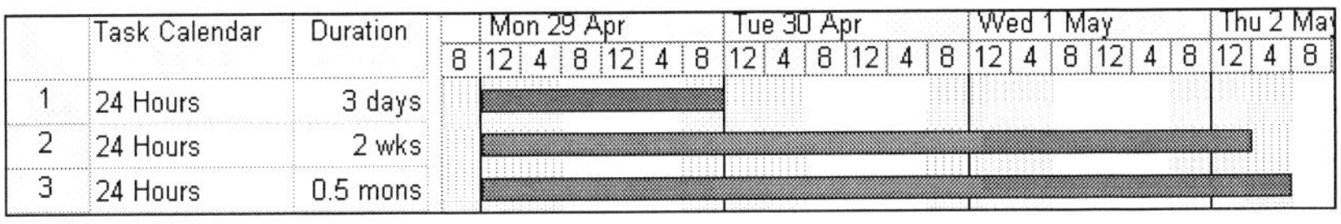

Task Calendar
Duration
Mon 29 Apr
Tue 30 Apr
Wed 1 May
8 12 4 8 12 4 8 12 4 8 12 4 8 12 4 8 12 4 8 12 4 8
1
24 Hours
3 days
2
24 Hours
2 wks
3
24 Hours
0.5 mons

  - **Set as Default:** – Sets the current calendar settings as the default in the Global.mpt project and is used thereafter as the calendar settings in all new projects.

Other scheduling software also exhibit this time conversion problem when using multi-calendars. This display can lead to a great deal of confusion. To avoid confusion when using multi-calendars, it is suggested that you only display durations in hours.

17.6 Schedule

The options below affect how your schedules calculate and **SHOULD** be understood.

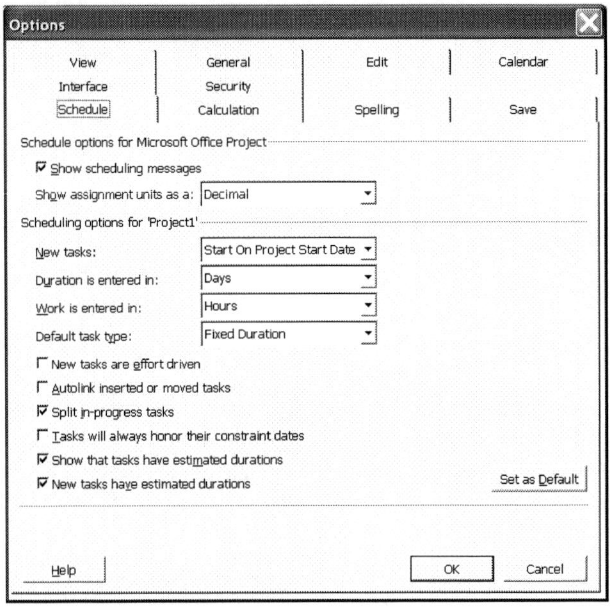

- **Schedule options for Microsoft Project**
  - **Show scheduling messages** – Uncheck this box to prevent the software from displaying advice on scheduling.
  - **Show assignment unit as a:** – The options are **Percentage** or **Decimal**. You have the option of assigning and displaying resource assignments as either a percentage, such as 50%, or a decimal, such as 0.5. Both these values would assign a person to work half time on that task.
- **Scheduling options for 'Project 1'**
  - **New tasks:** – When adding new tasks, you have the option of the new task either:
    - Scheduled to **Start On Project Start Date** and the task will be scheduled **As Soon As Possible**, or
    - **Start on Current Date** and the task will be assigned (without any warning dialog box) a **Start No Earlier Constraint** equal to the current date.
  - **Duration entered in:** – You should select the units you will most frequently use for your task duration units. When "days" are selected, you will only need to type "**5**" in order to enter a 5-day task duration. Similarly, you will have to enter "**5w**" for a task that is 5 weeks long or "**5h**" for a task that is 5 hours long.
  - **Work entered in:** – This is similar to the **Durations entered in:** function, but applies to assigning work to a task. When hours are selected, you will only need to type "**48**" to enter 48 hours of work. If 5 days of work are to be assigned then "**5d**" needs to be entered.

Note: the options above are the author's preferred defaults.

The next two subjects, **Default task type:** and **Effort Driven**, are **EXTREMELY** important for understanding how resources are used and applied.

- **Default task type:** – There is a relationship between the **Duration** of a task, the **Work** (the number of hours required to complete a task) and the **Units per Time Period** (the rate of doing the work or the number of people working on the task). The relationship is:

  Duration x Units per time Period = Work

  The three options for the **Default task type:** are:

  Fixed Duration – The **Duration** will stay constant if either **Units per time Period** or **Work** is changed. If you change the **Duration**, then **Work** changes. This is the **RECOMMENDED** option for new users as the duration will not change as resources are manipulated.

  Fixed Units – The **Units** stay constant if either **Duration** or **Work** is changed. If you change the **Units per time Period**, then **Duration** changes.

  Fixed Work – The **Work** stays constant if either **Duration** or **Units per time Period** is changed. Your estimate will not change when you change **Duration** or **Units per time Period**. If you change the **Work**, then **Duration** changes.

- Check any of the six boxes at the bottom of the form:
  - **New tasks are effort driven:** – An **Effort Driven** task keeps the total work constant as resources are added and removed from a task. When a task is not **Effort Driven** then the addition of resources to a task will increase the total work assigned to a task. This option is not available with a **Fixed Work** task since, otherwise, it would not be fixed work. New users should consider **NOT** checking this option, then resources may be added independently without the Work other resource's changing.
  - **Autolink inserted or moved tasks:** – With this box checked, new or moved tasks are automatically linked to the tasks above and below in the new or inserted location. Moved tasks that had relationships to tasks immediately above and below in the old location are deleted and the original predecessors and successors are now joined with an FS relationship. This function may confuse you since the logic is changed automatically without warning when you add or move tasks. It is suggested that the option is **NEVER** switched on as dragging an activity to a new location may completely change the logic of a schedule.
  - **Split in-progress tasks:** – This option allows the splitting of in-progress tasks. This option must be checked for the following functions to operate:

    1. **Tools**, **Tracking**, **Update Project...**, **Update work as completed through**,
    2. **Tools**, **Tracking**, **Update Project...**, **Reschedule uncompleted work to start after**, and
    3. **Tools**, **Options...**, **Calculation** tab, **Status Date Calculation Options** which are the found option found under **Calculation options for 'Project 1'** allowing incomplete or completed parts of tasks to be moved to the Status Date.

- **Tasks will honor their constraint dates:** – This option will make all constraints override relationships. For example, a task with a **Must Start On** constraint, which is prior to a predecessor's Finish Date, will have an Early Start on the constraint date and not the scheduled date. (This is similar to converting all P3 and SureTrak **Must Start On** constraints to **Mandatory** constraints.) When checked, the **Total Slack** may not calculate as the difference between Late Start and Early Start. Examine the following two examples with the option box checked and unchecked:

Tasks will honor their constraint dates: option box checked.

Start	Finish	Late Finish	Total Float	Constraint Date	Constraint Type
10 Jan '06	17 Jan '06	12 Jan '06	-3 days	NA	As Soon As Possible
18 Jan '06	18 Jan '06	13 Jan '06	-3 days	NA	As Soon As Possible
14 Jan '06	14 Jan '06	14 Jan '06	-3 days	14 Jan '06	Finish No Later Than

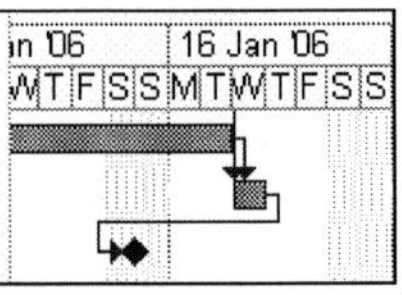

Note: the third task starts before the predecessor finishes and the total slack of the second task is calculated at –3 days, which is not the difference of the early and late dates, thus the conventional Total Float calculation is also incorrect.

Tasks will honor their constraint dates: option box NOT checked.

Start	Finish	Late Finish	Total Float	Constraint Date	Constraint Type
10 Jan '06	17 Jan '06	12 Jan '06	-3 days	NA	As Soon As Possible
18 Jan '06	18 Jan '06	13 Jan '06	-3 days	NA	As Soon As Possible
18 Jan '06	18 Jan '06	14 Jan '06	-3 days	14 Jan '06	Finish No Later Than

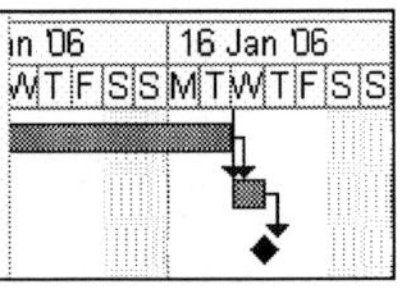

It is suggested that the option is **NEVER** switched on as schedule may appear to be achievable when it is not.

- **Estimated Durations** – These two options do not affect the calculation of projects. A new task is assigned an estimated duration. Once a duration is entered, this assignment is removed. The **General** tab on the **Task** form has a check box which establishes when a task has an estimated duration.
  - **Show that tasks have estimated durations:** – A task with an estimated duration is flagged with a "?" after the duration in the data columns when this option is checked.
  - **New tasks have estimated durations:** – When a task is added, it will have the estimated check box checked.

17.7 Calculation

The following options **SHOULD** to be understood to confidently progress your schedule.

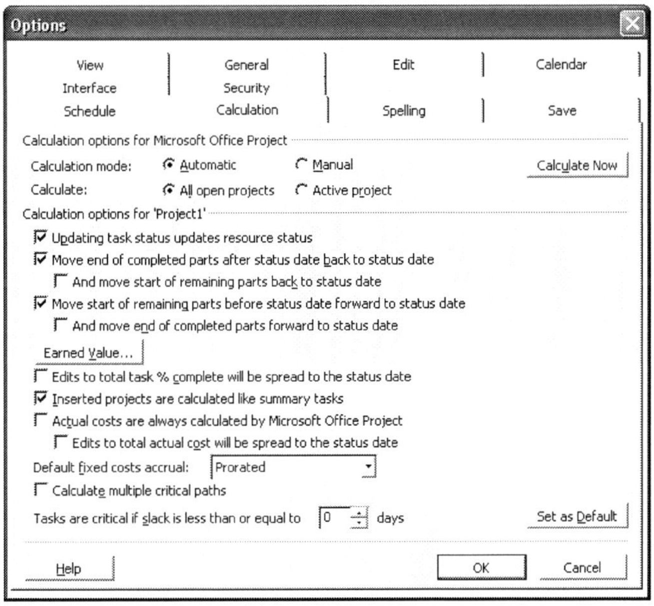

- **Calculation options for Microsoft Project**:– Select one of the following two options:
  - **Automatic** – This option recalculates the schedule every time a change is made to data.
  - **Manual** – This option requires you to press the **F9 Key** to recalculate the schedule.

- Also, select one of the two following options:
  - **Calculate all open projects** – Calculates all open projects when the recalculation takes place.
  - **Calculate the active project** – Calculates only the active project (the one that is displayed on your screen) when recalculation takes place.
  - Calculate Now – Recalculates the schedule(s) in accordance with the selected option above.

  To recalculate a project when calculation option is set to manual, press **F9**.

- **Calculate options for 'Project 1'**
  - **Updating task status updates resource status:** – This option links **% Complete** to **% Work** and in turn **%Work** is linked to **Actual Work**. Therefore when this option is checked:

    Actual Work = **% Work x Work** and
    Remaining Work = **Work – Actual Work**

    When unchecked the **% Complete** and **% Work** may have different values and **% Work** or **Actual Work** have to be entered separately from the **% Complete**.

The example below shows the calculations with this option checked. If you uncheck this option, you may enter actual work and remaining work independently of the **% Complete**.

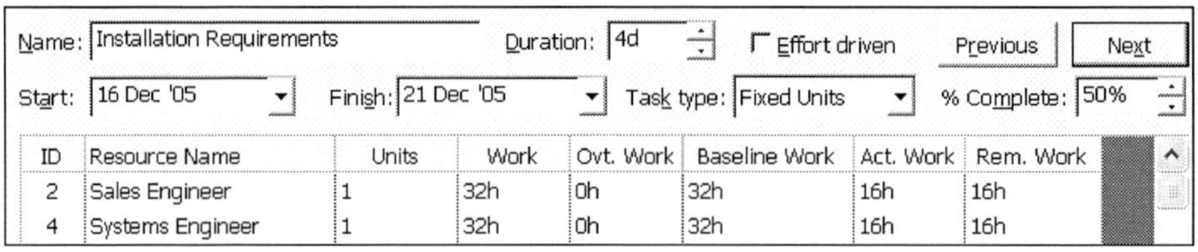

Summary Duration % Completes
A Summary Task **% Complete** is calculated by the sum of **Actual Durations** of all the **Detailed Tasks** divided by the sum of the **Durations**.

A summary task **% Work** is calculated based on the sum of **Actual Work** divided by the sum of the **Work** and will gives an indication of progress based on hours of effort.

This option also determines the manner in which a % Complete that is assigned to a Parent Task gets spread to detailed tasks. See the following two examples where 60% is applied to the summary task entitled "Research":

- With the **Updating task status updates resource status CHECKED**.

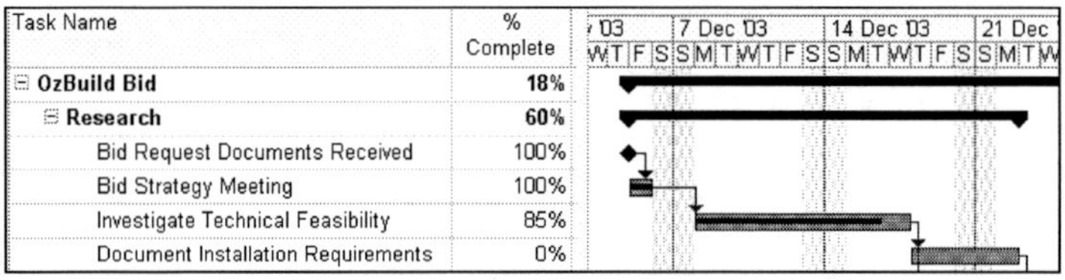

- With the **Updating task status updates resource status UNCHECKED**.

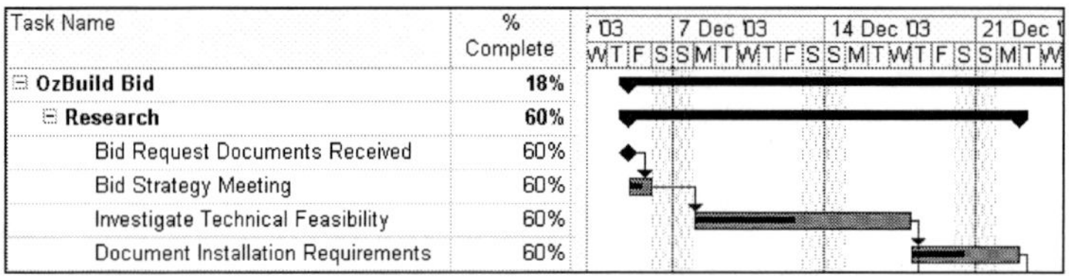

- New functions have been introduced in Microsoft Project 2002, which are intended to assist schedulers to place new tasks in a logical position with respect to the **Status Date**. The following four options are new to Microsoft Project 2002 and are covered in detail in the **Status Date Calculation Options** section of **Chapter 15 TRACKING PROGRESS**.
  - **Move end of completed task parts after status date back to status date**
  - **And move start of remaining parts back to status date**
  - **Move start of remaining parts before status date forward to status date**
  - **And move end of completed parts forward to status date**

  These functions do not work intuitively and the instructions in the Microsoft documentation do not outline clearly how these functions work. It is recommended that you review these options carefully before applying them.

- The [Earned Value...] button opens the **Earned Value** form:

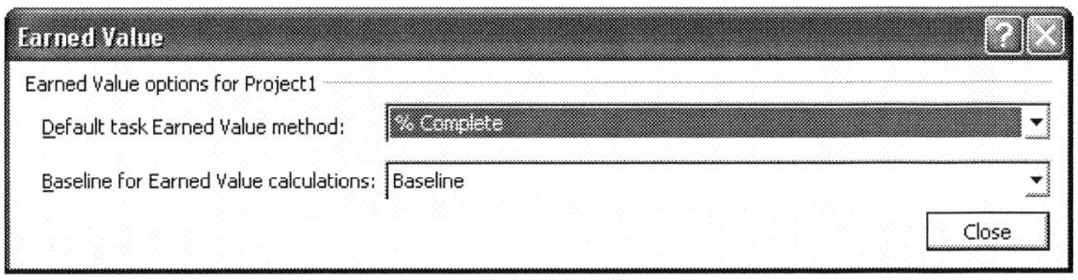

  - The **Default task Earned Value method:** may be set to use the task **% Complete** or the new field in Microsoft Project 2002 – **Physical % Complete**. This is useful when the Earned Value is not progressing in proportion to the duration.
  - The **Baseline for Earned Value calculations:** allows the selection of any of the baselines for the calculation of the Earned Value.
- **Edits to total task % complete will be spread to the status date** – When this option is selected and a **Status Date** is set in the **Project Information** form, then the % Complete is spread from the Actual Start to the Status Date and may be viewed in the **Task** or **Resource Usage** forms. (Right-click to add a % Complete column.)
- **Inserted projects are calculated like summary tasks** – When checked, an inserted project task's Total Float is calculated to the end of all Projects like a group of related projects. When unchecked, each Project's Total Float is calculated to the end of that Project only.
- **Actual costs are always calculated by Microsoft Project** – With this option checked, the resource **Actual Cost** is calculated by Microsoft Project from the resource **Work** and **Rates**. When unchecked, you have to directly enter the **Actual Cost** and **Actual Work**.
- **Edits to total actual costs will be spread to the status date** – When this option is selected and a **Status Date** is set in the **Project Information** form, then Actual Costs are spread from the Actual Start to the Status Date. Actual Costs should not be entered in a time-phased form such as the Task or Resource Usage forms, but the results may be viewed in these forms.
- **Default fixed costs accrual:** – This option sets the default accrual method for **Fixed Costs** (covered in **Chapter 19 ASSIGNING RESOURCES AND COSTS TO TASKS**). Fixed cost may be accrued at the **Start**, **End**, or **Prorated** over the duration of the task.
- **Calculate multiple critical paths** – When checked Tasks without successors have their Late Dates set to equal their Early Dates, and are calculated with zero Total Slack, and are on the critical path. (This is the same function as the Primavera SureTrak and P3 **Open Ends** options.)
- **Tasks are critical if slack is less than or equal to ? days** – This option will flag tasks as being critical that have **Total Slack** (Total Float) less than or equal to the value as established in this form. Critical tasks may be displayed in the **Critical** column and on bars when the "critical bar" is displayed. This is useful for flagging "near critical" tasks.

17.8 Spelling

The spell check options are self explanatory and are applied to all projects:

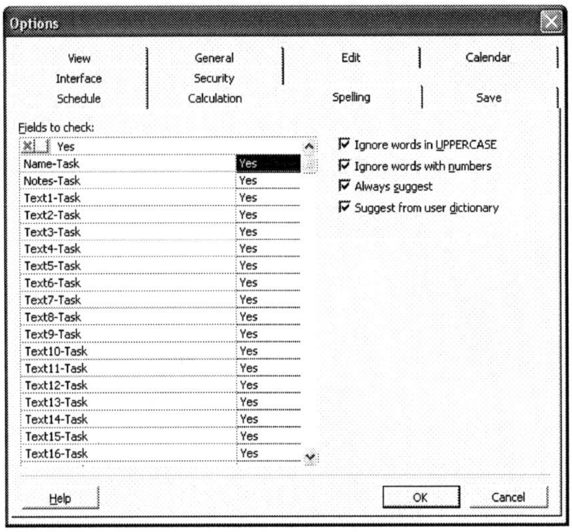

17.9 Security

This is a new feature in Microsoft Project 2003 allowing:

- The properties of Author, Manager, Company and Last Saved by are removed from a file when it is saved, and
- There are security options that may be set to nominate which Macros may be used or not used. This is aimed at preventing Macros with a virus being activated.

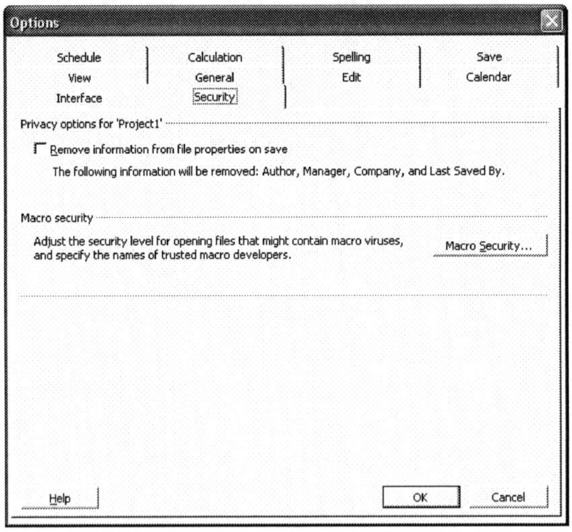

17.10 Save

This Options tab specifies where and how your files are saved. It does not affect how your project calculates.

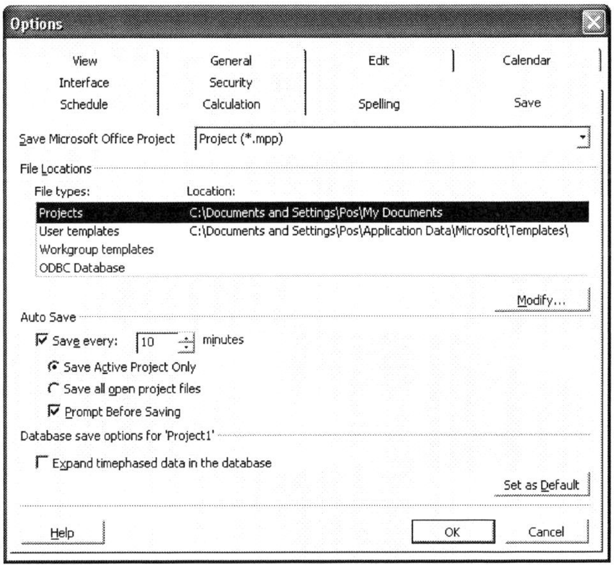

- **Save Microsoft Project files as:** – There are several formats that your project may be saved in. The default file type may be defined here but the file type may be changed at the time a project file is saved. This book only covers **Project (*.mpp)** and **Template (*.mpt)** formats in detail:
  - **Project (*.mpp)** – The normal format to save Microsoft Project 2003 files.
  - **Template (*.mpt)** – Use this format to save projects that you want to use as project templates. Save them in the **User templates** directory that is specified in the **Tools**, **Options…**, **Save** tab.
  - **Microsoft Project 98 (*.mpp)** – This format will allow users of Microsoft Project 98 to read your files. If you have Microsoft Project 98, you may save the file in MPX format for transfer to other scheduling programs.
  - **Project Database (*.mpd)** – This is a database format, which allows multiple projects to be saved in one database. It may be used for exporting data to other programs.
  - **Microsoft Access Database (*.mdb)** – A part or all of a project may be saved in Microsoft Assess format and may be read and modified by Access.
  - **ODBC Database** – your project may be saved as an ODBC compliant database.
- **File Locations** – This is where you set the default locations for your project data files, template files, workgroup templates, and ODBC database files.
- **Auto Save** – These options are self-explanatory.
- **Database save options for 'Project 1'**
  - **Expand time-phased data in the database** – This function creates time phased data in external databases.

17.11 Interface

This form, new to Microsoft Project 2003, has options for:

- The **Show indicator and Options buttons for:** settings, and
- The Project Guide, which are intended to assist the user but do not provide further scheduling options.

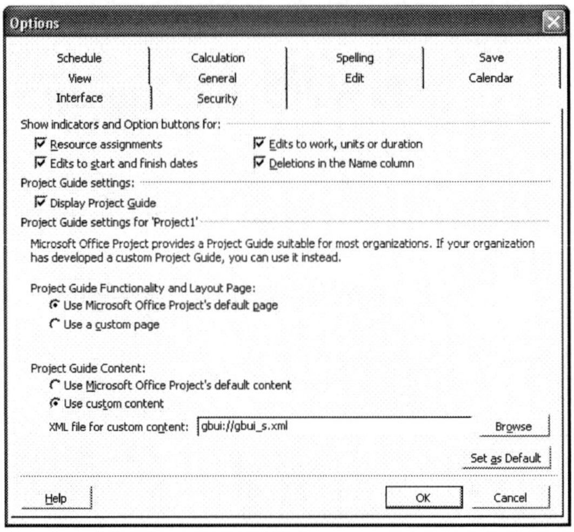

17.11.1 Graphical Indicators

The Indicator buttons give advice to the scheduler when certain functions are used which have more than one potential outcome.

- Below is an example of how the **Show indicators and buttons for:** options work when a task description is deleted:

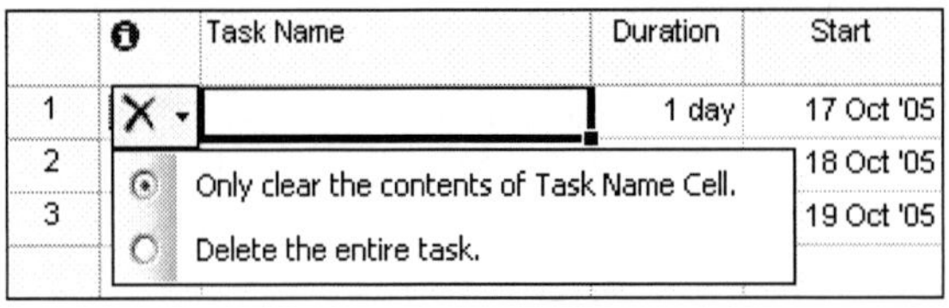

and when a date is edited:

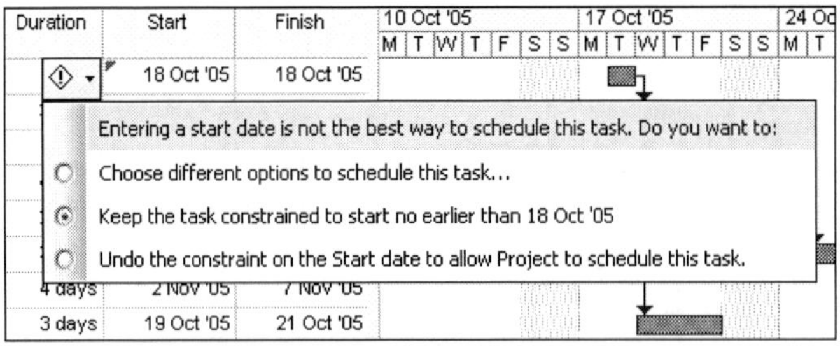

17.11.2 Project Guide

The **Project Guide** is a wizard type function that assists the user to create a new schedule. It is self explanatory and not covered in this book.

18 CREATING RESOURCES

A resource may be defined as something or someone that is assigned to a task and is required to complete the task. This includes people or groups of people, materials, equipment and money.

There are a large number of resource functions available in Microsoft Project 2003. Without getting into too much detail, this book will outline the important resource-related functions to enable you to create and assign resources to your schedules.

It is recommended that the minimum number of resources are assigned to tasks when it is planned to status a schedule. Avoid cluttering the schedule with resources that are in plentiful supply or are of little importance. Every resource added to the schedule will need to be statused. In this way, the workload will go up as more resources are allocated to schedule.

When you create your resources, you should consider them within the context of the following headings:

- **Input Resources** – Those that are required to complete the work:
  - Individual people by name.
  - Groups of people by trade or skill.
  - Individual equipment or machinery by name.
  - Groups of equipment or machinery by type.
  - Groups of resources such as Crews or Teams made up of equipment and machinery.
  - Materials.
  - Money.
- **Output Resources** – Things that are being delivered or produced:
  - Specifications completed.
  - Bricks laid.
  - Lines of code written.
  - Tests completed.

The following steps should be followed to create and use resources in a Microsoft Project 2003 schedule:

- Create your resources in the **Resource Sheet**.
- Assign the resources to tasks.
- Manipulate the resource calendar if resources have special timing requirements.

This chapter will concentrate on:

- The creation of Resources in the **Resource Sheet**,
- Understanding **Task Type** and **Effort-Driven Tasks**,
- Assign the Resources to Tasks, and
- Editing **Resource Calendars**.

Resources may be shared between projects using the **Tools**, **Resource Sharing** function.

18.1 Creating Resources in the Resource Sheet

Select **View**, **Resource Sheet** to add resources to the Resource Sheet.

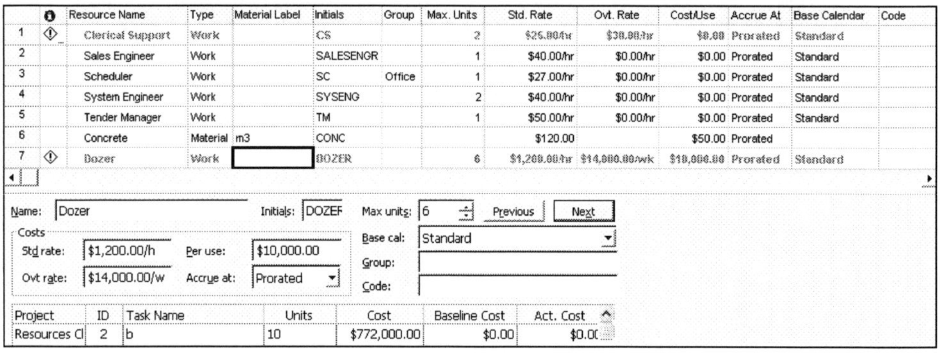

- To add a new resource:
  - ➢ Click on the first blank line, or
  - ➢ Highlight the line where you would like to insert your resource and select **Insert**, **New Resources**, or
  - ➢ Use the **Ins Key**.
- The first column, with a [icon] in the header, is the indicator column; [icon] is the Notes indicator; and [icon] is the resource over-allocation indicator.
- Enter the **Resource Name**. Microsoft Project 2003 allows duplicate Resource Names to be entered into the Resource Table, but this situation should be avoided by visually checking all your Resource Names.
- Enter one of the two Resource **Types**: **Work** or **Materials**. They function differently:
  - ➢ **Work** resources, such as people, often have a limit to the number that are available. This type of resource may have a maximum number assigned, an overtime rate and a calendar. It is not available with a **Materials** resource.
  - ➢ **Materials** resources have a **Material Label** which is not available with a **Work** resource.
- A **Material** resource may be assigned a **Material Label**, which may be the unit of measurement, such as m^3 or feet.
- The **Initials** are filled out by Microsoft Project 2003 with the first character of the **Resource Name**. This Initial field should be unique and is displayed in some Views and should be edited so it makes sense to you. For example, you could make this a person's initials, log-in name or payroll number.
- You may enter any text in the **Group** field if you want to group, filter or sort your resources by this field. This could be used for Department, Branch or Skill.

- Enter the **Max Units**. 100% (or 1.00) would represent one person and 400% (or 4.00) would represent four people. The method of entering the **Resource Units** may be formatted as a percentage or as a decimal. Select the **Tools**, **Options...**, **Schedule** tab to set your option:

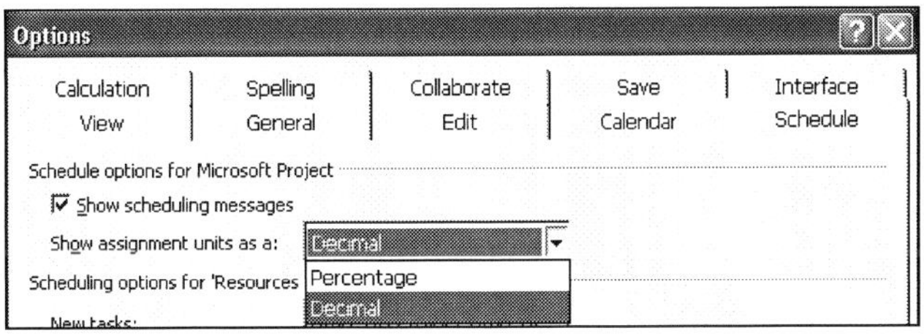

- Edit the **Standard Rate**, if required, for the resource. The default units are "hours" (hr), but this may be edited to be the cost per:
  - minute – min
  - hour – hr
  - day – day
  - week – wk
  - month – mon
- Enter the **Overtime Rate** for the resource. A salaried person may not be paid overtime and would have a zero rate for overtime.
- Enter the **Cost per Use**. This could represent a mobilization cost and is applied each time a resource is assigned to a task. The **Cost per Use** is accrued when an actual start date is entered.
- **Accrue At**:
  - **Start** – The costs for this resource are incurred when the task commences.
  - **Prorated** – The costs for this resource are spread over the duration of the task.
  - **End** – The costs for this resource are incurred when the task is complete.
- **Base Calendar** – This is the calendar assigned to the resource and may be edited to suit each individual resource.

18.2 Grouping Resources in the Resource Sheet

Resources may be Grouped on any data fields, such as Custom Fields and Custom Outline Codes, using the **Project**, **Group by:** function. The example below shows resources grouped by **Resource Group**:

Resource Name	Type	Materia Label	Initials	Group	Max. Units	Std. Rate	Ovt. Rate	Cost/Use	Accrue At	Base Calendar
⊟ **Group: Equipment**				**Equipment**	**7**			**$1,700.00**		
Dozer	Work		DOZER	Equipment	6	$2,500.00/wk	$0.00/hr	$1,000.00	Prorated	Standard
Excavator	Work		EXCAV	Equipment	1	$400.00/day	$0.00/hr	$700.00	Prorated	Standard
⊟ **Group: Material**				**Material**				**$50.00**		
Concrete	Material	m3	CONC	Material		$120.00		$50.00	Prorated	
Crushed Rock	Material	ton	CRCK	Material		$80.00		$0.00	Prorated	
⊟ **Group: Office**				**Office**	**7**			**$0.00**		
Clerical Support	Work		CS	Office	2	$25.00/hr	$30.00/hr	$0.00	Prorated	Standard
Sales Engineer	Work		SALESENGI	Office	1	$40.00/hr	$0.00/hr	$0.00	Prorated	Standard
Scheduler	Work		SC	Office	1	$27.00/hr	$0.00/hr	$0.00	Prorated	Standard
System Engineer	Work		SYSENG	Office	2	$40.00/hr	$0.00/hr	$0.00	Prorated	Standard
Tender Manager	Work		TM	Office	1	$50.00/hr	$0.00/hr	$0.00	Prorated	Standard

18.3 Resources Information Form

The **Resource Information** form is opened by double-clicking on a specific row with in the **Resource Sheet** view. There five four tabs in this form. These will not be described in detail:

18.3.1 General

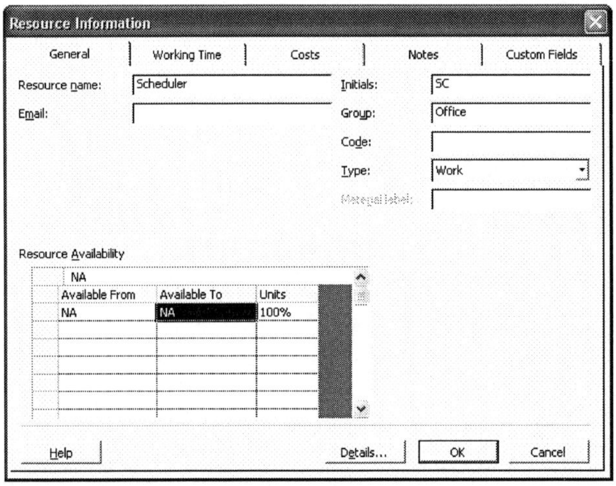

- The **Resource Availability** section in this form allows you to decide the availability of all resources.
- The **Email** field is used in email communication and not covered in this book.

18.3.2 Working Time

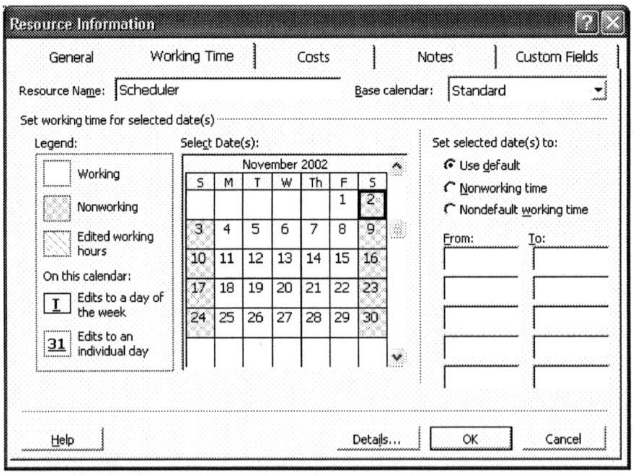

- This is where you may select each resource's **Base calendar** and edit the calendar to suit the resource's availability.

18.3.3 Costs

The **Cost rate** table allows you to adjust the resource rate over time and assign up to five different rates to a resource. The rate is assigned in the **Resource Information** form. This form is opened by double-clicking on the **Resource Name** for any assigned resource in either the **Resource Sheet**, **Task Usage** view and the **Resource Usage** view.

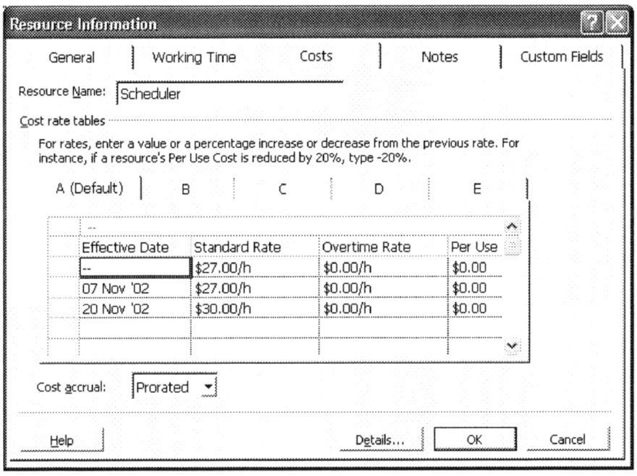

18.3.4 Notes

This is where notes are made about the resource.

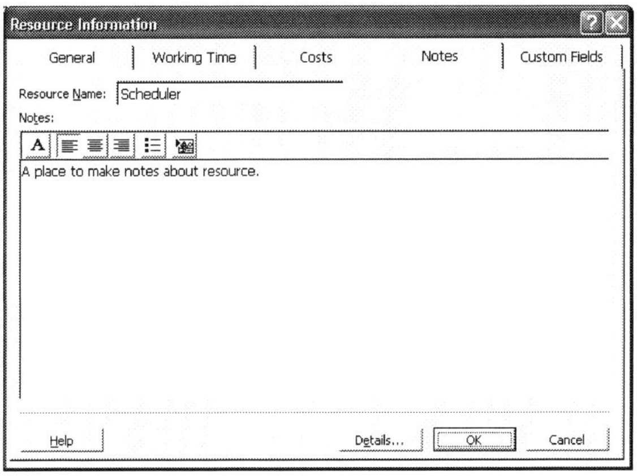

18.3.5 Custom Fields

Custom Fields are for assigning user defined costs to resources and not covered in detail in this book.

- Custom Resources fields may be created in the **Customize Fields** form by selecting **Tools, Customize, Fields….**
- After the fields have been defined in the **Customize Fields** form then these fields may be used to assign custom defined costs to resources from the **Custom Fields** tab of the **Resource Information** form.

18.4 Editing and Using Resource Calendars

Base Calendars are applied to tasks but they may not accommodate specific resource requirements. For example, a base calendar would not reflect when a person goes on vacation or is occupied on another project. Resource Calendars, on the other hand, can be used to schedule this resource-specific non-work time. Resource Calendars should be used when specific resources have a unique availability.

When a resource is created, a unique resource calendar is created automatically and is a copy of the **Base Calendar** selected when the resource is created. This Resource Calendar may be modified, if required.

The Resource Calendar may be edited by either:

- Clicking on **Resource Calendar** in the **Resources Sheet**, or
- Opening the **Resource Information** form by double-clicking on a **Resource Name** in most forms and selecting the **Working Time** tab.

All edits to the calendars are performed in the same manner as the Base Calendars.

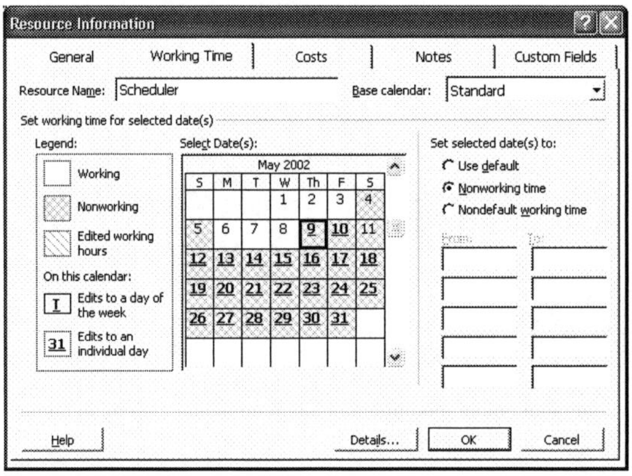

A task will always be calculated using a **Resource Calendar** except when a **Task Calendar** has been assigned. In this case, the **Scheduling ignores resource calendars** option in the **Task Information** form becomes available.

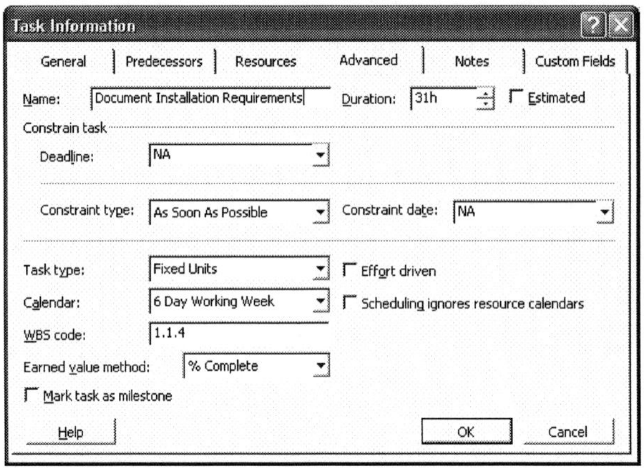

WORKSHOP 15

Defining Resources

Preamble

The resources must now be added to this schedule. Since we have statused our project, we need to revert to the un-statused schedule that we saved prior to statusing the current schedule.

Assignment

1. Save and close your current project.
2. Open the **OzBuild No Resources** project and save it as **OzBuild With Resources**.
3. Select **Tools**, **Options…** from the menu. From the **Schedule** tab, set **Show assignment Units as a: Decimal**.
4. Open the **Resource Sheet** view and add the following resources to the project:

	Resource Name	Type	Material Label	Initials	Group	Max. Units	Std. Rate	Ovt. Rate	Cost/Use	Accrue At	Base Calendar
1	Clerical Support	Work		CS	Office	2	$25.00/hr	$30.00/hr	$0.00	Prorated	Standard
2	Sales Engineer	Work		SAE	Office	1	$40.00/hr	$0.00/hr	$0.00	Prorated	Standard
3	Scheduler	Work		SC	Office	2	$27.00/hr	$0.00/hr	$0.00	Prorated	Standard
4	Systems Engineer	Work		STE	Office	2	$40.00/hr	$0.00/hr	$0.00	Prorated	Standard
5	Bid Manager	Work		BM	Office	1	$50.00/hr	$0.00/hr	$0.00	Prorated	Standard
6	Report Binding	Material	Each Folder	BIND			$50.00		$0.00	Prorated	
7	Contract Consultant	Work		CC		1	$0.00/hr	$0.00/hr	$1,000.00	Prorated	Standard

5. Save your project file.

19 ASSIGNING RESOURCES AND COSTS TO TASKS

When a resource is assigned to a task it has three principal components:

- **Quantity**, in terms of **Work** or **Material** required to complete the task,
- **Units** which represents the number of people working on a task and
- **Cost**, calculated from the **Standard Rate**, **Overtime Rate** and **Cost per Use**.

Microsoft Project 2003 also has a function titled **Fixed Costs**, these are costs assigned to tasks without resources.

The resource data may be assigned and displayed in a number of Microsoft Project 2003 forms. You may either enter the **Units** and then the **Work** will be calculated, or enter the **Work** and then the **Units** will be calculated.

Some forms allow the **Units** only, some allow the **Work** only, and some allow both **Units** and **Work** to be entered and displayed in the one form.

Resources may be assigned to tasks using a number of methods including:

Method	Menu Command
• **Fixed Costs** assignment	Display the **Fixed Costs** and **Fixed Costs Accrual** columns.
• **Resource Assignment** form	Click on the **Assign Resource** icon. You may assign **Units** only.
• **Task Information** form	Double-click on a **Task** name or click on the **Task Information** icon You may assign **Units** only.
• The **Resource Task Details** forms	In the bottom window, select the **Task Details Form**, **Task Form** or **Task Name Form**, and then select the appropriate option from the **Format**, **Details** option. Each form has different options.
• The **Resources Names** and **Resource Initials** columns	Insert appropriate columns by right-clicking on the column heading where you want to insert a column and then assign a Resource.
• **Contoured** Resource assignment	Open the **Assignment Information** form by: • Double-clicking on a resource in the **Task Usage** or **Resource Usage** view, or • Right-clicking on a resource and selecting **Assignment Information** from the menu.
• **Shared Resources**	Select **Tools**, **Resource**, **Share Resources...** to share resources with other projects.

It is important that you understand these options before you embark on developing and statusing some complex schedules.

19.1 Task Type and Effort-Driven

These **Task type** and **Effort-driven** options operate after the first **Work** resource has been assigned to a task. The relationship amongst the following four variables are controlled by these functions:

- The number of **Resources**,
- The **Task duration**,
- The resource **Units per time Period**, and
- The amount of **Work** to be performed.

The **Task type** and **Effort-driven** functions control which variable changes when one of these four parameters is changed or when **Work** resources are added or removed from a task. These options may be set in a number of forms including the **Task** and the **Task Details** form. **Material** resources do not change values after they have been assigned.

19.1.1 Task Type – Fixed Duration, Fixed Units, Fixed Work

There is a relationship between the **Duration** of a task, the **Work** (the number of hours required to complete a task) and the **Units per Time Period** (the rate of doing the work or number of people working on the task). The relationship is:

Duration x Units per time Period = Work

There are three options for the **Default task type:** which decide how this relationship operates. They are:

- **Fixed Duration** The **Duration** stays constant if either **Units per time Period** or **Work** are changed. If you change the **Duration**, then the **Work** changes.
- **Fixed Units** The **Units per time Period** stay constant if either **Duration** or **Work** is changed. If you change the **Units per time Period**, then the **Duration** changes.
- **Fixed Work** The **Work** stays constant if either **Duration** or **Units per time Period** are changed. Your estimate will not change when you change **Duration** or **Units per time Period**. If you change the **Work**, then the **Duration** changes.

19.1.2 Effort-Driven

Once a resource has been assigned to a task, the **Effort** is the number of resource hours assigned to a task. The **Effort-driven** option decides how the effort is calculated when a resource is added or when a resource removed to a **Fixed Units** or **Fixed Work** task. There are two options:

- **Effort-driven** When a resource is added or removed from a task, the total effort assigned against a task remains constant. Adding or removing resources leaves the total effort assigned to a task as a constant unless all resources are removed.
- **Non Effort-driven** When a resource is added or removed from a task, the total effort against the other resources remains constant. Adding or deleting resources increases or decreases the total task effort.

It is recommended that you consider setting your default options to:

- The **Default task type:** as **Fixed Work** in the **Tools**, **Options...**, **Schedule** tab.
- This will make new tasks as **NOT** effort driven. The **New tasks are effort driven** box in the **Tools**, **Options...**, **Schedule** tab appears gray.

Then set these as your default options by clicking on the Set as Default icon and all new projects will have these options set. With this scenario:

- When you change the durations of your tasks your estimate of hours and cost will not change, and
- You will be able to enter resources against a task and you will be able to assign the hours against each resource as you assign the resource.
- You may then change tasks to different options as required by exception.

19.2 Fixed Costs

Fixed costs are a function where you may assign costs to a task without creating resources. It is a useful function if you require a cash flow only.

A fixed cost may be assigned directly to a task without a Resource Assignment.

- A fixed cost is assigned using the **Fixed Cost** column.
- The fixed cost may be accrued at the **Start**, **End** or **Prorated** over the duration of the task. This option is selected from the **Fixed Cost Accrual** column. The default for the **Fixed Cost Accrual** is set in the **Tools**, **Options...**, **Calculation** tab:

	Task Name	Duration	Fixed Cost	Fixed Cost Accrual	24 Oct '05 M	T	W	T	F
1	**Summary**	**5 days**	**$0.00**	**Prorated**	$240.00	$40.00	$40.00	$40.00	$240.00
2	$200.00 Fixed cost accrued at Start	5 days	$200.00	Start	$200.00				
3	$200.00 Fixed cost Prorated	5 days	$200.00	Prorated	$40.00	$40.00	$40.00	$40.00	$40.00
4	$200.00 Fixed cost accrued at End	5 days	$200.00	End					$200.00

- Fixed Costs will not be displayed in the **Resource Sheet** or **Resource Usage** views but are available in the **Task Usage** view and in a **Report**. The **Cash Flow** report will only give you the cash flow in weeks.
- Fixed costs are added to resource costs and the total of the two is shown in the **Cost** column.

Task Name	Fixed Cost	Fixed Cost Accrual	Resource Initials	Cost
Summary	**$0.00**	**Prorated**		**$8,800.00**
$200.00 Fixed cost accrued at Start	$200.00	Start		$200.00
$200.00 Fixed cost Prorated	$200.00	Prorated		$200.00
$200.00 Fixed cost accrued at End	$200.00	End		$200.00
Activity With Resources Only	$0.00	Prorated	PEH	$4,000.00
Activity With Costs & Resources	$200.00	Prorated	PEH	$4,200.00

When there are both **Fixed Costs** and **Resource Costs** assigned to a task, there is no column to display **Resource Costs** only.

When a Baseline is set, the Fixed costs and the Resource costs are added together in the Baseline costs value.

19.3 Assigning Resources Using the Resource Assignment Form

Highlight one or more tasks that you want to assign resources. Click the **Assign Resource** icon on the **Standard** toolbar to display the **Assign Resource** form.

The **Units** may be the number of people working on a task and may be displayed as Units shown in the picture on the left, or as a Percentage shown on the right. This option is changed in the **Schedule** tab of the **Options** form.

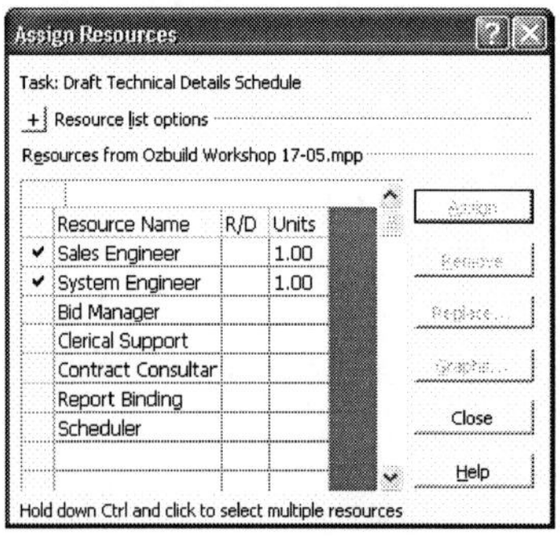

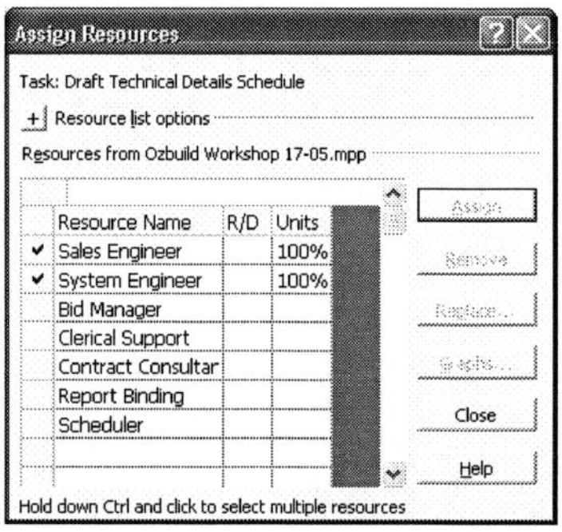

- Select the **Resource** from the form.
  - Type in the number of resources you want to assign under **Units**. This may be displayed in % or whole numbers depending on how your options are set. 1.00 is the same value as 100%.
  - Click on the Assign button to assign a resource to a task.
- You may then assign another resource to a task.
- Select a resource and click on the Remove button to remove a resource.
- Select a resource and click on the Replace... button to open the Replace Resource form to replace a resource.

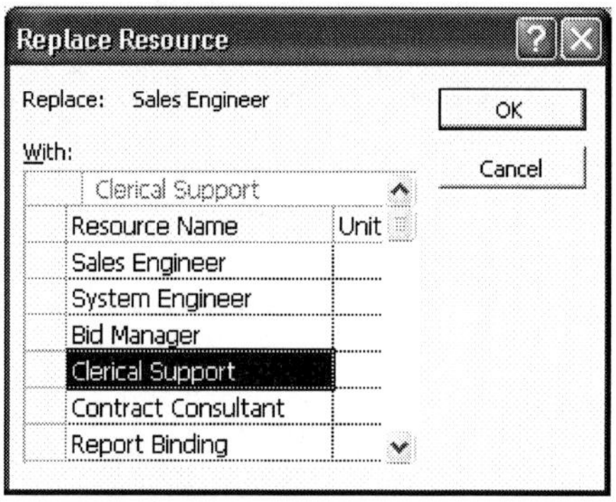

- The Resource list options at the top of the screen allows the resource list to be reduced by the use of filters and is very useful when there is a large number of resources and new resources to be created from your address book.

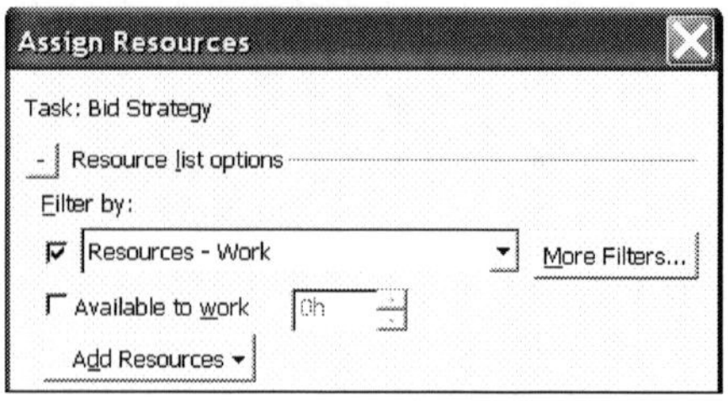

19.4 Assigning Resources using the Task Details Form

In the bottom window select the **Task Detail Form**, **Task Form** or **Task Name Form**. Then select the appropriate option from the **Format**, **Details** option. See below:

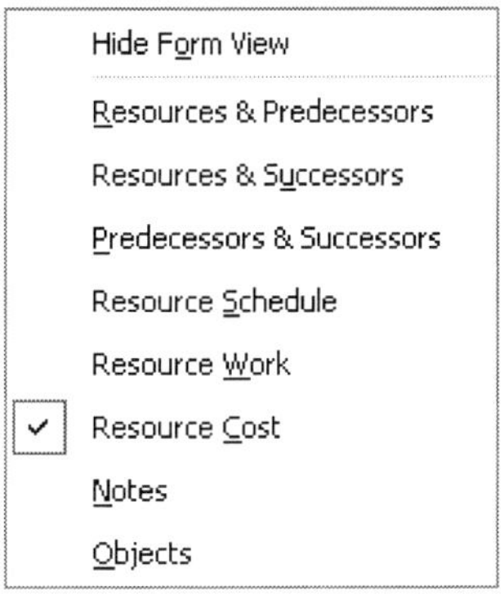

Form	Assignment Option
• **Resource & Predecessors**	**Units** and **Work**
• **Resource & Successors**	**Units** and **Work**
• **Resource Schedule**	**Work** only
• **Resource Work**	**Units** and **Work**
• **Resource Cost**	**Units** only with calculated costs

The **Task Entry** form with the **Resource Work** details form is displayed below.

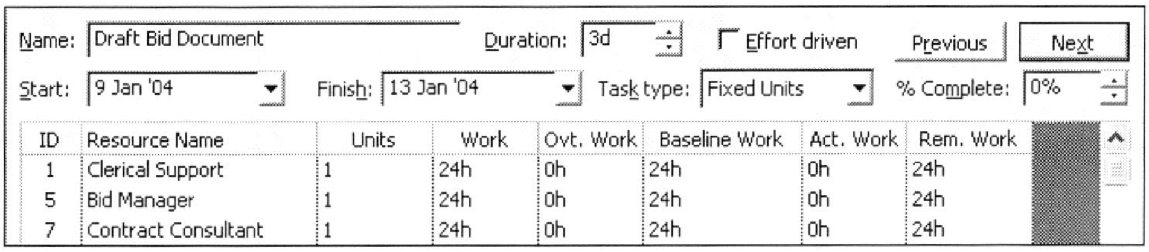

- Select the required resource in the drop-down box under the **Resource Name** heading.
- Enter the number of resources under **Units**, or
- Enter the amount of **Work** – The number of hours may be entered under this heading.
- **Ovt. Work** – Overtime Work may be assigned to reduce the duration of the task. The **Work** remains the same, but the **Overtime** value represents the amount of work being done on overtime.
- **Baseline Work** is copied from **Work** when a **Baseline** is set.
- **Act. Work** – **Actual Work** is entered when work is in-progress.
- **Rem. Work – Remaining Work** is calculated by subtracting **Actual Work** from **Work**.

19.5 Assigning Task Information Form

Double-click on a **Task** name or click on the **Task Information** icon to open the **Task Information** form.

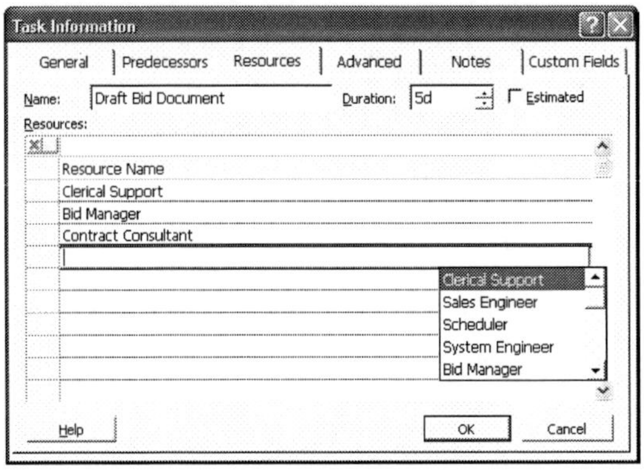

- Select the required resource from the drop-down box.
- Type in the number of required resources in the **Units** box.

19.6 Assignment of Resources to Summary Tasks

Summary tasks may be assigned **Fixed Costs**, **Work Resources** and **Material Resources**.

You must also be aware that when a Work resource is assigned to a summary task the task type is set to **Fixed Duration** and that setting may not be changed. Thus, any change in duration of a summary task due to rescheduling of associated detailed tasks will result in a change to the work assignment and the calculated costs of a Summary task.

It is recommended that unless a Summary task Work resource assignment is required to vary in proportion to the Summary task duration, then Work Resources should not be assigned to a Summary task. You should consider using Fixed Costs or a Material resource if appropriate.

19.7 Sharing Resources From With Other Projects

Microsoft Project allows resources to be shared with multiple projects. This feature is not covered in detail in this book. Select **Tools**, **Resource**, **Share Resources…** to open the **Share Resources** form where the project that it is wished to share resources from may be nominated and calculations options for leveling may be set.

19.8 Rollup of Costs and Hours to Summary Tasks

The task **Cost** and **Work** fields are calculated from the sum of the costs and work assigned to the related detailed tasks and those of the summary task.

It becomes difficult for a scheduler to check the total cost of a Summary task after costs and work have been assigned to both the Summary task and associated Detailed tasks. It is recommended that you consider only assigning costs and work to Detailed tasks.

Summary Tasks have the costs and work rolled up to give you a cost at any Outline level.

	Task Name	Duration	Start	Finish	Cost	Work
1	**OzBuild Bid**	**33 days**	**5 Dec '05**	**23 Jan '06**	**$26,046.00**	**656 hrs**
2	**Research**	**13 days**	**5 Dec '05**	**21 Dec '05**	**$8,720.00**	**216 hrs**
3	Bid Request Documents Received	0 days	5 Dec '05	5 Dec '05	$0.00	0 hrs
4	Bid Strategy Meeting	1 day	5 Dec '05	5 Dec '05	$1,040.00	24 hrs
5	Investigate Technical Feasibility	8 days	6 Dec '05	15 Dec '05	$5,120.00	128 hrs
6	Document Installation Requirements	4 days	16 Dec '05	21 Dec '05	$2,560.00	64 hrs
7	**Estimation**	**15 days**	**15 Dec '05**	**9 Jan '06**	**$8,416.00**	**232 hrs**
8	Request Component Tenders	3 days	3 Jan '06	5 Jan '06	$960.00	24 hrs
9	Develop Project Schedule	4 days	22 Dec '05	29 Dec '05	$864.00	32 hrs
10	Draft Technical Details Schedule	11.25 days	15 Dec '05	4 Jan '06	$5,952.00	160 hrs
11	Compile Cost from Component Bids	2 days	6 Jan '06	9 Jan '06	$640.00	16 hrs
12	**Proposal**	**10 days**	**10 Jan '06**	**23 Jan '06**	**$8,910.00**	**208 hrs**
13	Draft the Bid Document	3 days	10 Jan '06	12 Jan '06	$2,800.00	72 hrs
14	Meeting to Review the Draft Bid Document	1 day	13 Jan '06	13 Jan '06	$1,040.00	24 hrs
15	Design Presentation	1 day	16 Jan '06	16 Jan '06	$720.00	16 hrs
16	Edit Proposal Draft Bid Document	1 day	16 Jan '06	16 Jan '06	$600.00	16 hrs
17	Negotiate Component Work Packages	4 days	17 Jan '06	20 Jan '06	$3,030.00	64 hrs
18	Final Review of Bid Document	1 day	23 Jan '06	23 Jan '06	$720.00	16 hrs
19	Submit Bid	0 days	23 Jan '06	23 Jan '06	$0.00	0 hrs

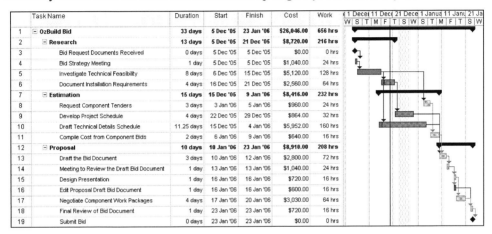

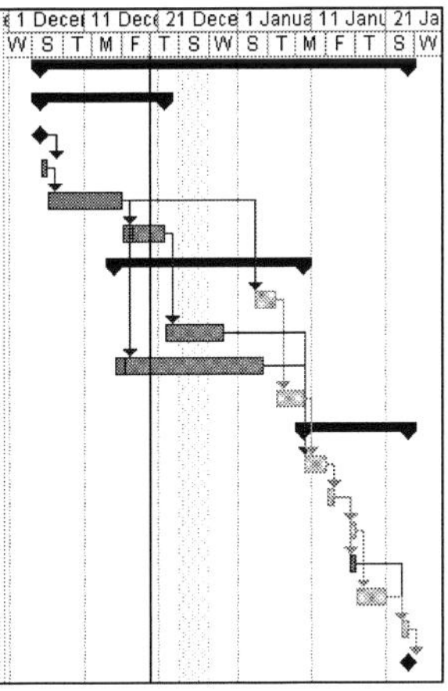

The Baseline Costs and Work are copied from the Cost and Work fields at the time that the Baseline is set and are not calculated from their child tasks. Thus, when tasks are added, deleted or moved to different Summary tasks then the Baseline Costs and Work are no longer the sum of their child tasks.

There is a function titled **Summary Task Interim Baseline Calculation**, which is covered in the **STATUSING PROJECTS WITH RESOURCES** chapter. This will allow the Summary Tasks Baseline Dates and Costs to be recalculated.

19.9 Contour the Resource Assignment

A Resource Assignment may be assigned to a task with a non-linear profile. This function is titled **Work Contour** and is similar to the Resource Curve function in Primavera P3 and SureTrak software. To assign a contour to a resource assignment:

- Open the **Assignment Information** form by:
  - ➢ Double-clicking on a resource in the **Task Usage** or **Resource Usage** view, or
  - ➢ Right-clicking on a resource and selecting **Assignment Information** from the menu.
- The picture below shows the Sales Engineer being assigned as **Back Loaded**.

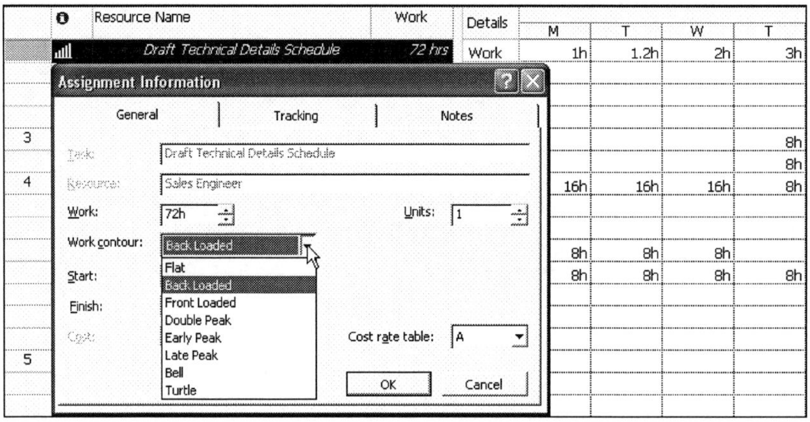

- From the **Work contour:** drop-down box, select the **Work Contour** type.
- The picture below is the **Resource Usage** form. It shows one task titled **Example Task** with six resources. Each resource name is the same as the assigned **Work** contour. The picture shows the effect of:
  - ➢ The assignment for each of the **Work Contour** types, and
  - ➢ The icon displayed in the **Task Information** column is a graphical representation of the contour.

		Task Name	Work	Details	10 June M	T	W	T	F
1		Example Task	320 hrs	Work	39.68h	95.27h	68.1h	78.27h	38.68h
		Flat	40 hrs	Work	4h	25h	0h	8h	3h
		Back Loaded	40 hrs	Work	1.67h	5h	8.33h	11.67h	13.33h
		Front Loaded	40 hrs	Work	13.33h	11.67h	8.33h	5h	1.67h
		Double Peak	40 hrs	Work	6h	12h	4h	12h	6h
		Early Peak	40 hrs	Work	6h	16h	10h	6h	2h
		Late Paek	40 hrs	Work	2h	6h	10h	16h	6h
		Bell	40 hrs	Work	2.4h	9.6h	16h	9.6h	2.4h
		Turtle	40 hrs	Work	4.28h	10h	11.43h	10h	4.28h

The **Flat** option allows you to type in any value into each on the cells. A resource assignment is automatically set to **Flat** when one entry is made in the table.

WORKSHOP 16

Assigning Resources to Tasks

Preamble
The resources must now be assigned to their specific tasks.

Assignment
Open the **OzBuild With Resources** project and complete the following steps:

1. Select the **Gantt Chart** view.
2. Split the pane by selecting **Window**, **Split**.
3. Display the **Task Details Form** in the lower pane window, select **Format**, **Details**, and choose **Resource Work**.
4. Using the **Resources Work** and the **Assign Resources** form assign the following resources:

Note: Once you have entered the **Resource Name** in the **Resources Work** form and assigned the **Units**, Microsoft Project will calculate the worked hours automatically after you click out of the form. Ensure the tasks are **NOT Effort Driven** as you enter resources, otherwise the Total Work may stay constant as you add additional resources.

Task No	Task Name	Resource	Units	Work
4	Bid Strategy Meeting	Sales Engineer	1	8 hrs
		System Engineer	1	8 hrs
		Bid Manager	1	8 hrs
5	Investigate Technical Feasibility	System Engineer	2	128 hrs
6	Document Installation Requirements	Sales Engineer	1	32 hrs
		System Engineer	1	32 hrs
8	Request Component Bids	Sales Engineer	1	24 hrs
9	Develop Project Schedule	Scheduler	1	32 hrs
10	Draft Technical Details Schedule	Sales Engineer	1	72 hrs
		System Engineer	1	72 hrs
11	Compile Costs from Component Bids	Sales Engineer	1	16 hrs
13	Draft Bid Document	Clerical Support	1	24 hrs
		Bid Manager	1	24 hrs
		Contract Consultant	1	24 hrs
14	Meeting to Review the Draft Bid Document	Sales Engineer	1	8 hrs
		System Engineer	1	8 hrs
		Bid Manager	1	8 hrs

continued over…

WORKSHOP 16 CONTINUED

5. Use the **Task Information** form to assign the remaining resource to each task. You may display this form by double-clicking the Task Name item on your Gantt Chart.

Task No	Task Name	Resource	Units	Work
15	Design Presentation	System Engineer	1	8 hrs
		Bid Manager	1	8 hrs
16	Edit Proposal Draft Bid Document	Clerical Support	1	8 hrs
		Bid Manager	1	8 hrs
17	Negotiate Component Work Packages	Sales Engineer	1	32 hrs
		Bid Manager	1	32 hrs
		Report Binding	3 Folders	
18	Final Review of Bid Document	System Engineer	1	8 hrs
		Bid Manager	1	8 hrs

6. Insert the **Resource Names** and **Resources Initials** columns to the right of the Task Name. Align the data in the columns to the left.

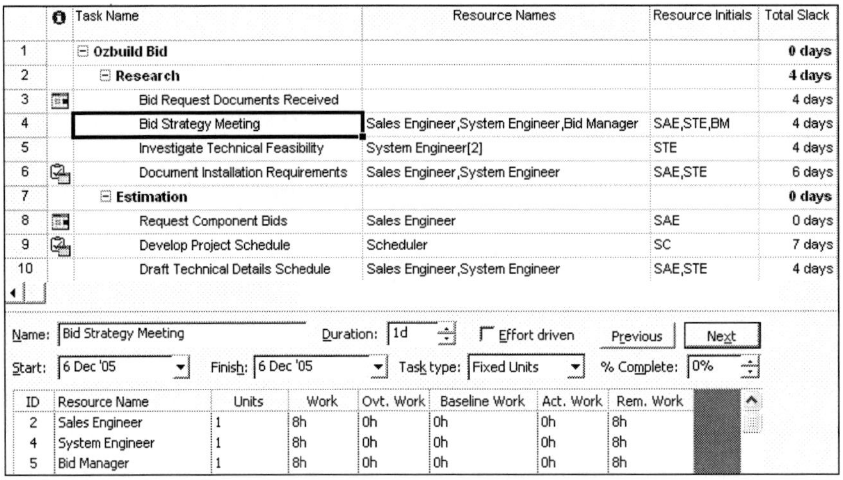

	Task Name	Resource Names	Resource Initials	Total Slack
1	Ozbuild Bid			0 days
2	Research			4 days
3	Bid Request Documents Received			4 days
4	Bid Strategy Meeting	Sales Engineer,System Engineer,Bid Manager	SAE,STE,BM	4 days
5	Investigate Technical Feasibility	System Engineer[2]	STE	4 days
6	Document Installation Requirements	Sales Engineer,System Engineer	SAE,STE	6 days
7	Estimation			0 days
8	Request Component Bids	Sales Engineer	SAE	0 days
9	Develop Project Schedule	Scheduler	SC	7 days
10	Draft Technical Details Schedule	Sales Engineer,System Engineer	SAE,STE	4 days

Name: Bid Strategy Meeting Duration: 1d Effort driven Previous Next
Start: 6 Dec '05 Finish: 6 Dec '05 Task type: Fixed Units % Complete: 0%

ID	Resource Name	Units	Work	Ovt. Work	Baseline Work	Act. Work	Rem. Work
2	Sales Engineer	1	8h	0h	0h	0h	8h
4	System Engineer	1	8h	0h	0h	0h	8h
5	Bid Manager	1	8h	0h	0h	0h	8h

7. Now display the **Task Usage** view, insert the **Assignment Units** column and check that your Assignment Units and Work match the table above by scrolling down.

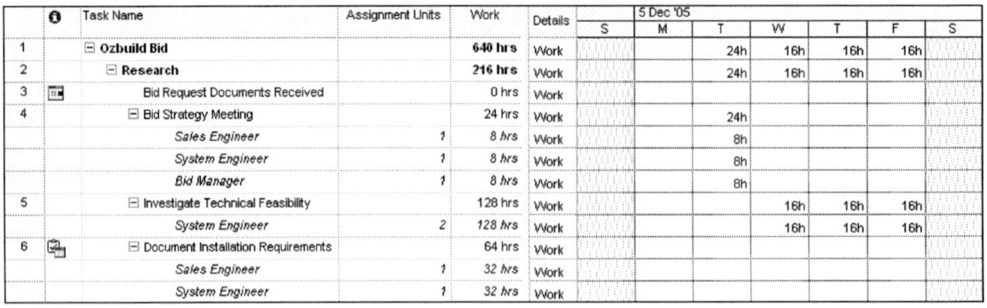

	Task Name	Assignment Units	Work	Details	5 Dec '05 S	M	T	W	T	F	S
1	Ozbuild Bid		640 hrs	Work			24h	16h	16h	16h	
2	Research		216 hrs	Work			24h	16h	16h	16h	
3	Bid Request Documents Received		0 hrs	Work							
4	Bid Strategy Meeting		24 hrs	Work			24h				
	Sales Engineer	1	8 hrs	Work			8h				
	System Engineer	1	8 hrs	Work			8h				
	Bid Manager	1	8 hrs	Work			8h				
5	Investigate Technical Feasibility		128 hrs	Work				16h	16h	16h	
	System Engineer	2	128 hrs	Work				16h	16h	16h	
6	Document Installation Requirements		64 hrs	Work							
	Sales Engineer	1	32 hrs	Work							
	System Engineer	1	32 hrs	Work							

20 STATUSING PROJECTS WITH RESOURCES

Statusing a project with resources uses a number of features which are very interactive. It is suggested that after you have read this chapter and before you work on a live project that you create a simple schedule with a couple of tasks and assign two or three resources against each task. Set the **Options** to reflect the information that you want to enter and what you want Microsoft Project 2003 to calculate. Go through the statusing process with dummy data representing all the data you want to collect. Then enter and check that the results are as you expected. The options to consider are:

- Have you linked **% Complete** and **% Work** with the **Updating task status updates resource status:** option? If unlinked then the **% Work** may be different to **% Complete**.
- How is the measure of progress at summary task level being displayed? The summary **% Work** is based hours of efforts which is more meaningful than the summary **% Complete** which based on durations.
- Are the resource **Actual Costs** calculated by Microsoft Project 2003 with setting the **Actual costs are always calculated by Microsoft Project** option checked?
- Are your tasks scheduled to start after a date using the **Reschedule uncompleted tasks to start after:** in conjunction with the **Split** task option?

Statusing a project with resources takes place in two distinct steps:

- The dates are statused using the methods outlined in the **TRACKING PROGRESS** chapter, and
- The resources' hours and costs are updated.

This chapter covers the following topics:

- Understanding **Baseline Dates** (Target Dates), **Baseline Costs** (Budget) and **Baseline Work**.
- Understanding the **Status Date**, **Work After Date** and **Current Date** with respect to resources.
- Information Required to Update a Resourced Schedule.
- Updating Resources.
- Splitting Tasks.
- Summary Task Interim Baseline Calculation.
- How to **Contour the Resource Assignment**.

Microsoft Project 2003 does not provide the capability for producing S-Curves. The data may be exported or copied and pasted to products such as Excel where the S-Curves may be created.

20.1 Understanding Baseline Dates, Duration, Costs and Hours

Baseline Dates are also known as Target Dates and are normally the original Project Early Start and Early Finish dates. These are the dates against which project progress is measured.

Baseline Duration is the original planned duration of a task.

Baseline Costs are also known as Budgets and represent the original project cost estimate. These are the figures against which the expenditures and Cost at Completion (or Estimate at Completion) are measured.

Baseline Work is also known as Budgeted Quantity and represents the original estimate of the project quantities. These are the quantities against which the consumption of resources are measured.

The **Baseline Costs** and **Work** of resources are **NOT** automatically recorded in the Baseline fields of each resource as the resource is assigned to a task. All **Baseline** information **(Dates**, **Costs** and **Work)** are saved when the project or selected Tasks have the Baseline set.

If resources have been assigned then the Baseline Costs and Work are recorded at the same time as the Baseline dates.

Setting the **Baseline Dates** is covered in the **TRACKING PROGRESS** chapter.

Baseline Dates may be displayed by using the following methods:

- Baseline Date Start and Finish columns, or
- Bar Chart Bars, or
- Dates on the Bars.

	Task Name	Start	Finish	Baseline Start	Baseline Finish
1	**Ozbuild Bid**	**5 Dec '05**	**23 Jan '06**	**5 Dec '05**	**23 Jan '06**
2	**Research**	**5 Dec '05**	**20 Dec '05**	**5 Dec '05**	**20 Dec '05**
3	Bid Request Documents Recieved	5 Dec '05	5 Dec '05	5 Dec '05	5 Dec '05
4	Bid Strategy Meeting	5 Dec '05	5 Dec '05	5 Dec '05	5 Dec '05
5	Investigate Technical Feasibility	6 Dec '05	15 Dec '05	6 Dec '05	15 Dec '05
6	Document Installation Requirements	16 Dec '05	20 Dec '05	16 Dec '05	20 Dec '05

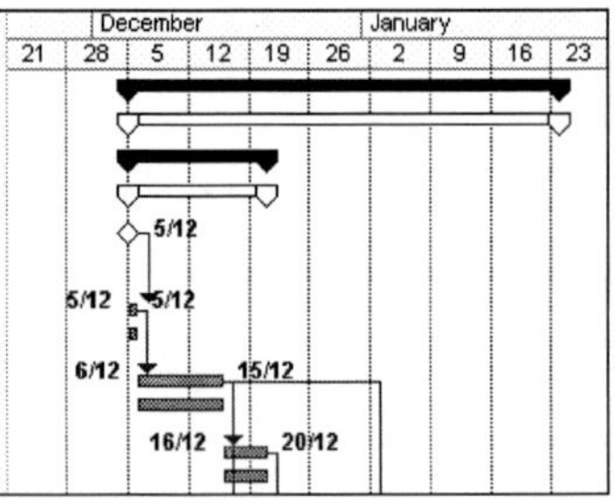

20.2 Understanding the Data Date

The **Data Date** is a standard scheduling term. It is also known as the **Review Date**, **Status Date**, **As of Date** and **Update Date**.

- The **Data Date** is the date that divides the past from the future in the schedule. The **Data Date** is not normally in the future but is often in the recent past due to the time it may take to collect the information to status the schedule.
- **Actual Costs** and **Quantities/Hours** or **Actual work** occur before the data date.
- **Costs** and **Quantities/Hours To Complete** or **Work to Complete** occur after the Data Date.
- **Remaining duration** is the duration required to complete a task. It is calculated forward from the **Data Date**.

Microsoft Project 2003 has four dates associated with updating a schedule. These are covered in the **TRACKING PROGRESS** chapter. In summary, these date fields are:

- **Current Date** – This date is set to the computer's system date each time a project file is opened. It is used to calculate **Earned Value** data when a **Status Date** has not been set.
- **Status Date** – By default this is blank. After assigning this date it will not change (as the **Current Date** does) when the project is saved or reopened at a later date. When set, this date overrides the **Current Date** for calculating **Earned Value** data.
- **Update work as completed through:** – When a project is updated using the **Update Project** form (the project is statused as if it were progressed exactly according to plan), the **Status Date** is set to the same date as **Project Update Date.**
- **Reschedule uncompleted work to start after:** date – This function is used to move the **Incomplete Work** of **In-Progress** tasks into the future.
  - ➢ **In-Progress** tasks must be able to **Split** for this function to operate. The option to split tasks is found on the **Tools**, **Options...**, **Schedule** tab.
  - ➢ The **Status Date** is **NOT** set to the **Reschedule uncompleted work to start after:** date when this function is invoked.
  - ➢ This function will not move the incomplete portions of tasks back in time when a task is completed ahead of schedule. This causes the task's **Remaining Duration** to occur sometime in the future and not immediately after the selected date. The example below shows a schedule with the **Reschedule uncompleted work to start after:** date set to 10 June. The top task's Remaining Duration commences on 11 June, the task is split and the Remaining Duration occurs immediately after 10 June. The lower task's Remaining Duration does not commence until 18 June. This is not realistic and manual intervention is required to correct this.

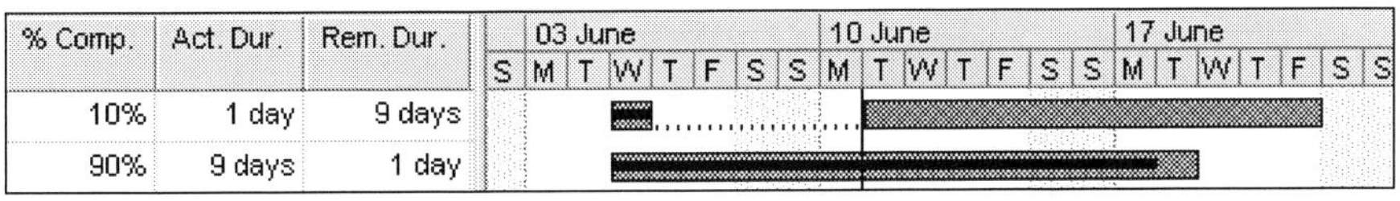

NEITHER the **Current Date** nor the **Status Date** is used to calculate the **Early Finish** of an **In-Progress** task when a schedule is calculated using **F9**, or when the **Calculation options for Microsoft Project**'s option of **Automatic Scheduling** is enabled. The end date of an in-progress task in Microsoft Project 2003 is normally calculated from the **Actual Start Date** plus the **Duration**. This is a different method of calculation than employed by some other scheduling software, which calculate the end date of a task from a single **Data Date** plus the **Remaining Duration**.

20.3 Formatting the Current Date and Status Date Lines

To format the display of the **Current Date** and **Status Date** lines on the Bar Chart, select **Format**, **Gridlines…** to display the **Gridlines** form:

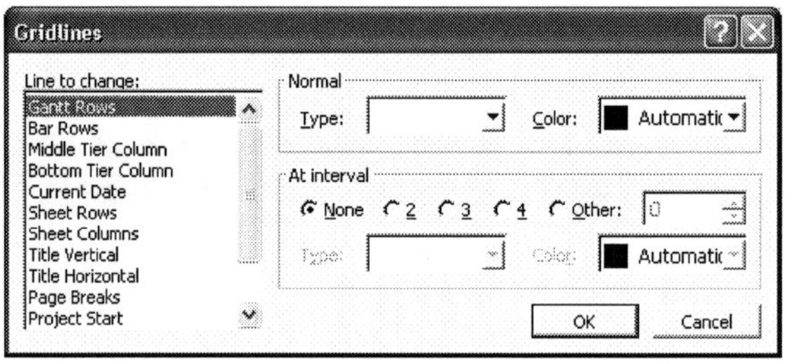

- These **Options** allow the selection of colors and line types for all sight lines as shown in the Lines to Change box on the left of the example above.
- All other sight lines may also be formatted in this form.

20.4 Information Required to Update a Resourced Schedule

A project schedule is usually updated at the end of a period, such as each day, week or month. One purpose of updating a schedule is to establish differences between the plan and the current schedule.

Microsoft Project 2003 may calculate task Actual Costs from the rates entered in the Resource Table or the Costs may be entered manually. If the Actual Costs are to be calculated by Microsoft Project 2003 then the Actual Costs do not need to be collected.

The following information is required to status a resourced schedule:

Tasks completed in the update period:

- **Actual Start** date of the task.
- **Actual Finish** date of the task.
- **Actual Costs** spent, **Actual Resource Hours** spent, and/or **Actual Material Quantities**.

Tasks commenced in the update period:

- **Actual Start** date of the task.
- **Remaining Duration** or **Expected Finish** date.
- **Actual Costs** and **Actual Resource Hours** and/or **Actual Material Quantities**.
- **Hours** or **Quantities** to complete. Costs to complete are always calculated using the resource rates in the Resource Table.
- **Suspend** and **Resume** dates for tasks that have had their work suspended. These are used for splitting tasks.

Tasks Not Commenced:

- Changes in Logic or date constraints.
- Changes in estimated **Costs**, **Hours** or **Quantities**.

The schedule may be updated once this information is collected.

You have the option of allowing Microsoft Project 2003 to calculate many of these fields from the **% Complete** by selecting the appropriate option found in the **Tools**, **Options...**, **Calculation** tab.

20.5 Updating Dates and Percentage Complete

The schedule should be first updated as outlined in the **TRACKING PROGRESS** chapter. In summary, this is completed by entering:

- The **Actual Start** and **Actual Finish** dates of **Complete** tasks.
- The **Actual Start**, **% Complete** and **Remaining Duration** or **Expected Finish** of **In-Progress** tasks.
- Adjust Logic and **Durations** of **Unstarted** tasks.

Before you do this, you should set the **Tools, Options..., Calculation** tab to ensure that the actual costs and hours calculate the way you want.

A simple process to update a project is to:

- Use the **Tools, Tracking, Update Project...** and select **Update work as complete through:** to update the project through to the data date.
- Adjust the **Actual Start** and **Actual Finish** dates for **Complete** tasks.
- Drag the finish date of **In-Progress** tasks to the required date.
- Adjust the **Actual Start** dates of **In-Progress** tasks and drag the **% Complete** to the required value.

The option of **Tools, Tracking, Update Project...** will only work when the following Project Options are enabled. Then, **Work** and **Costs** are calculated by Microsoft Project 2003.

- **Updating task status updates resource status**, and
- **Actual costs are always calculated by Microsoft Project**.

20.6 Entering a % Complete Against Summary Tasks

A **% Complete** may be entered against a summary task and all the child tasks and their resources will be updated automatically to reflect the Summary % Complete. Full details of this function are outlined in the **TRACKING PROGRESS** chapter. In the picture below 75% was entered against the **Research** task:

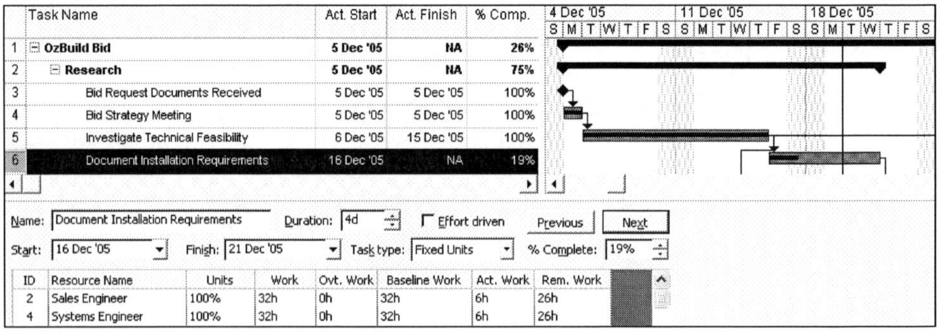

20.7 Updating Resources

There are many permutations available in the **Tools**, **Options…** form for calculating resource data. Due to the number of resource options and numerous forms available in Microsoft Project 2003, it is not feasible to document all the combinations available for resource calculation.

This book, therefore, outlines some typical scenarios and examples of entering the status data that you may want to try on your projects.

Material resources are updated in the same way as **Work** resources and are not covered separately.

20.7.1 Updating Tasks with Fixed Costs Only

A project with fixed costs only is the simplest option for managing costs. The example below displays:

- A 10-day **Duration** task which is 40% Complete.
- A **Baseline Cost** of $100.00 which was created by an original **Fixed Cost** of $100.00 when the **Baseline** was set.
- A **Fixed Cost** of $80.00 representing a revised estimate which has reduced the original estimate by $20.00 to $80.00.
- An **Actual Cost** of $32.00 and **Remaining Cost** of $48.00, which are calculated from the 40 **% Complete**.
- The **Current Bar** (the upper bar in the picture below) shows that the task started two days late and is scheduled to end two days late as compared to the **Baseline Bar** (the black lower bar).

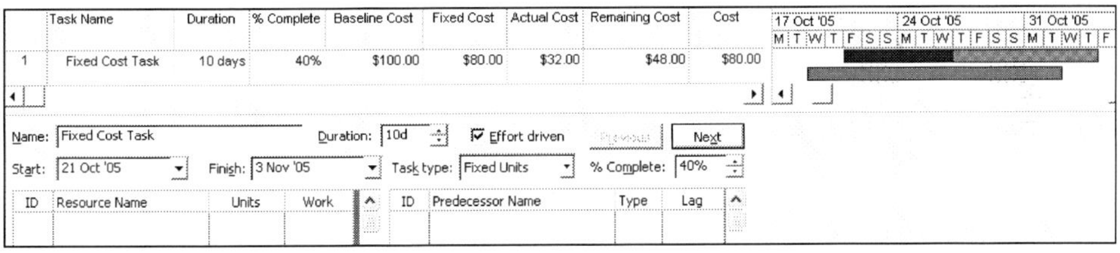

You will notice that the **Task Details** view with the **Resource Cost** details form does not show any resources or costs when **Fixed Costs** are used.

20.7.2 Forecasting Resource Hours

The next level of complexity usually occurs when a schedule is used for the management of resource hours but not costs.

One or more resources may be applied to a task and you may want to enter both the **Actual Work** and the **Remaining Work** independently. In this situation you will need to unlink **% Complete** and **Actual Work** with the **Updating task status updates resource status:** option in the **Tools**, **Options…**, **Calculation** tab. Now the field **% Work** field will be linked to the **Work, Actual Work** and **Remaining Work** fields and will now operate independently of the **% Complete** field.

The example below uses the **Task Details** form.

- A 10-day **Duration** task which is 40% Complete.
- The task's **Baseline Work** of 240hrs is from the addition of the two resource allocations when the **Baseline** was set.
  - ➢ The **Bid Manager** was originally assigned at 100%, or full-time, giving a **Baseline Work** of 80 hours.
  - ➢ The **Clerical Support** was originally assigned at 200%, or 2 full-time people, giving a **Baseline Work** of 160 hours.
- The task's **Actual Work** of 90 hours is the addition of the two resources' **Actual Work** of 20 hours and 70 hours.
- The task's **Remaining Work** of 220 hours is the addition of the **Remaining Work** of 70 hours and 150 hours.
- The **% Work** of 29% is calculated by dividing the **Actual Work** by the **Work.** This is different than the **% Complete**, which represents the elapsed time.
- Again, the **Current Bar** shows the task started two days late and is scheduled to end two days late.

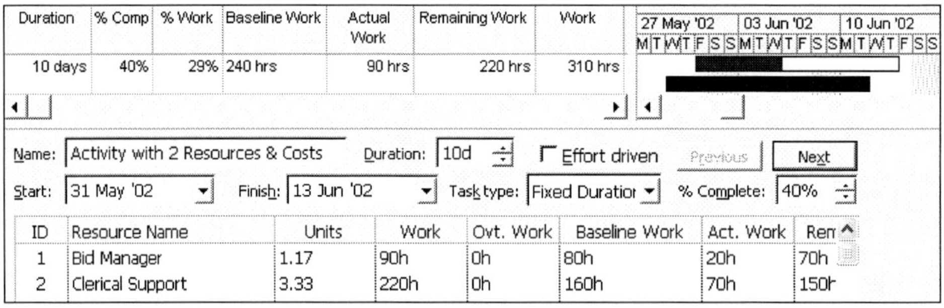

- The **Actual Work** and **Remaining Work** were entered manually and have not been calculated by Microsoft Project 2003 from the **% Complete**.
- The duration of the task is still 10 days long because it is a **Fixed Duration** task. The **Units** have increased from 1 to 1.17 and 2 to 3.33 for the Bid Manager and Clerical Support, respectively, as the **Work** is now greater than the **Baseline Work** due to the increase in the number of hours required to complete the task.

20.7.3 Forecasting Resource Hours and Costs Form

The example below is similar to the previous example, but it now displays the **Resource Costs** (not the **Work**) calculated by Microsoft Project 2003 with **% Complete** and **Actual Work** unlinked.

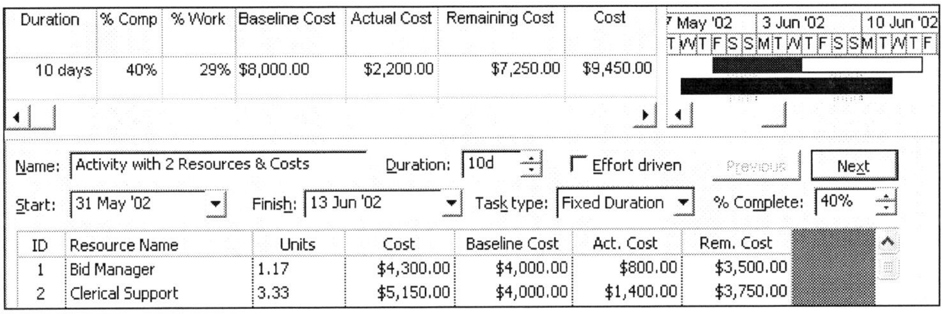

The next level of complexity occurs usually when a schedule is used for the management of resource hours and costs.

You may want to enter both the **Actual Work** and **Actual Costs** separately. For example, you may want to take the costs and hours from a timesheet and/or an accounting system that may have a different resource rate than your schedule. Input your data in the lower pane of the window, which is known as the **Resource Details** form.

In this situation you will need to:

- Unlink **% Complete** and **Actual Work** with the **Updating task status updates resource status:** option in the **Tools, Options..., Calculation** tab, and
- Unlink the **Actual Work** and **Actual Costs** by disabling the **Actual costs are always calculated by Microsoft Project** option in the **Tools, Options..., Calculation** tab.

The example below shows how the costs are calculated with:

- **Work** and **Costs** unlinked, and
- Updated hours as per the previous example.

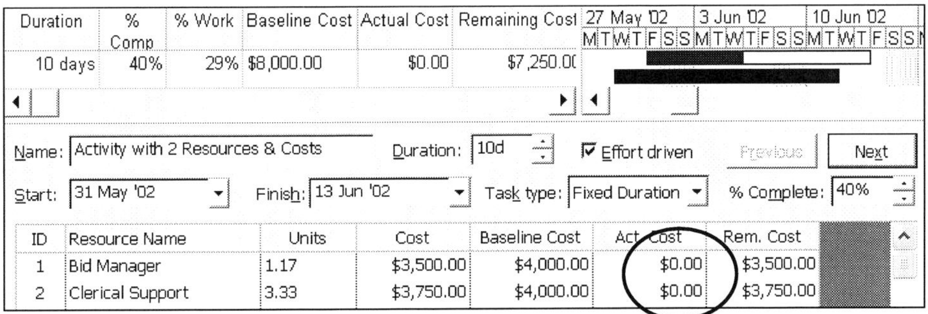

The **Actual Costs** have not been calculated by Microsoft Project 2003 and remain at zero. The **Remaining Costs** are based on the resources' **Standard Rate**.

Now that the Actual Costs and Actual Work are unlinked, you may type in the **Actual Costs** per resource in the lower pane of the window (i.e. the **Resource Details** form) without affecting the **Actual Work.**

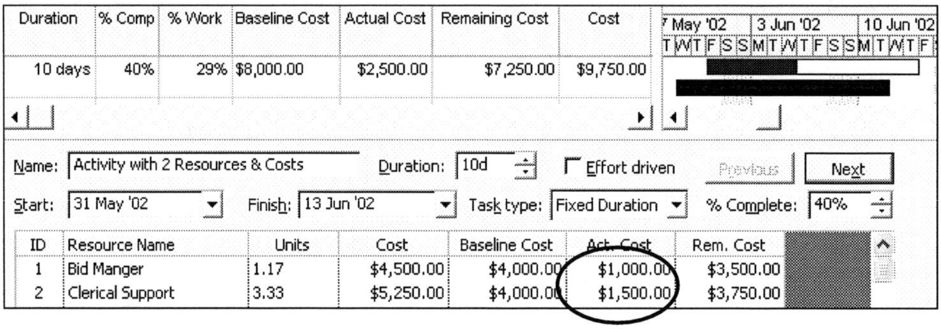

20.7.4 Using the Task Usage and Resource Usage Views

The Task Usage and Resource usage forms allow the most flexibility when entering resource quantities to-date and to go. The picture below is the **Task Usage** form from the OzBuild workshops showing the **Work**, **Actual Work** and **Baseline Work** rows. Additional rows of information may be obtained by right-clicking to display a menu or selecting **Format**, **Details**.

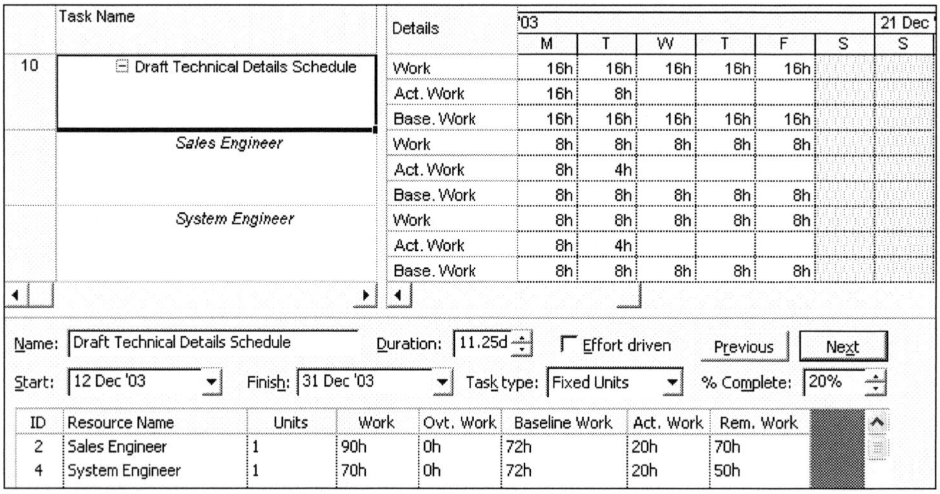

A timesheet style view may be created with the **Resource Usage** view. This view may be printed or used by team members to directly update their hours to date in the **Act. Work** (Actual Work) rows and their estimate hours in the **Work** rows. The advantage of this view is that all activities are grouped together in one band under each name.

Calculated cost fields that may not have data entered into are shown as gray.

20.8 Splitting Tasks

When the **Split in-progress tasks** option is enabled, a task may be **Split** by:

- Dragging the incomplete portion of a task in the bar chart, or
- Clicking on the icon and then moving your cursor over the point on the task bar where you want a split and drag the task, or
- Using the **Tools**, **Tracking**, **Update Project...**, **Reschedule uncompleted work to start after:** function, or
- Commencing a task before its predecessor finishes.

In the picture below the upper task was split using the icon and the lower task was split because it commenced before its predecessor.

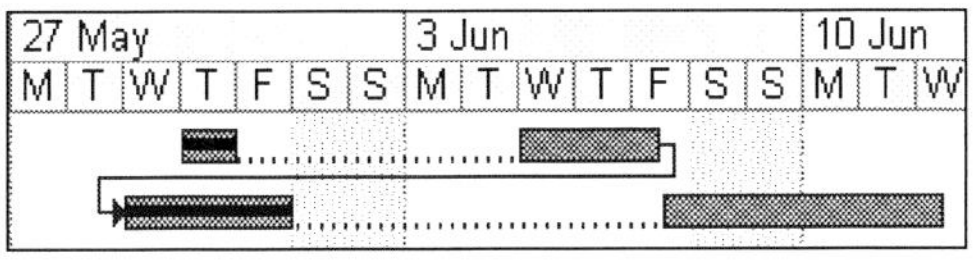

A **Split** may be reversed by dragging the split portion back in the bar chart if there is no predecessor pushing the split out. The first task in the example below has been dragged back. The second task may not be dragged back due to the FS relationship.

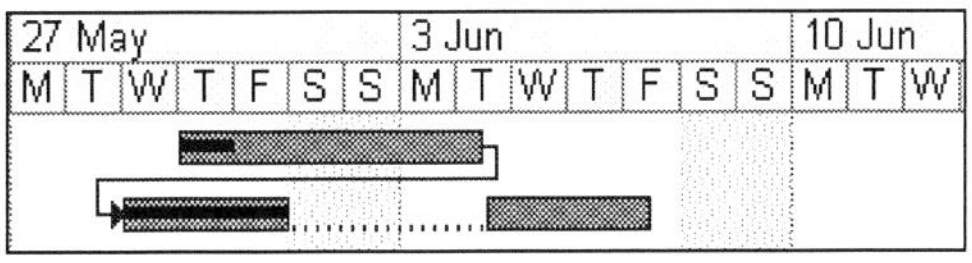

Once a task is **Split**, then the resources are not scheduled to work during the split period. This concept is demonstrated in the **RESOURCE HISTOGRAMS S-CURVES AND TABLES** chapter.

20.9 Summary Task Interim Baseline Calculation

Microsoft Project 2002 introduced new function titled **Summary Task Interim Baseline Calculation** that enables the Baseline dates and costs of summary activities to be recalculated when tasks are added, deleted or moved. The date aspect of this function is covered in the **TRACKING PROGRESS** chapter.

There are a number of permutations that may be used with this option and a little experimentation with a simple schedule, such as the one below, will enable you to understand this function.

Often it is required to recalculate the Baseline Costs Summary Tasks based on the Original Baseline Costs after activities have been moved to different Summary Tasks. This function may recalculate the Summary Task Baseline Costs from the detailed tasks Baseline Costs when the costs for a project have been revised and are showing a forecast with a deviation from the Baseline. The example below demonstrates this process:

- Original schedule with Baseline Costs and Dates:

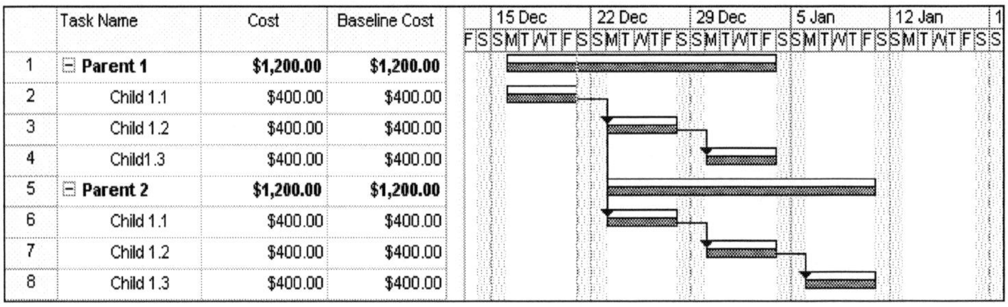

	Task Name	Cost	Baseline Cost
1	Parent 1	**$1,200.00**	**$1,200.00**
2	Child 1.1	$400.00	$400.00
3	Child 1.2	$400.00	$400.00
4	Child1.3	$400.00	$400.00
5	Parent 2	**$1,200.00**	**$1,200.00**
6	Child 1.1	$400.00	$400.00
7	Child 1.2	$400.00	$400.00
8	Child 1.3	$400.00	$400.00

- After the project has been delayed and the task Child 1.3 moved under Parent 1:

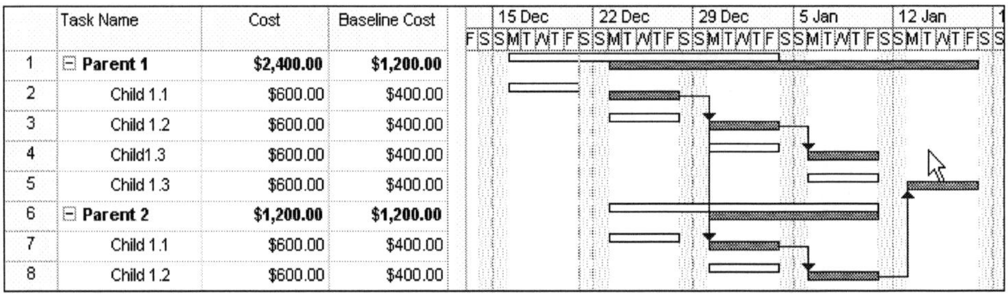

	Task Name	Cost	Baseline Cost
1	Parent 1	**$2,400.00**	**$1,200.00**
2	Child 1.1	$600.00	$400.00
3	Child 1.2	$600.00	$400.00
4	Child1.3	$600.00	$400.00
5	Child 1.3	$600.00	$400.00
6	Parent 2	**$1,200.00**	**$1,200.00**
7	Child 1.1	$600.00	$400.00
8	Child 1.2	$600.00	$400.00

You will observe that neither the summary bars nor the Baseline Costs of Parent 1 or Parent 2 are now correct.

- Now highlight the parent tasks only and then select **Tools Tracking**, **Save Baseline....** and select the options shown in the Save Baseline form below:

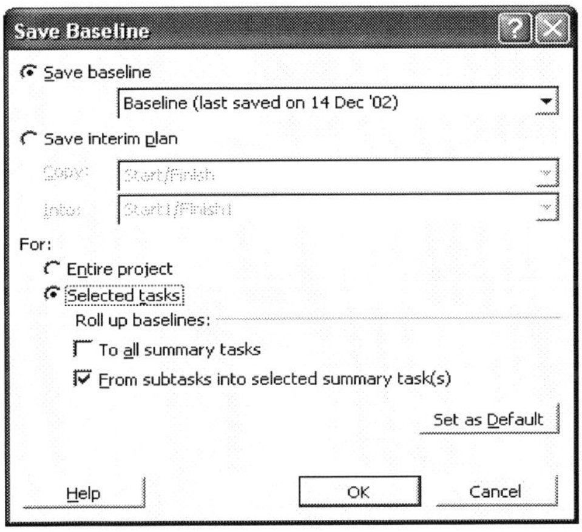

You will see from the picture below that now both the Baseline Dates displayed as the upper bar and the Baseline Costs are recalculated from the original baseline data.

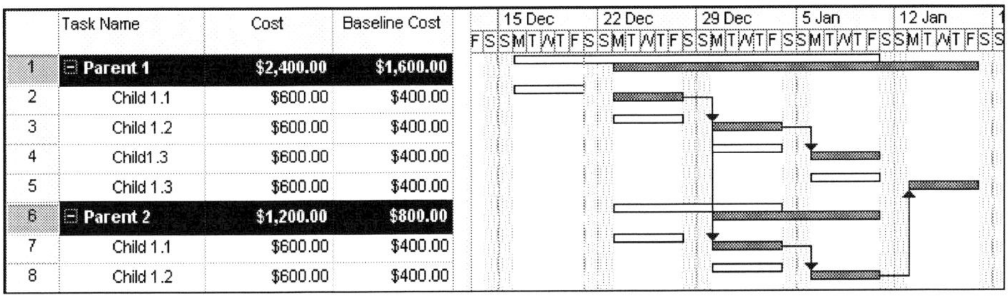

	Task Name	Cost	Baseline Cost
1	Parent 1	$2,400.00	$1,600.00
2	Child 1.1	$600.00	$400.00
3	Child 1.2	$600.00	$400.00
4	Child1.3	$600.00	$400.00
5	Child 1.3	$600.00	$400.00
6	Parent 2	$1,200.00	$800.00
7	Child 1.1	$600.00	$400.00
8	Child 1.2	$600.00	$400.00

20.10 Summary Tasks and Earned Value

Actual Costs and **Work** may be summarized at any level in the same way as **Work** and **Costs**. The picture below is showing the **Earned Value** table and the costs have been summarized up to the Project Level:

	Task Name	BCWS	BCWP	ACWP	SV	CV	EAC	BAC	VAC
1	**Ozbuild Bid**	**$15,664.00**	**$12,608.00**	**$9,574.13**	**($3,056.00)**	**$3,033.87**	**$17,860.13**	**$25,134.00**	**$6,048.00**
2	**Research**	**$8,720.00**	**$8,720.00**	**$6,640.13**	**$0.00**	**$2,079.87**	**$6,640.13**	**$8,720.00**	**$2,079.87**
7	**Estimation**	**$6,944.00**	**$3,888.00**	**$2,934.00**	**($3,056.00)**	**$954.00**	**$3,030.00**	**$8,224.00**	**$2,017.93**
8	Request Componei	$320.00	$0.00	$320.00	($320.00)	($320.00)	$640.00	$960.00	$320.00
9	Develop Project Sc	$864.00	$432.00	$864.00	($432.00)	($432.00)	$0.00	$864.00	($864.00)
10	Draft Technical De	$5,760.00	$3,456.00	$1,750.00	($2,304.00)	$1,706.00	$1,750.00	$5,760.00	$2,843.33
11	Compile Costs fron	$0.00	$0.00	$0.00	$0.00	$0.00	$640.00	$640.00	$0.00
12	**Proposal**	**$0.00**	**$0.00**	**$0.00**	**$0.00**	**$0.00**	**$8,190.00**	**$8,190.00**	**$0.00**
13	Draft Bid Documen	$0.00	$0.00	$0.00	$0.00	$0.00	$2,800.00	$2,800.00	$0.00
14	Meeting to Review	$0.00	$0.00	$0.00	$0.00	$0.00	$1,040.00	$1,040.00	$0.00
15	Design Presentatio	$0.00	$0.00	$0.00	$0.00	$0.00	$720.00	$720.00	$0.00
16	Edit Proposal Draft	$0.00	$0.00	$0.00	$0.00	$0.00	$600.00	$600.00	$0.00
17	Negotiate Compon(	$0.00	$0.00	$0.00	$0.00	$0.00	$2,310.00	$2,310.00	$0.00
18	Final Review of Bi(	$0.00	$0.00	$0.00	$0.00	$0.00	$720.00	$720.00	$0.00
19	Submit Bid	$0.00	$0.00	$0.00	$0.00	$0.00	$0.00	$0.00	$0.00

The terminology below is used by a large number of companies and organizations. It provides standard terms to describe Earned Value calculations, which many people understand. The method that Microsoft Project calculates the Earned Value Fields is documented in the help file. Below are some of the terms that are in common use:

- AC or ACWP — Actual Cost or Actual Cost of Work Performed
- EV or BCWP — Earned Value or Budget Cost of Work Performed
- PV or BCWS — Planned Value or Budget Cost of Work Scheduled
- BAC — Budget At Completion
- C/SCSC — Cost/Schedule Control Systems Criteria (CS^2)
- CV — Cost Variance to date, BCWP – ACWP
- EAC — Estimate at Completion
- ETC Time — Estimate To Complete expressed in Time
- ETC — Estimate To Complete
- FAC $ — Forecast at Completion
- FC CV — Forecast Cost Variance at Completion (Budget – Forecast)
- FC SV — Forecast Schedule Variance at Completion (Baseline End Date – Scheduled End Date)
- FTC CT — Forecast To Complete Calendar Time
- SV — Schedule Variance to date, BCWP – BCWS

The **Physical % Complete** field may be used to calculate the Earned Value (Budget Cost of Work Performed) independently from the value in the task **% Complete** field. This function is useful for measuring the progress of work that is not progressing linearly. This may be set for each individual task in the **General** tab of the **Task Information** form and the default for new tasks may be set with the [Earned Value...] button in the **General** tab of the Options form, where also the Baseline to be used for the Earned Value calculations may be selected.

WORKSHOP 17

Updating Costs

Preamble

We need to status the tasks and resources.

Assignment

Note: If your settings are not exactly the same as the computer on which this exercise was undertaken you may end up with different results.

Open your project file and complete the following steps:

1. We will initially allow Microsoft Project to calculate costs and hours from the % Complete. Adjust the options using **Tools, Options…, Calculation** tabs to:
   - **CHECK** the **Updating task status updates resource status option**. This will link % Complete and Actual Work. (This option also needs to be checked to allow Summary % Completes to be spread correctly to detailed tasks.)
   - **CHECK** the **Actual costs are always calculated by Microsoft Project** option. With this option checked the resource Actual Cost is calculated by Microsoft Project 2002 from the resource Work and Rates.
   - **UNCHECK** both **Move end of completed task parts after status date back to status date** and **Move start of remaining parts before status date forward to status date** options.
   - Select the **Schedule** tab and **UNCHECK** the **Split in-progress tasks** option. This will prevent the splitting of tasks.
2. Save the Baseline using the **Tools, Tracking, Save Baseline….**
3. Split the screen. Display the **Gantt Chart** with **Tracking** table in the upper screen and the **Task Details Form** with **Resource Work** details form in the lower screen.
4. Run the **Format, Gantt Chart Wizard…** to format the bars. Display the **Baseline** without text and display logic links lines.

Continued Over…

WORKSHOP 17 CONTINUED

5. The **Data Date** is 19 Dec 05:
   - Update the project using **Tools, Tracking, Update Project…**, as per the picture below:

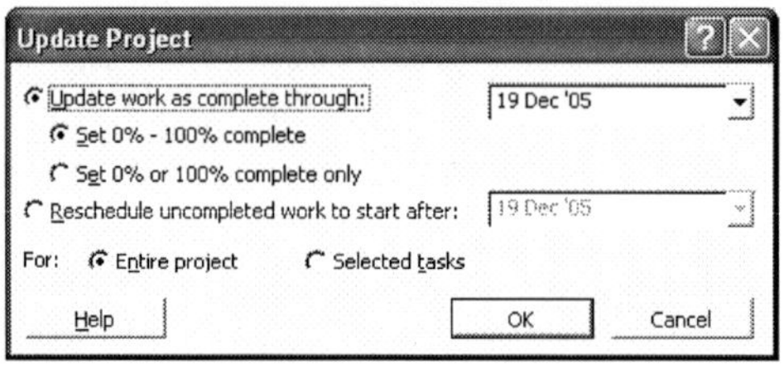

   - Enter 75% Complete against the summary task **Research**.
   - Observe how the % Complete fields of the Detailed Tasks of Research are calculated.
   - Open the **Resource Cost** form in the lower pane. Observe how the resource costs have been updated in the **Document Installation Requirements** task.
   - Enter a 20% Complete against task **Draft Technical Details Schedule** and enter an actual Start Date of 15 Dec 05.
   - Check the **Status Date** in the **Project Information** form it should be 19 Dec 05.
   - Use the **Format, Gridlines…** command to format the **Status Date** as a black dashed line.

Your schedule should look like this:

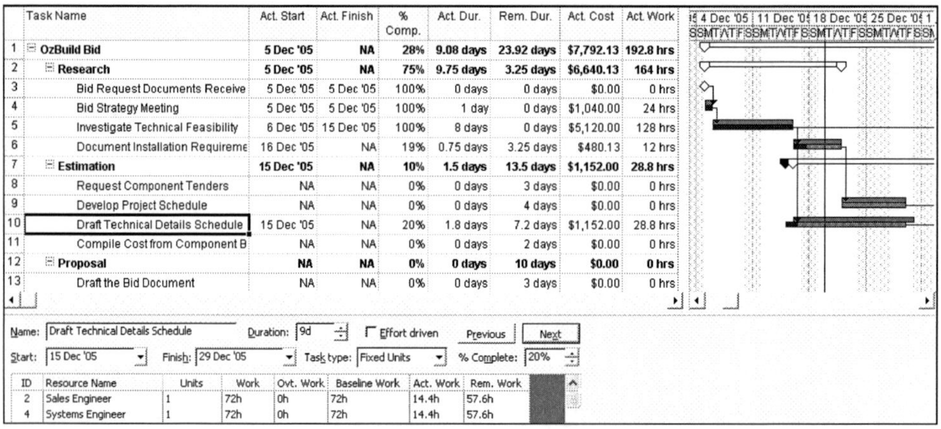

	Task Name	Act. Start	Act. Finish	% Comp.	Act. Dur.	Rem. Dur.	Act. Cost	Act. Work
1	**OzBuild Bid**	**5 Dec '05**	**NA**	**28%**	**9.08 days**	**23.92 days**	**$7,792.13**	**192.8 hrs**
2	**Research**	**5 Dec '05**	**NA**	**75%**	**9.75 days**	**3.25 days**	**$6,640.13**	**164 hrs**
3	Bid Request Documents Receive	5 Dec '05	5 Dec '05	100%	0 days	0 days	$0.00	0 hrs
4	Bid Strategy Meeting	5 Dec '05	5 Dec '05	100%	1 day	0 days	$1,040.00	24 hrs
5	Investigate Technical Feasibility	6 Dec '05	15 Dec '05	100%	8 days	0 days	$5,120.00	128 hrs
6	Document Installation Requireme	16 Dec '05	NA	19%	0.75 days	3.25 days	$480.13	12 hrs
7	**Estimation**	**15 Dec '05**	**NA**	**10%**	**1.5 days**	**13.5 days**	**$1,152.00**	**28.8 hrs**
8	Request Component Tenders	NA	NA	0%	0 days	3 days	$0.00	0 hrs
9	Develop Project Schedule	NA	NA	0%	0 days	4 days	$0.00	0 hrs
10	Draft Technical Details Schedule	15 Dec '05	NA	20%	1.8 days	7.2 days	$1,152.00	28.8 hrs
11	Compile Cost from Component B	NA	NA	0%	0 days	2 days	$0.00	0 hrs
12	**Proposal**	**NA**	**NA**	**0%**	**0 days**	**10 days**	**$0.00**	**0 hrs**
13	Draft the Bid Document	NA	NA	0%	0 days	3 days	$0.00	0 hrs

Name: Draft Technical Details Schedule   Duration: 9d   Effort driven   Previous   Next
Start: 15 Dec '05   Finish: 29 Dec '05   Task type: Fixed Units   % Complete: 20%

ID	Resource Name	Units	Work	Ovt. Work	Baseline Work	Act. Work	Rem. Work
2	Sales Engineer	1	72h	0h	72h	14.4h	57.6h
4	Systems Engineer	1	72h	0h	72h	14.4h	57.6h

6. We will now enter our hours and costs to date and hours to go. To prevent Microsoft Project from calculating the costs, open the **Options** form, **Calculation** tab and:
   - **UNCHECK** the **Updating task status updates resource status**. This option is to prevent the updated % Complete from calculating **Actual Work** and **Remaining Work**.
   - **UNCHECK** the **Actual costs are always calculated by Microsoft Project**. This option is to prevent Microsoft Project from calculating the **Actual Costs** from the **Actual Work**.

WORKSHOP 17 CONTINUED

- Update the **Draft Technical Details Schedule** with the following information:
  20 hours of **Actual Work** and 50 hours of **Remaining Work** against the **Sales Engineer**, and
  20 hours of **Actual Work** and 70 hours of **Remaining Work** against the **System Engineer**.

Click the **OK** button to accept the changes.

- You will notice that the task duration extends. This is because the task is **Fixed Units**, so the **Units** stayed the same and duration was extended. Your schedule should look like this:

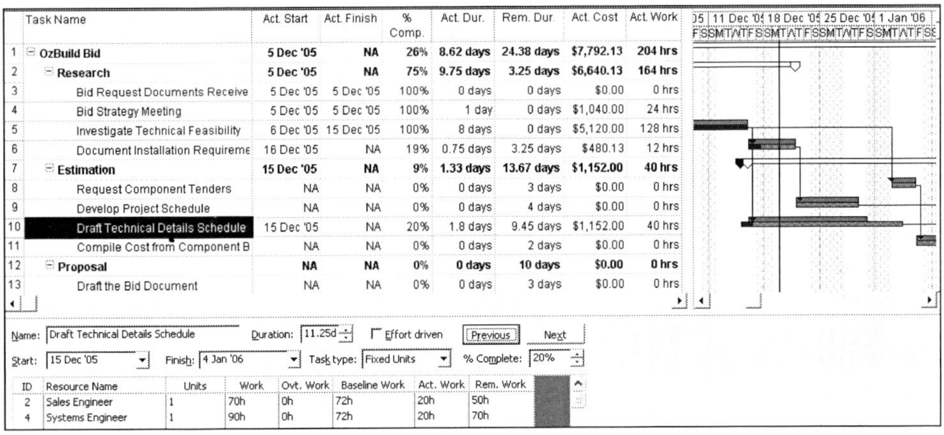

	Task Name	Act. Start	Act. Finish	% Comp.	Act. Dur.	Rem. Dur.	Act. Cost	Act. Work
1	**OzBuild Bid**	**5 Dec '05**	**NA**	**26%**	**8.62 days**	**24.38 days**	**$7,792.13**	**204 hrs**
2	**Research**	**5 Dec '05**	**NA**	**75%**	**9.75 days**	**3.25 days**	**$6,640.13**	**164 hrs**
3	Bid Request Documents Receive	5 Dec '05	5 Dec '05	100%	0 days	0 days	$0.00	0 hrs
4	Bid Strategy Meeting	5 Dec '05	5 Dec '05	100%	1 day	0 days	$1,040.00	24 hrs
5	Investigate Technical Feasibility	6 Dec '05	15 Dec '05	100%	8 days	0 days	$5,120.00	128 hrs
6	Document Installation Requireme	16 Dec '05	NA	19%	0.75 days	3.25 days	$480.13	12 hrs
7	**Estimation**	**15 Dec '05**	**NA**	**9%**	**1.33 days**	**13.67 days**	**$1,152.00**	**40 hrs**
8	Request Component Tenders	NA	NA	0%	0 days	3 days	$0.00	0 hrs
9	Develop Project Schedule	NA	NA	0%	0 days	4 days	$0.00	0 hrs
10	Draft Technical Details Schedule	15 Dec '05	NA	20%	1.8 days	9.45 days	$1,152.00	40 hrs
11	Compile Cost from Component B	NA	NA	0%	0 days	2 days	$0.00	0 hrs
12	**Proposal**	**NA**	**NA**	**0%**	**0 days**	**10 days**	**$0.00**	**0 hrs**
13	Draft the Bid Document	NA	NA	0%	0 days	3 days	$0.00	0 hrs

Name: Draft Technical Details Schedule Duration: 11.25d Effort driven Previous Next
Start: 15 Dec '05 Finish: 4 Jan '06 Task type: Fixed Units % Complete: 20%

ID	Resource Name	Units	Work	Ovt. Work	Baseline Work	Act. Work	Rem. Work
2	Sales Engineer	1	70h	0h	72h	20h	50h
4	Systems Engineer	1	90h	0h	72h	20h	70h

7. Apply the **Resource Cost** details form:
   - Update the **Draft Technical Details Schedule** with the following information:
     Actual Cost of $1,000.00 against the **Sales Engineer**, and
     Actual Cost of $750.00 against the **System Engineer**.
   - Click the **OK** button to accept the changes.
   - Note: The **Remaining Costs** have been calculated by Microsoft Project.
   - You schedule should look like this:

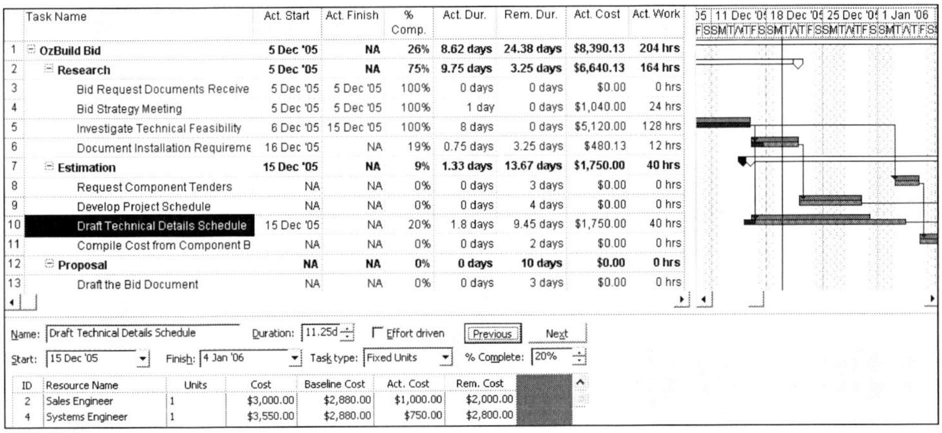

	Task Name	Act. Start	Act. Finish	% Comp.	Act. Dur.	Rem. Dur.	Act. Cost	Act. Work
1	**OzBuild Bid**	**5 Dec '05**	**NA**	**26%**	**8.62 days**	**24.38 days**	**$8,390.13**	**204 hrs**
2	**Research**	**5 Dec '05**	**NA**	**75%**	**9.75 days**	**3.25 days**	**$6,640.13**	**164 hrs**
3	Bid Request Documents Receive	5 Dec '05	5 Dec '05	100%	0 days	0 days	$0.00	0 hrs
4	Bid Strategy Meeting	5 Dec '05	5 Dec '05	100%	1 day	0 days	$1,040.00	24 hrs
5	Investigate Technical Feasibility	6 Dec '05	15 Dec '05	100%	8 days	0 days	$5,120.00	128 hrs
6	Document Installation Requireme	16 Dec '05	NA	19%	0.75 days	3.25 days	$480.13	12 hrs
7	**Estimation**	**15 Dec '05**	**NA**	**9%**	**1.33 days**	**13.67 days**	**$1,750.00**	**40 hrs**
8	Request Component Tenders	NA	NA	0%	0 days	3 days	$0.00	0 hrs
9	Develop Project Schedule	NA	NA	0%	0 days	4 days	$0.00	0 hrs
10	Draft Technical Details Schedule	15 Dec '05	NA	20%	1.8 days	9.45 days	$1,750.00	40 hrs
11	Compile Cost from Component B	NA	NA	0%	0 days	2 days	$0.00	0 hrs
12	**Proposal**	**NA**	**NA**	**0%**	**0 days**	**10 days**	**$0.00**	**0 hrs**
13	Draft the Bid Document	NA	NA	0%	0 days	3 days	$0.00	0 hrs

Name: Draft Technical Details Schedule Duration: 11.25d Effort driven Previous Next
Start: 15 Dec '05 Finish: 4 Jan '06 Task type: Fixed Units % Complete: 20%

ID	Resource Name	Units	Cost	Baseline Cost	Act. Cost	Rem. Cost
2	Sales Engineer	1	$3,000.00	$2,880.00	$1,000.00	$2,000.00
4	Systems Engineer	1	$3,550.00	$2,880.00	$750.00	$2,800.00

WORKSHOP 17 CONTINUED

8. Now apply the **Earned Value** table to review how your project is progressing:

	Task Name	BCWS	BCWP	ACWP	SV	CV	EAC	BAC	VAC
1	**OzBuild Bid**	**$8,720.00**	**$7,803.47**	**$8,390.13**	**-$916.53**	**-$586.67**	**$27,797.71**	**$25,854.00**	**-$1,943.71**
2	**Research**	**$7,440.00**	**$6,651.47**	**$6,640.13**	**-$788.53**	**$11.33**	**$8,705.14**	**$8,720.00**	**$14.86**
3	Bid Request Docum	$0.00	$0.00	$0.00	$0.00	$0.00	$0.00	$0.00	$0.00
4	Bid Strategy Meeting	$1,040.00	$1,040.00	$1,040.00	$0.00	$0.00	$1,040.00	$1,040.00	$0.00
5	Investigate Technica	$5,120.00	$5,120.00	$5,120.00	$0.00	$0.00	$5,120.00	$5,120.00	$0.00
6	Document Installatio	$1,280.00	$491.47	$480.13	-$788.53	$11.33	$2,500.97	$2,560.00	$59.03
7	**Estimation**	**$1,280.00**	**$1,152.00**	**$1,750.00**	**-$128.00**	**-$598.00**	**$12,493.06**	**$8,224.00**	**-$4,269.06**
8	Request Componen	$0.00	$0.00	$0.00	$0.00	$0.00	$960.00	$960.00	$0.00
9	Develop Project Sch	$0.00	$0.00	$0.00	$0.00	$0.00	$864.00	$864.00	$0.00
10	Draft Technical Detai	$1,280.00	$1,152.00	$1,750.00	-$128.00	-$598.00	$8,750.00	$5,760.00	-$2,990.00
11	Compile Cost from C	$0.00	$0.00	$0.00	$0.00	$0.00	$640.00	$640.00	$0.00
12	**Proposal**	**$0.00**	**$0.00**	**$0.00**	**$0.00**	**$0.00**	**$8,910.00**	**$8,910.00**	**$0.00**
13	Draft the Bid Docum	$0.00	$0.00	$0.00	$0.00	$0.00	$2,800.00	$2,800.00	$0.00

9. You will notice that the project is scheduled to be $1,943.71 over cost. See the **VAC** (Variance at Completion) column.

10. To demonstrate the splitting of tasks, assume that the data date is 23 Dec 05 and no more work will continue until after the New Year. Apply the **Tracking** table and update the **% Complete** column as follows:
   - **Document Installation Requirements** as 100%,
   - **Develop Project Schedule** as 50%, and
   - **Draft Technical Details Schedule** as 60%.

11. We will not update the Work and Costs, but ensure that splitting in-progress tasks is checked in the **Tools**, **Options…**, **Schedule** tab.

12. Use the **Tools**, **Tracking**, **Update Project…** to reschedule all incomplete work after 03 Jan 06, allow scheduling conflict.

13. After rescheduling all incomplete work after 03 Jan 06, your schedule should look like the following:

	Task Name	Act. Start	Act. Finish	% Comp.
1	**OzBuild Bid**	**5 Dec '05**	**NA**	**49%**
2	**Research**	**5 Dec '05**	**21 Dec '05**	**100%**
3	Bid Request Documents Recieved	5 Dec '05	5 Dec '05	100%
4	Bid Strategy Meeting	5 Dec '05	5 Dec '05	100%
5	Investigate Technical Feasibility	6 Dec '05	15 Dec '05	100%
6	Document Installation Requirements	16 Dec '05	21 Dec '05	100%
7	**Estimation**	**15 Dec '05**	**NA**	**43%**
8	Request Component Bids	NA	NA	0%
9	Develop Project Schedule	22 Dec '05	NA	50%
10	Draft Technical Details Schedule	15 Dec '05	NA	60%
11	Compile Costs from Component Bids	NA	NA	0%
12	**Proposal**	**NA**	**NA**	**0%**
13	Draft Bid Document	NA	NA	0%
14	Meeting to Review the Draft Bid Document	NA	NA	0%
15	Design Presentation	NA	NA	0%
16	Edit Proposal Draft Bid Document	NA	NA	0%
17	Negotiate Component Work Packages	NA	NA	0%
18	Final Review of Bid Document	NA	NA	0%
19	Submit Bid	NA	NA	0%

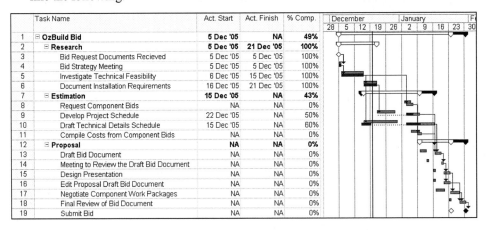

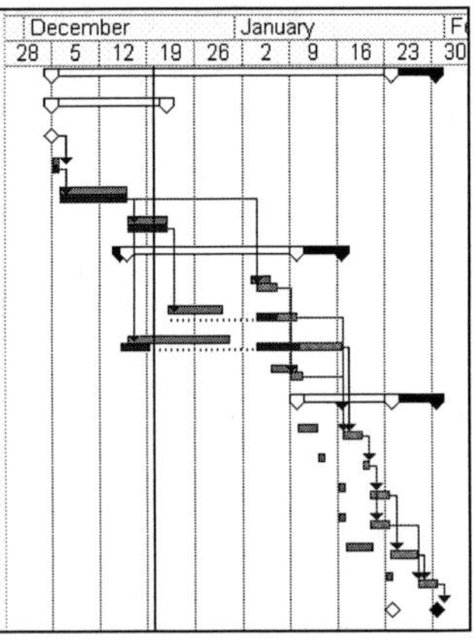

14. You will notice there will be negative float on Submit Bid which is generated by the Must Finish By constraint.

21 RESOURCE HISTOGRAMS, TABLES AND S-CURVES

This chapter will briefly cover the following topics:

- **Resource Histograms** – Allow the display of each resource on a time-phased vertical bar graph. These are termed **Resource Graphs** in Microsoft Project 2003.
- **Resource Tables** – Allow the display of one or more resource requirements in a table on a time-phased basis. This information is displayed by the **Task Usage** and **Resource Usage** views.
- **S-Curves** – Allow the display of the planned, earned or actual consumption of resources or costs as a line graph. A single S-Curve may be produced for one or selected resource costs or hours using the Resource Graph View. The display of multiple S-Curves showing the Planned, Actual and Earned cost or hours of resources is not possible. It is also not possible to display Fixed Costs on the S-Curves, but the source data for resources may be exported to a spreadsheet where the S-Curves may be readily created for multiple S-Curves.

Function	Command Menu
• To display a **Resource Graph**	Select the **Resource Graph** View.
• To display a **Resource Tables**	Select the **Task Usage** or **Resource Usage** Views.
• To print a **Resource Profile** or **Resource Table**	Select the appropriate **View**, make the view active and use the normal print commands.

21.1 Resource Graph

A **Resource Graph** may be displayed with the **Resource Graph** view:

- The **Resource Graph** may be viewed in the top or bottom pane. The view below displays the Gantt Chart in the top pane and Resource Graph in the bottom pane and shows the Sales Engineer is overloaded:

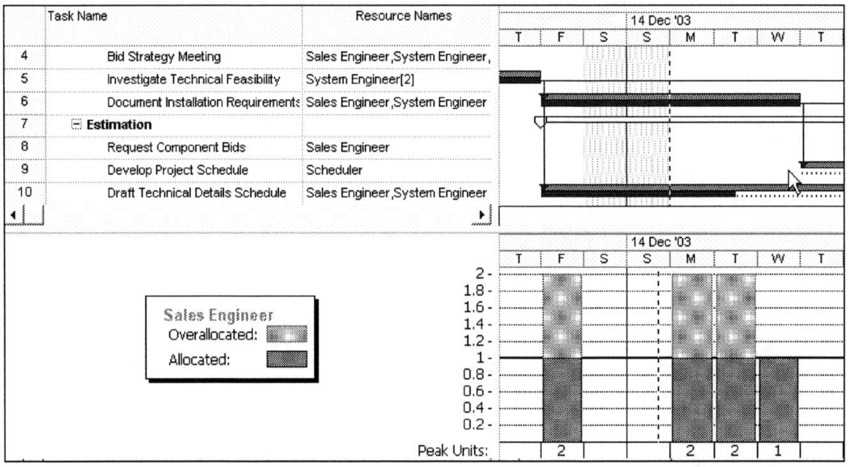

- Only one resource is displayed at a time. You may scroll through the resources by:
  - ➢ Striking the **Page Up** and **Page Down Keys**, or
  - ➢ Clicking on the scroll bar in the lower left-hand side of the **Resource Graph** view.
- The **Resource Graph Units** options may display the vertical scale by using any of the units shown in the gray box on the right. The **Units** to be displayed may be selected by:
  - ➢ Selecting **Format**, **Details**, or
  - ➢ Right-clicking in the graph area of the screen.

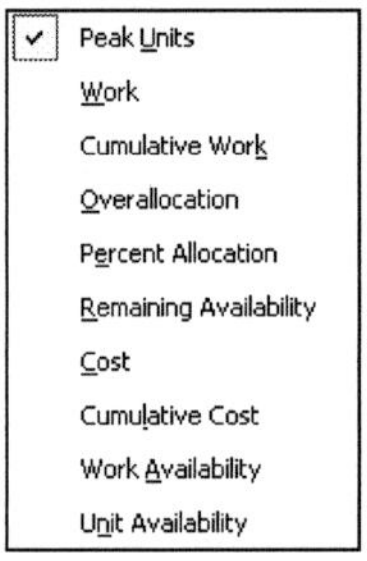

- The **Resource Graph** gridlines and colors may be formatted by opening the **Bar Styles** form and:
  - ➢ Selecting **Format**, **Bar Styles…** when the **Resource Graph** is the active pane, or
  - ➢ Right-clicking in the graph area of the screen and selecting the **Bar Styles** option.
- The **Timescale** of the graph is adopted from the timescale setting. You will find the **Zoom** icons useful for changing the timescale.

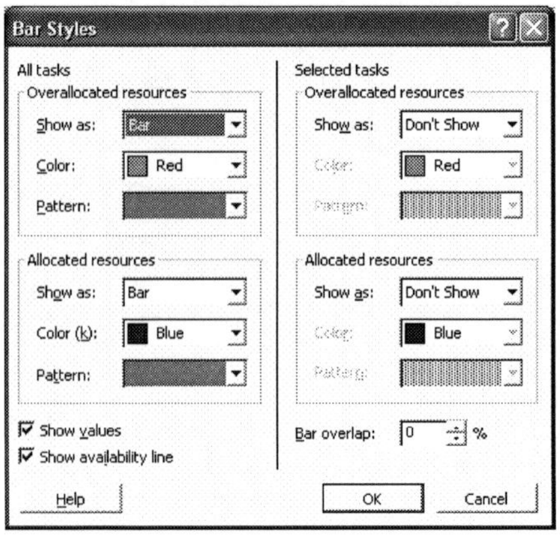

21.2 Resource Tables

Resource tables are displayed using the **Task Usage** or **Resource Usage** views.

- The **Task Usage View** organizes the schedule by **Task** and then by **Resource**:

	Task Name	Assignment Units	Work	Details	4 Dec '05 S	M	T	W	T	F	S	11 Dec '05 S	M	T	W	T	F
1	⊟ OzBuild Bid		656 hrs	Work		24h	16h	16h	16h	16h			16h	16h	16h	32h	32h
2	⊟ Research		216 hrs	Work		24h	16h	16h	16h	16h			16h	16h	16h	16h	16h
3	Bid Request Documents Received		0 hrs	Work													
4	⊟ Bid Strategy Meeting		24 hrs	Work		24h											
	Sales Engineer	100%	8 hrs	Work		8h											
	Systems Engineer	100%	8 hrs	Work		8h											
	Bid Manager	100%	8 hrs	Work		8h											
5	⊟ Investigate Technical Feasibility		128 hrs	Work			16h	16h	16h	16h			16h	16h	16h	16h	
	Systems Engineer	200%	128 hrs	Work			16h	16h	16h	16h			16h	16h	16h	16h	
6	⊟ Document Installation Requirements		64 hrs	Work													16h
	Sales Engineer	100%	32 hrs	Work													8h
	Systems Engineer	100%	32 hrs	Work													8h

- The **Resource Usage** view organizes the schedule by **Resource** and then by **Task**. Note the Sales Engineer is overloaded as indicated by the icon in the indicators column:

	ℹ	Resource Name	Work	Details	1 Jan '06 S	M	T	W	T	F	S	8 Jan '06 S	M	T	W
2	⟐	⊟ Sales Engineer	190 hrs	Work			8h	16h	16h	16h			16h	8h	8h
		Bid Strategy Meeting	*8 hrs*	Work											
		Document Installation Requirements	*32 hrs*	Work											
		Request Component Tenders	*24 hrs*	Work			8h	8h	8h						
		Draft Technical Details Schedule	*70 hrs*	Work			0h	8h	8h	8h			8h	8h	8h
		Compile Cost from Component Bids	*16 hrs*	Work						8h			8h		
		Meeting to Review the Draft Bid Document	*8 hrs*	Work											
		Negotiate Component Work Packages	*32 hrs*	Work											
3		⊟ Scheduler	32 hrs	Work			0h	8h	8h	8h			8h		
		Develop Project Schedule	*32 hrs*	Work			0h	8h	8h	8h			8h		
4		⊟ Systems Engineer	282 hrs	Work			0h	8h	8h	8h			8h	8h	8h
		Bid Strategy Meeting	*8 hrs*	Work											
		Investigate Technical Feasibility	*128 hrs*	Work											

- When these views are displayed in the top **Pane** they display all resources and all tasks.
- When these views are displayed in the bottom **Pane** they display values for resources or tasks that have been selected in the top **Pane**.

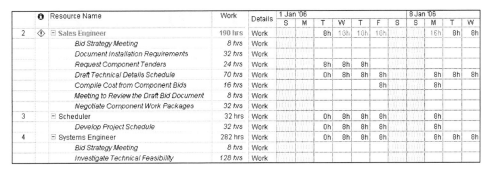

21.3 Printing Resource Profiles and Tables

To **Print** a **Task Usage**, **Resource Usage**, or **Resource Graph**, make the appropriate **Pane** active and use the print functions as described in the **PRINTING AND REPORTS** chapter.

21.4 Exporting Table Data to Create S-Curves and Histograms in Excel

Most people are familiar with copying and pasting data from one software program to another. Many people are also comfortable with creating S-Curves in spreadsheets. In this section, we will cover the basics of transferring the data from Microsoft Project 2003 to a spreadsheet for the creation of S-Curves.

21.4.1 Export Using Copy and Paste

A data table may be highlighted, copied, and then pasted into spreadsheets to create S-Curves. The headings are not copied, and the data on the left-hand side of the **Pane** and the data on the right-hand side of the **Pane** have to be copied separately and aligned with the headings in Excel.

To transfer the data into Excel:

- Organize the **Task Usage** or **Resource Usage** view and **Details** so it displays the rows, columns, units and timescale you require in your graph.
- Type the column headings from both sides of the pane into the top row of your spreadsheet.
- Highlight the rows on the left-hand side of your Pane that you want to copy to the spreadsheet. Position the cursor under the spreadsheet heading and paste. If your headings do not line up, move either the data or the headings so that they do line up.
- Follow the same process with the data on the right-hand side of the Pane.
- You may now create your S-Curves in the spreadsheet using its built-in graphing function.

21.4.2 Export Using Analysis Toolbar

The **Analysis** toolbar is designed to allow you to export time-phased data to Excel. Excel may be used to export data and create graphs. This is a Wizard-style function and the instructions will step you through the process.

21.4.3 S-Curve Example

The example below has:

- An Early Baseline curve shown in diamonds as the upper left curve,
- A Late Baseline curve shown in squares as the lower right curve, and
- A Progress Curve shown in triangles as the center incomplete curve.

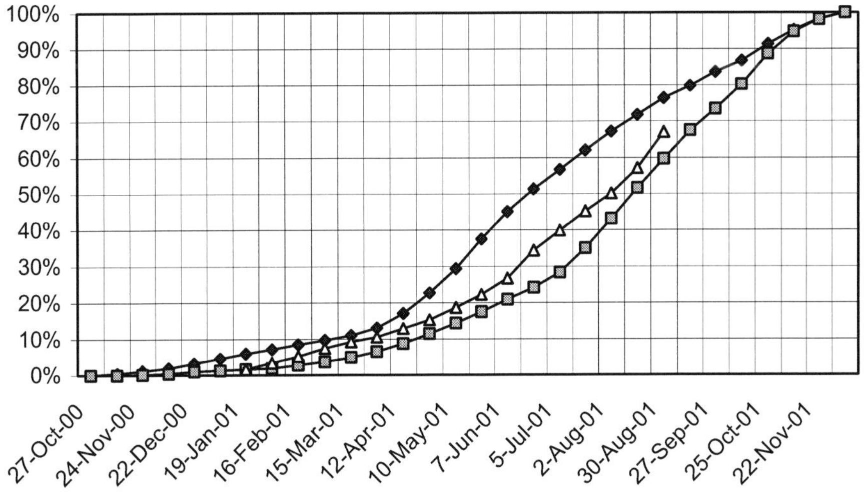

You may try the following process to create S-Curves in Excel:

- Assign a resource with a cost value to each task and export the total project hours or costs on a time-phased basis to Excel.
- To Create the **Late Baseline**, copy the project and change all the tasks to **As Late As Possible**. This will only calculate correctly if the network is a **Closed Network**. Copy and paste the data into the next line of the spreadsheet.
- As the project progresses, copy the Actuals into the third line. This may be achieved by creating an **Export Status View**, applying the view to the schedule and exporting the Actuals after each update.
- The actual costs are reduced to % by a formula before creating the Excel Chart.

When the actual line has crossed the late baseline curve there is a decreasing chance of bringing the project in on time with the current plan. It is important that you monitor the angle of the late curve; if it is flatter that the baseline then your project is in trouble as your progress is too slow.

You may create more that one curve, one for each phase, for example, to isolate scheduling problems.

The single S-Curve displayed below was produced by:

- Displaying the **Resource Sheet** in the top pane and selecting all the resources,
- Displaying the **Resource Graph** in the bottom pane,
- Displaying the **Cumulative Costs** and
- Selecting the **Show as:** option as **Line**.

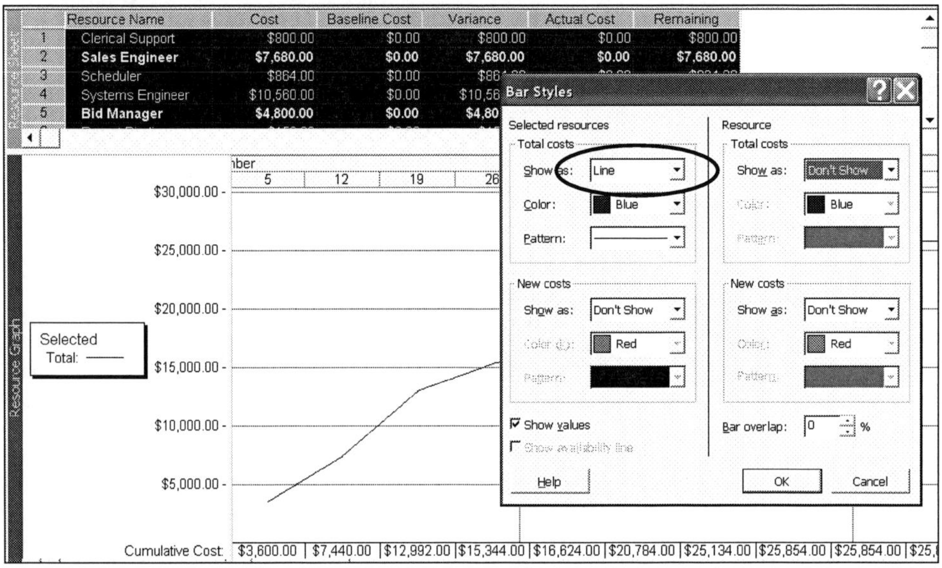

22 TOOLS AND TECHNIQUES FOR SCHEDULING

22.1 Understanding Menu Options

You will find that the menu options change as you select rows, cells, all rows and cells, or views. Therefore, not all menu options are always available. The following topics will be covered in this chapter:

- Menu items sometimes found under **Edit**:
  - **Cut**, **Copy** and **Paste Task**
  - **Cut**, **Copy** and **Paste Cell**
  - **Copy Picture**
  - **Fill**
  - **Clear**
  - **Find…** and **Replace…**
  - **Go To…**
- **Insert**, **Recurring tasks…**
- **Splitting** a Task
- **Copy**, **Cut** and **Paste** from Spread Sheets
- **Unique Task**, **Resource** and **Assignment ID**
- **Organizer**

22.2 Cut, Copy and Paste Row

This function allows you to select one or more consecutive or non-consecutive tasks (using Ctrl-Click) and either copy them, or cut them and paste as a group to a new location that you select with your mouse.

22.3 Cut, Copy and Paste Cell

You may cut or copy information and then paste into one or more cells from:

- One cell, or
- Adjacent cells by dragging in rows and/or columns, or
- A non-contiguous group of cells by Ctrl-clicking.

This function operates in a similar way to Excel's cut-and-paste operation:

- Highlight the cell(s) you want to copy or cut.
- Select **Edit**, **Copy Cell** or **Ctrl+C** to copy a cell or **Edit**, **Cut Cell** or **Ctrl+X** to cut a cell.
- Position the mouse where you want to paste the data and select **Edit**, **Paste** or **Ctrl+V**.

This is an interesting function since it allows you to copy from one column to another when the format is compatible. This function may also be used for transferring and/or updating your schedule from other software such as Excel and is covered later in this chapter.

22.4 Copy Picture

This allows a section or total screen to be copied to the clipboard and then either pasted into a document or saved as a **gif** file. Select **Edit**, **Copy Picture...** to open the **Copy Picture** form:

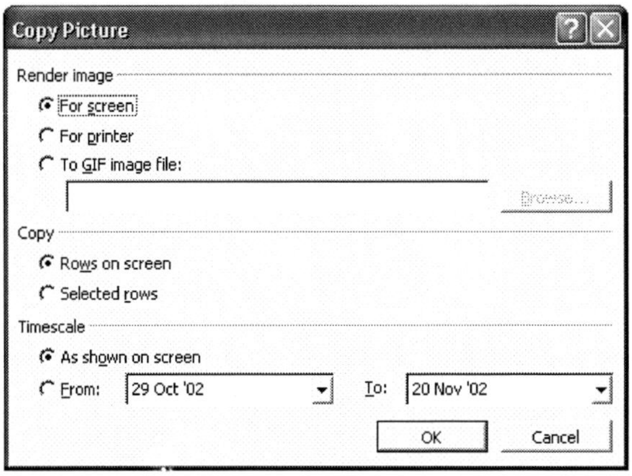

The options are self-explanatory and when pasted into a document the picture looks like the example below.

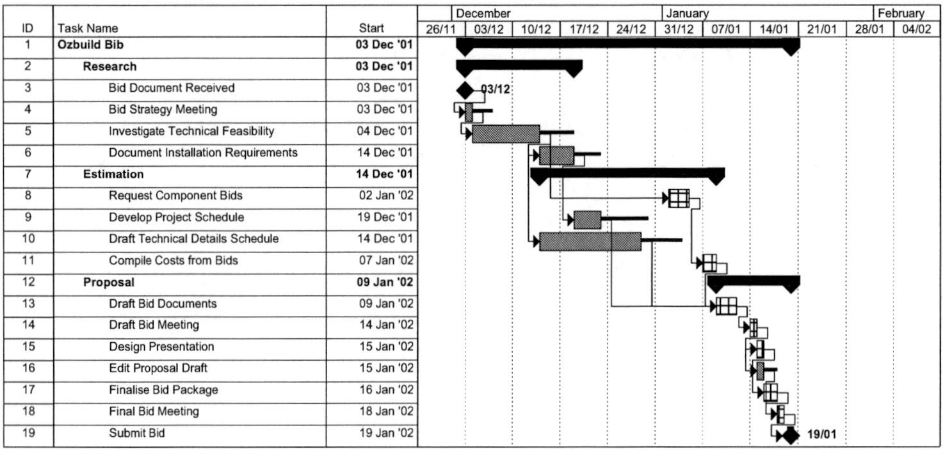

ID	Task Name	Start
1	**Ozbuild Bib**	**03 Dec '01**
2	**Research**	**03 Dec '01**
3	Bid Document Received	03 Dec '01
4	Bid Strategy Meeting	03 Dec '01
5	Investigate Technical Feasibility	04 Dec '01
6	Document Installation Requirements	14 Dec '01
7	**Estimation**	**14 Dec '01**
8	Request Component Bids	02 Jan '02
9	Develop Project Schedule	19 Dec '01
10	Draft Technical Details Schedule	14 Dec '01
11	Compile Costs from Bids	07 Jan '02
12	**Proposal**	**09 Jan '02**
13	Draft Bid Documents	09 Jan '02
14	Draft Bid Meeting	14 Jan '02
15	Design Presentation	15 Jan '02
16	Edit Proposal Draft	15 Jan '02
17	Finalise Bid Package	16 Jan '02
18	Final Bid Meeting	18 Jan '02
19	Submit Bid	19 Jan '02

22.5 Fill

The **Edit**, **Fill** command allows you to select one cell or a range of cells and then copy them up, down, left, or right without the need to copy and paste.

22.6 Clear

The **Edit**, **Clear** command allows you to clear the following data from a task:

- **All** – This clears all information from the cell you have highlighted.
- **Format** – This sets the formats back to the default.
- **Notes** – Clears the Notes contents on a task.
- **Contents** – This clears the contents of the highlighted cells but leaves the formatting.
- **Entire task** – This deletes the entire task but leaves blank rows.

22.7 Find and Replace

The **Edit**, **Find…** and **Edit**, **Replace…** functions allow you to find any task by matching data with your defined criteria. This will also allow you to find the data and then replace it with another piece of information.

22.8 Go To

The **Edit**, **Go To…** function allow you to quickly find a date and move the timescale horizontally to that date or a task if you know the task ID number:

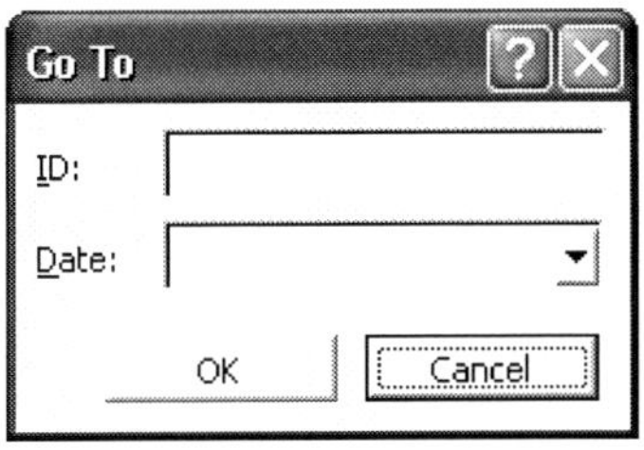

22.9 Insert Recurring Task

The **Insert**, **Recurring tasks…** function allows the insertion of more than one task that occurs on a regular basis. Select **Insert**, **Recurring Task…** to display the **Recurring Task Information** form. The options are self-explanatory.

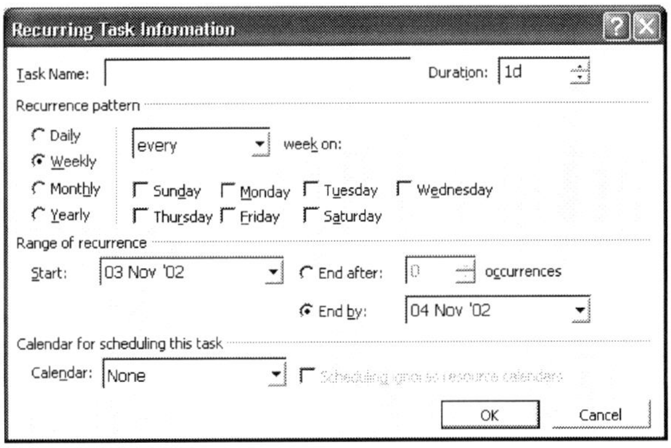

22.10 Splitting a Task

A task may be manually split by clicking on the icon, moving your cursor over the point on the task bar where you want a split, and dragging the task. Splitting a task effectively increases the duration of the task but work associated with a split task is scheduled intermittently.

The Tools, Options, Schedule **Split in-progress tasks** option must be checked to allow the splitting of a task.

An in-progress task may also be split by:

- Dragging the incomplete portion of a task in the bar chart, or
- Using the **Tools**, **Tracking**, **Update Project...**, **Reschedule uncompleted work to start after:** function, or
- Commencing a task before its predecessor finishes as shown by the second task in the diagram below:

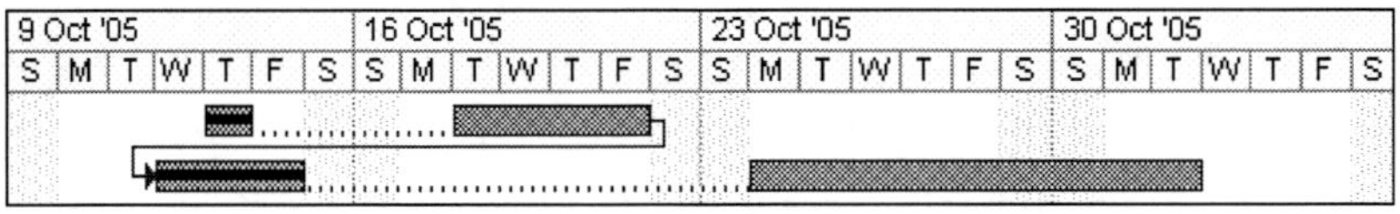

22.11 Copy or Cut-and-Paste to and from Spreadsheets

Microsoft Project 2003 will allow **Copy** or **Cut**-and-**Paste** to and from spreadsheets and other software packages. This may be useful for a number of purposes:

- Importing Tasks from other applications,
- Assigning Codes and Resources to Tasks,
- Statusing a schedule, and
- Exporting Data to other packages.

The **Copy** or **Cut** function copies the task data from highlighted columns and rows in the schedule to the Windows clipboard. These may then be pasted into another software package:

- Highlight the data in your schedule that you want to transfer to another software package and cut or copy the data.
- Move to the spreadsheet application and paste the task information. You will find the column headers are **NOT** pasted into the application; only data from the schedule is pasted.
- The data may be edited or updated in the application, as required.
- The tasks may be selected and pasted back into the schedule to update it with the normal copy and paste functions.

22.12 Paste Link – Cell Values in Columns

A cell may be copied and the Paste Linked to another cell. Thus when the copied cell is updated the linked cell is also updated with the same value. This could be used to update a number of activities that have the same % Complete. A linked cell has a small triangle in the corner and the linking may be removed by overtyping the cell value.

22.13 Unique Task, Resource and Assignment ID

The Task IDs change for all tasks below an inserted task. There are many contractual and management reasons to be able to identify each task by a unique task number.

22.13.1 Task Unique ID

When a new task is added to a blank schedule, it is assigned a **Unique ID** commencing with 1. When tasks are deleted these numbers are not reused and any new task is assigned a new sequential number. This **Unique ID** number may be displayed in a column titled **Unique ID**. It is also possible to display the **Unique Predecessor** and **Unique Successor** columns:

	Unique ID	Task Name	Unique ID Predecessors	Unique ID Successors
1	**18**	⊟ **OzBuild Bid**		
2	**17**	⊟ **Research**		
3	2	Bid Document Received		3
4	3	Bid Strategy Meeting	2	4
5	4	Investigate Technical Feasibility	3	5,6,8
6	5	Document Installation Requirements	4	7,9
7	**19**	⊟ **Estimation**		
8	6	Request Component Bids	4	9

22.13.2 Resource Unique ID

A resource **Unique ID** is created for each resource. Again, when a resource is deleted, the resource **Unique ID** is not re-used. This number may be displayed in the **Resource Sheet** or **Resources Usage** view column titled, **Unique ID**.

22.13.3 Resource Assignment Unique ID

When a resource is assigned to a task, the assignment is given a unique **Assignment ID**. This number may be displayed in the **Resource Sheet** or **Resources Usage** view column titled, **Unique ID**:

	ℹ	Task Name	Unique ID
1		⊟ **OzBuild Bid**	**18**
2		⊟ **Research**	**17**
3		Bid Document Received	2
4		⊟ Bid Strategy Meeting	3
		Sales Engineer	*2097170*
		System Engineer	*2097181*
		Tender Manager	*2097169*
5		⊟ Investigate Technical Feasibility	4
		System Engineer	*2097171*
6		⊟ Document Installation Requirements	5

22.14 Organizer

The **Global.mpt** file holds the schedule's default settings such as **Tables** and **Views**, which are inherited by new projects and not created from a template. The **Organizer** function is also used to copy information between projects or to update the **Global.mpt**.

Select **Tools**, **Organizer…**to open the **Organizer** form:

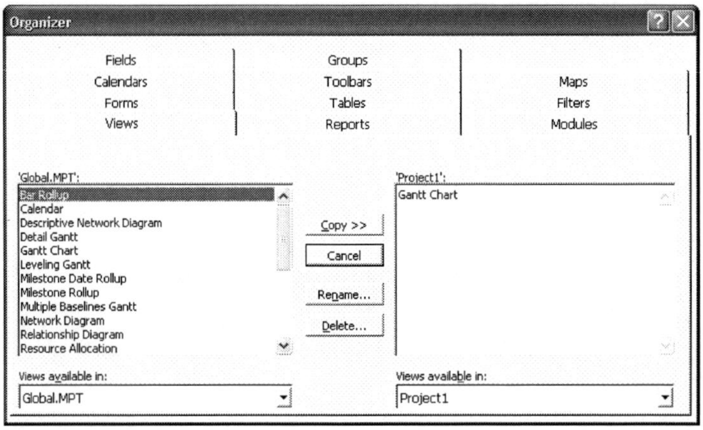

Except for the **Global.mpt** project, the projects you want to copy settings to and from will have to be opened in order to copy data.

- The Organizer function is used for renaming and deleting most items such as **Tables**, **Views and Calendars**.
- The two tab titles above that are self explanatory are:
  - **Maps** – These are predefined tables for exporting data, and
  - **Modules** – These are Visual Basic Macros.

23 WHAT IS NEW IN MICROSOFT PROJECT

There were very few changes made in functionality of Microsoft Project 2003 and for the benefit of users upgrading from Microsoft Project 2000 I have kept the new features from 2002 to 2003 in a separate section.

Earlier versions on Microsoft Project 2003 did not function correctly and it was not possible to outdent some tasks. A Hotfix Package (software patch) is available from the Microsoft website to correct this problem.

23.1 WHAT IS NEW IN MICROSOFT PROJECT 2003 STANDARD

23.1.1 Copy Picture to Office Wizard

There is a new function available on the **Analysis** toolbar that runs a wizard to copy a picture to PowerPoint, Word or Visio.

23.1.2 Print a view as a report

There is a new wizard available for displaying views and printings. Display the **Project Guide** toolbar and click on the Reports tab to run this function.

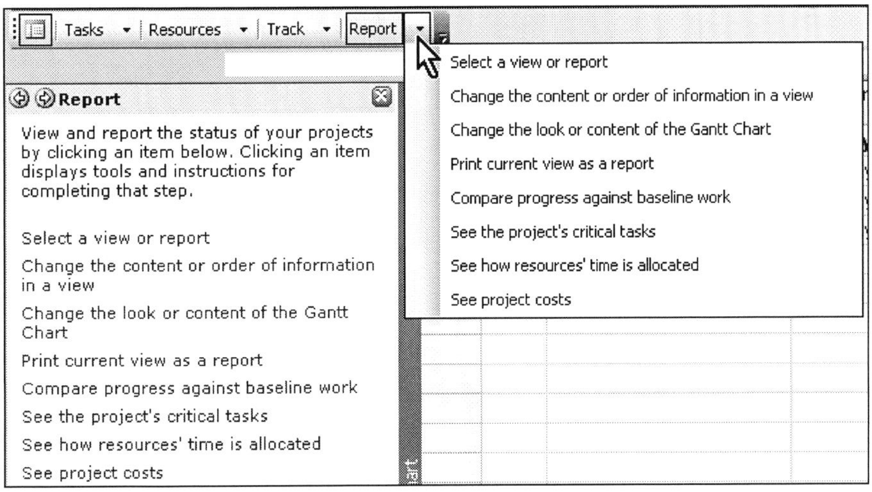

23.1.3 Options

In Microsoft Project 2000 there was an Option Titled **Workgroup**, this was replaced with **Collaborate** in 2002 and then removed in 2003 Standard.

Microsoft Project 2003 had a new option added titled **Security** allowing:

- The properties of Author, Manager, Company and Last Saved by are removed from a file when it is saved, and
- There are security options that may be set to nominate which Macros may be used or not used. This is aimed at preventing Macros with a virus being activated.

23.2 WHAT WAS NEW IN MICROSOFT PROJECT 2002 STANDARD

There were many features added or amended in the release of Microsoft Project Standard 2002. The following is a list of important features added that are considered to be useful for planning and scheduling projects. The additional enterprise functions of Microsoft Project Professional 2002 and Microsoft Project Server will not be discussed in this book.

Topic	Notes
• **Timescale**	The timescale, such as in Gantt Chart, may be displayed with one, two or three tiers with each tier displaying a different unit of time. By default, the timescale is shown in two tiers. The timescale may be adjusted by double-clicking the timescale or selecting **Format**, **Timescale…..**
• **Smart Tags**	This feature provides information about your scheduling results as well as some ease of use features to change your results. Microsoft Project 2003 does not allow user to create and customize the **Smart Tags**.
• **Multiple Baselines**	Microsoft Project 2003 can save 11 Baselines over the course of a project. To save a baseline, select **Tools**, **Tracking**, and then **Save Baseline….**
• **Status Date Calculation Options**	The Calculation Options form has additional options that allow user to automatically adjust where actual and remaining work of new in-progress activities are placed with respect to the **Status Date**. These options are found under **Tools**, **Option…**, **Calculation** tab.
• **Earned Value**	There is an additional field titled **Physical % Complete** which may be used for calculating the **Earned Value** of the Project based on any of the Baselines. Select **Tools**, **Option…**, select the **Calculation** tab and click on the Earned Value... button.
• **Project Guide**	The Project Guide allows user to define project parameters such as tasks, tracks, project calendar, resource, and report, etc. using a guide steps wizard. The Project Guide is activated by default and you may disable this features under **Tools**, **Option…**, **Interface** tab.
• **Summary Task Baseline Recalculation**	The baseline of Summary Tasks may not be valid when tasks are added, deleted or moved to another Summary Task. There are options found under **Tools**, **Tracking**, **Save Baseline…** to recalculate baseline data for the associated summary tasks.

23.2.1 Timescale

The timescale, such as in Gantt Chart, may be displayed with one, two or three tiers with each tier displaying a different unit of time. By default, the timescale is shown in two tiers. The timescale may be adjusted by double-clicking the timescale or selecting **Format**, **Timescale…**.. You may choose to display:

- One tier, the Middle,
- Two tiers, the Middle and Bottom, or
- All three tiers, as displayed below.

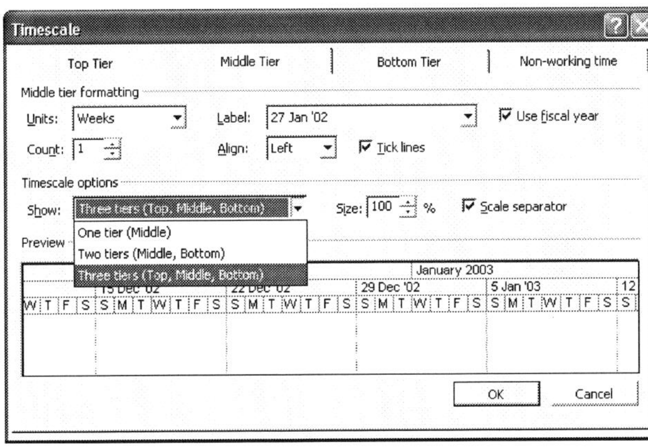

You may define and adjust the timeline (such as day, week, month and year) at each tier, depending on the level of detail you wish to present.

23.2.2 Graphical Indicators - Smart Tags

Microsoft Project 2003 uses smart tags to explain scheduling calculations and advise users of alternative options that may be more desirable. A green triangle in the edited cell indicates that the edit just performed has made a change to the task that has scheduling ramifications. The indicator only appears in the view types of Gantt, Sheet and Usage. A notice of a scheduling change hovers over the indicator, and clicking on the indicator displays alternate choices. The indicator in most cases displays as long as the edit may be undone; once a new edit is made the indicator disappears. Edits for which Feedback indicators display are:

- Additional resource assignment,
- Edits to Start and Finish,
- Edits to Work, Units or Duration, and
- Deletion in the Name column.

The example to the right has had a description highlighted and deleted.

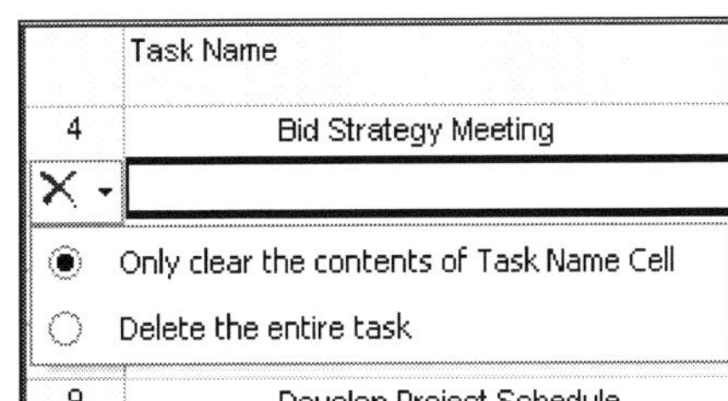

Select **Tools**, **Options…**, **Interface** tab to select which indications are displayed.

23.2.3 Multiple Baselines

This feature allows you to save all baseline fields up to a total of 11 times for one project. The **Tools, Tracking, Save Baseline…** dialog now has a dropdown list for each of the 11 baselines. Likewise, the Clear Baseline dialog has the same dropdown list. When a baseline is saved, the last date saved is stored with the baseline as well and is displayed in the dialog for future reference.

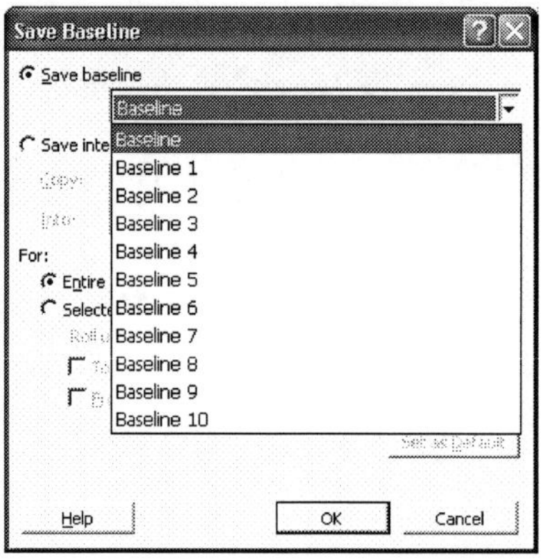

To accomplish this, 10 additional Baseline Start, Baseline Finish, Baseline Work, Baseline Duration, and Baseline Cost have been added.

The **Save interim plan** fields allow you to copy one baseline to another. You can now choose which baseline you would like to use in Earned Value calculations

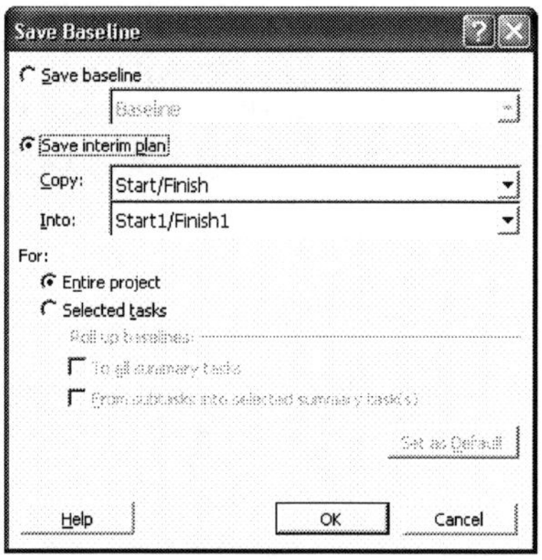

23.2.4 Status Date Calculation Options

New functions have been introduced in Microsoft Project 2003 intended to assist schedulers to place the new tasks as they are added in a logical position with respect to the **Status Date**. If the **Status Date** has not been set the **Current Date** is used.

Full details of this function are available in the **TRACKING PROGRESS** chapter, page 15-20.

23.2.5 Earned Value and Physical % Complete

There is an additional field titled **Physical % Complete**, which may be used for calculating the **Earned Value** of the Project based on any of the Baselines. Select the **Tools**, **Option...**, **Calculation** tab and click on the Earned Value... button to open the **Earned Value** form. You may now select:

- The **Default task Earned value method:**
  - **% Complete** or
  - **Physical % Complete**, which may be edited in a column.

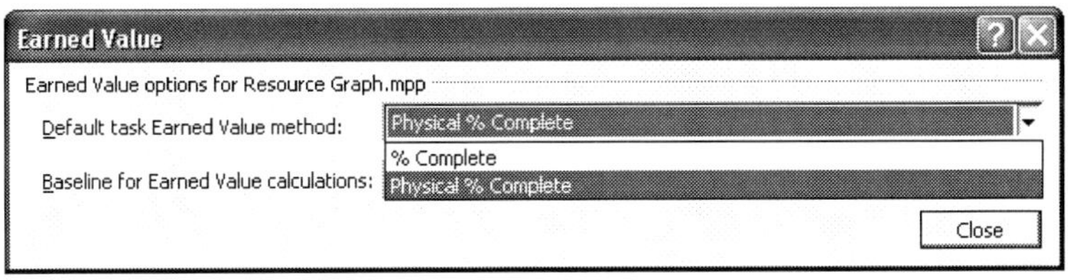

- The **Baseline for Earned Value Calculations:** and then select one of the 11 Baselines.

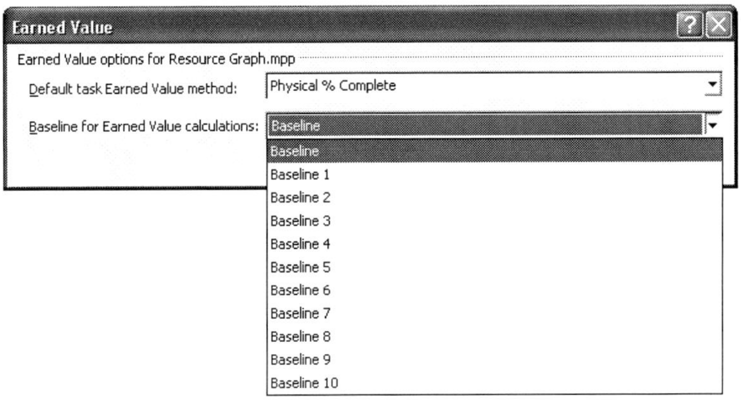

23.2.6 Project Guide

The Project Guide is accessed via a new toolbar titled **Project Guide**. The Project Guide includes options for **Tasks, Resources, Track, Report**, and **Next Steps** and **Related Activities**. The Project Guide will not be covered in detail as it is self-explanatory.

- When an option is selected from the Project Guide toolbar, a side-pane is displayed with a list of Help topics on the left of your screen.
- The Hide/Show option simply closes or opens the side-pane for the selected option. **Next Steps** and **Related Activities** display all the default Project Guide-based help available.
- The remaining options offer detailed information on each topic and include wizards to walk you through some specific goals.
- The View to the right of the side-pane updates to correspond with the action selected. Likewise the side-pane updates to display pertinent information as new data is entered into the project plan.

The picture below is an example of Project Guide in Microsoft Project 2003.

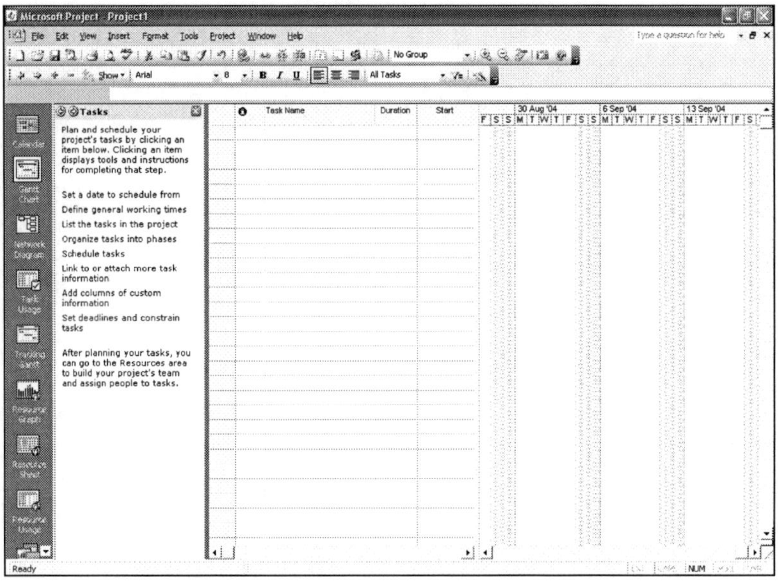

23.2.7 Summary Task Baseline Recalculation

By default, after the initial baseline is saved, a summary task's baseline is not updated when a subtask is modified, added or deleted. There are now two new options in the Save Baseline dialog that allow you to choose when summary task baseline information gets refreshed when saving a baseline for selected tasks. The options are:

- Roll up baseline to all summary tasks, and
- Roll up baseline from subtasks into selected summary task(s).

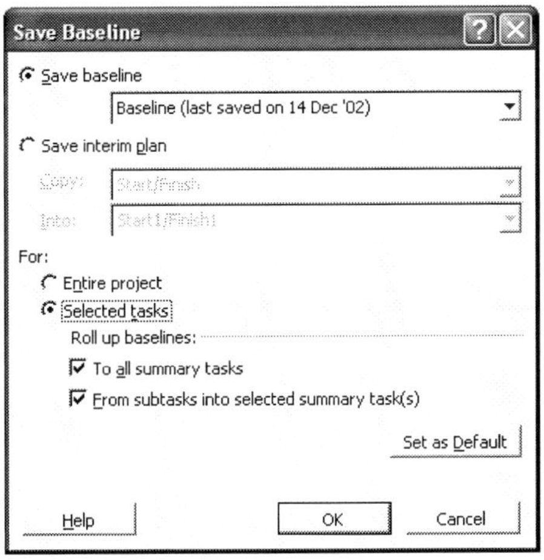

The first option allows baseline data from the immediate summary task to roll up into its ancestor summary tasks. The second option forces Microsoft Project to roll up the baseline details from its own subtasks. Details of this function are found in the **TRACKING PROGRESS** chapter. See page 15-4.

24 ITEMS NOT COVERED IN THIS BOOK

The following subjects are not covered in this book. This list is not exhaustive but is intended to give a guide to the other functions available in Microsoft Project 2003.

- **File, Save As Web Page…** – Saves the data in HTML format.
- **File, Save As Workspace…** – Saves and opens more than one file at a time.
- **File, Send To** – Emails your data to other people.
- **Format, Drawing** – Formats your drawings.
- **Edit, Paste Special…** – Allows options for pasting data or graphics into the schedule.
- **Edit, Paste as a Hyperlink** – Pastes data as a Hyperlink, the hyperlink is pasted into the Hyperlink columns.
- **Edit** and **Insert Objects** – Incorporates pieces of information from another application that may be displayed on the bar chart using OLE, Object Linking and Embedding.
- **Insert, Hyperlink…** – Allows you to save a Hyperlink to another file. The link need not be to another Microsoft Project file. You may click on the hyperlink icon in the **Task Information** column, and the function will take you to the file by opening the associated application to view the data. You may right-click on the Hyperlink to access the Hyperlink menu to edit, copy, etc.
- **Insert, Drawing** – Used to draw lines and other common geometrical objects (commonly found in graphics drawing software packages) on a Gantt Chart.
- **Tools, AutoCorrect Options…** – Specifies how the Spelling AutoCorrect function operates.
- **Tools, Import Outlook Tasks…** – Imports a list of tasks from Outlook.
- **Tools, Level Resources…** – This function delays tasks until resources become available and prevents overloading of resources.
- **Tools, Resource Sharing** – Shares resources with other projects.
- **Tools, Link Between Projects…** – Creates a link between two different projects.
- **Tools, Macro** – This function allows you to store a number of keystrokes and invoke them in one action. This is useful when executing repetitious keystroke operations. You may need to understand the Visual Basic programming language to fully implement this feature.
- **Collaborate** – These options are mainly used in conjunction with Project Server which is beyond the scope of this book
- **Help, Detect and Repair…** – Fixes problems with Microsoft Project 2003.
- **Compare Project Versions** – This function is found on the **Compare Project Versions** toolbar and allows the comparison of two project files to determine which data items have been changed.

- The **Analysis** toolbar has three options:
  - **Adjust Dates** – To change the start date of the project and associated task constraint dates.
  - **Analyze Timescale Data in Excel…** – Allows you to export time-phased data to Excel.
  - **PERT Analysis** –Functions like a Monte-Carlo Simulation. **PERT Analysis** also has its own toolbar.

Details of all these topics may be found in the User Manual and Help.

25 APPENDIX 1 – SCREENS USED TO CREATE VIEWS

Microsoft Project 2003 has a complex method of accessing the data on screen.

- The screen may be split with an **Upper** and **Lower Pane**.
  - ➢ The **Upper Pane** displays all project information, except when a filter is applied.
  - ➢ The **Lower Pane** displays information about the task or tasks that are highlighted in the upper pane.
- A **View** may be created and edited. A **View** is based on a **Screen**.
  - ➢ A **Combination View** sets the view for both the Upper and Lower Pane.
  - ➢ A **Single View** may be applied to either the **Upper** or **Lower Pane** as long as the **Screen** may be displayed in that **Pane**.
- There are 14 **Screens**, most of which may be applied to the **Upper** or **Lower Panes** through **Views**.
  - ➢ There are some restrictions. For example, the **Calendar** screen may not be applied to the bottom pane.
  - ➢ Some **Screens** are not useful when displayed in the incorrect pane. It is better to display the **Task** form in the bottom **Pane** and leave a **Gantt Chart** in the **Upper Pane**.
  - ➢ Some **Screens** are very similar to each other, such as the **Task** screen and the **Task Details** screen.
- Some **Screens** have **Details** forms where the content of a **View** may be changed. An example of this is a **Task Usage** view, which may be formatted to show **Work** or **Costs** and **Cumulative** or **Period**.
- Some Screens are split vertically and have a left-hand side and a right-hand side such as the **Gantt Screen**. Other **Screens** do not split vertically, like the **Task Screen** and the **Task Details Screen**, which are used in the **Task Form** view and the **Task Details Form** view.

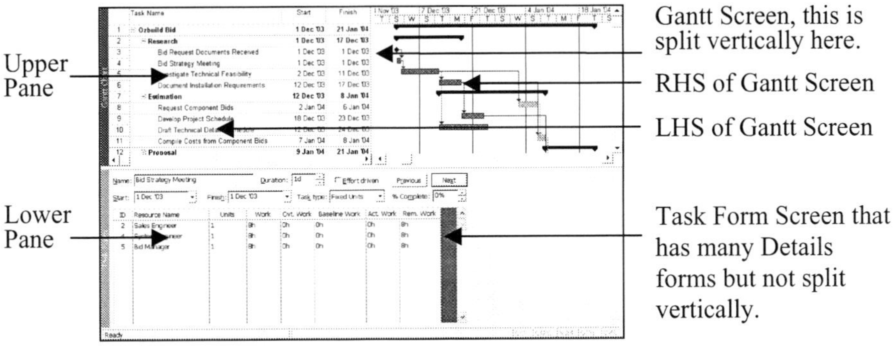

The following table summarizes the **Screen**, **View** and **Details** options:

	Screen Name	Panes	Split Vertically	Details Form Options	Format Column Options
1	Calendar	Upper Only	No	None	Right-click Format Options Task List... Go To... Timescale... Gridlines... Text Styles... Bar Styles... Zoom... Layout... Layout Now Remove Split
2	Gantt Chart	Best in Upper	Yes	Bars and Columns may be formatted	Yes
3	Network Diagram	Best in Upper	No	Task Boxes may be edited	No
4	Relationship Diagram	Either	No	None	No
5	Resource Form	Best in Lower	No	Hide Form View ✓ Schedule Cost Work Notes Objects	No
6	Resource Graph	Best in Lower	No	Gridlines... Bar Styles... Remove Split ✓ Peak Units Work Cumulative Work Overallocation Percent Allocation Remaining Availability Cost Cumulative Cost Work Availability Unit Availability	No
7	Resource Name Form	Best in Lower	No	Hide Form View ✓ Schedule Cost Work Notes Objects	No
8	Resource Sheet	Either	No	None	Yes

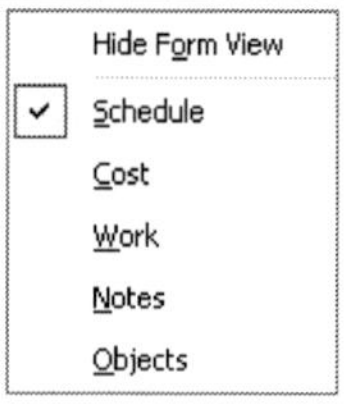

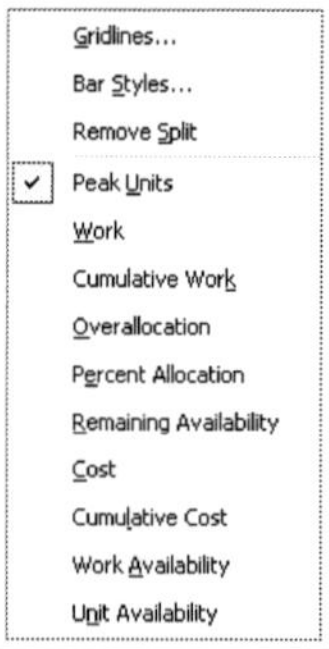

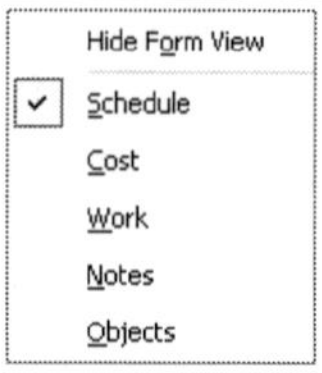

	Screen Name	Panes	Split Vertically	Details Form Options	Format Column Options
9	Resource Usage	Either	Yes	Detail Styles... ✓ Work Actual Work Cumulative Work Baseline Work Cost Actual Cost	No
10	Task Details Form	Best in Lower	Yes	Hide Form View ✓ Resources & Predecessors Resources & Successors Predecessors & Successors Resource Schedule Resource Work Resource Cost Notes Objects	No
11	Task Form	Best in Lower	No	Hide Form View Resources & Predecessors Resources & Successors ✓ Predecessors & Successors Resource Schedule Resource Work Resource Cost Notes Objects	No
12	Task Name Form	Best in Lower	No	Hide Form View Resources & Predecessors Resources & Successors Predecessors & Successors Resource Schedule ✓ Resource Work Resource Cost Notes Objects	No
13	Task Sheet	Best In Upper	No	None	Yes
14	Task Usage	Best in Lower	Yes	Detail Styles... ✓ Work Actual Work Cumulative Work Baseline Work Cost Actual Cost	Yes

The following table displays the Screens and provides some background on each:

Screen Name	Note and/or Screen Dumps
1. Calendar	May be displayed in top pane only. 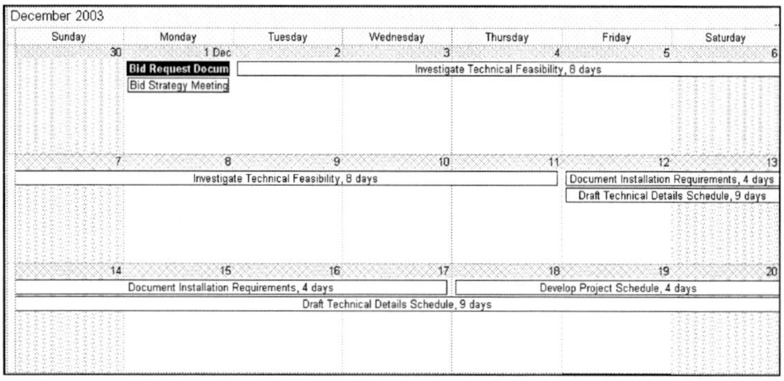 Some formatting is available through **Format**, **Layout…**: 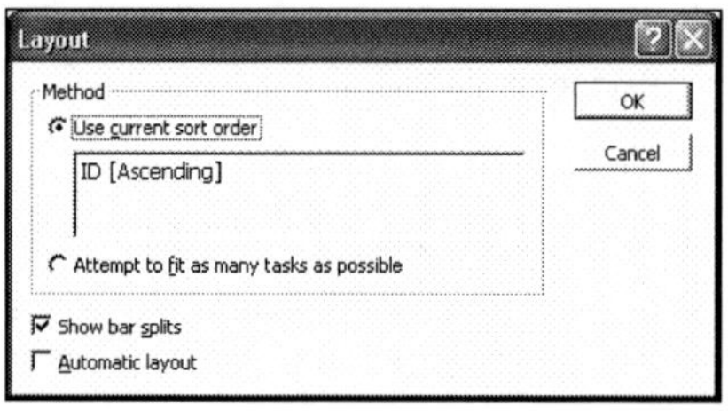
2. Gantt Chart	Designed to be displayed in the top pane: 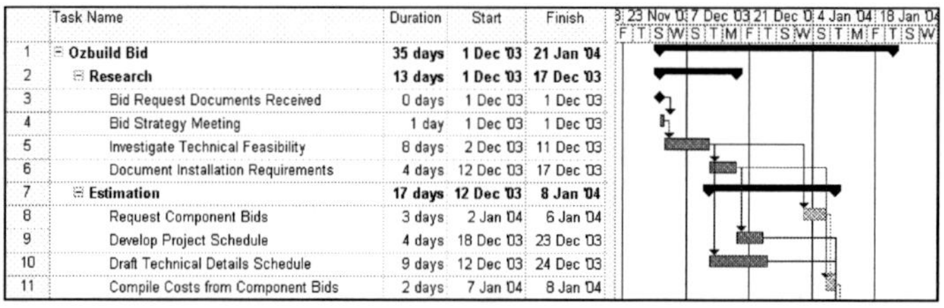 Columns and bars may be formatted.
3. Network Diagram	Designed to be displayed in the top pane: 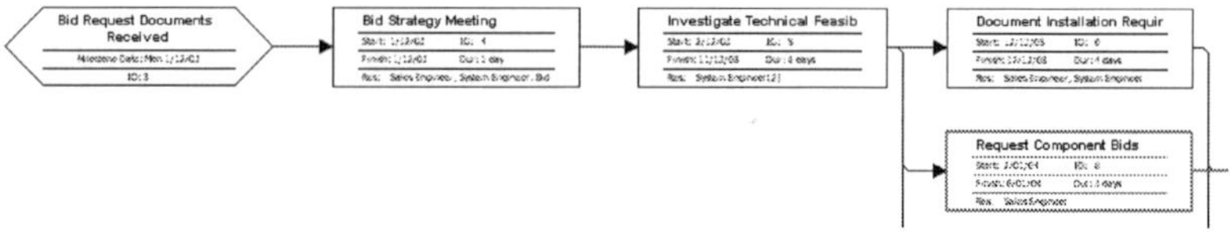

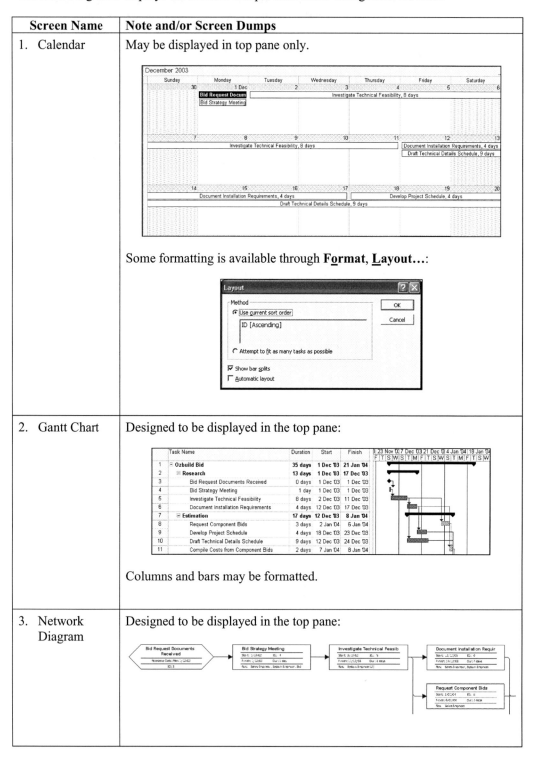

<table>
<tr><th>Screen Name</th><th>Note and/or Screen Dumps</th></tr>
<tr><td>4. Relationship Diagram</td><td>This displays relationships between tasks and may not be printed or edited.

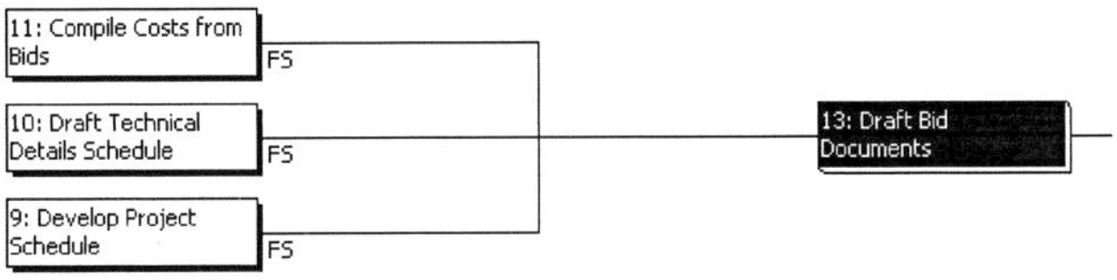

</td></tr>
<tr><td>5. Resource Form</td><td>Designed for the bottom pane:

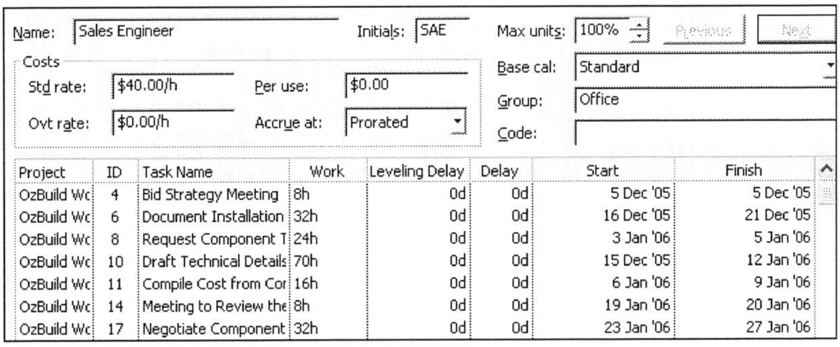

This View has the following options under **Format**, **Details**:

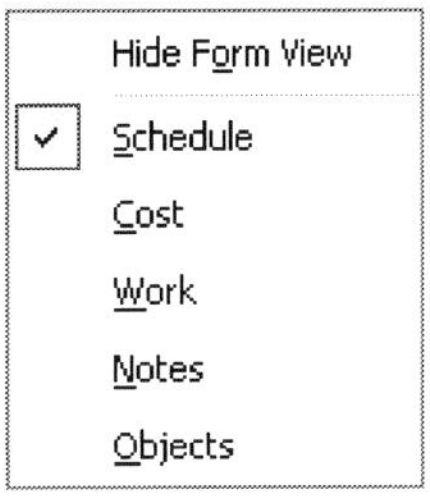

</td></tr>
</table>

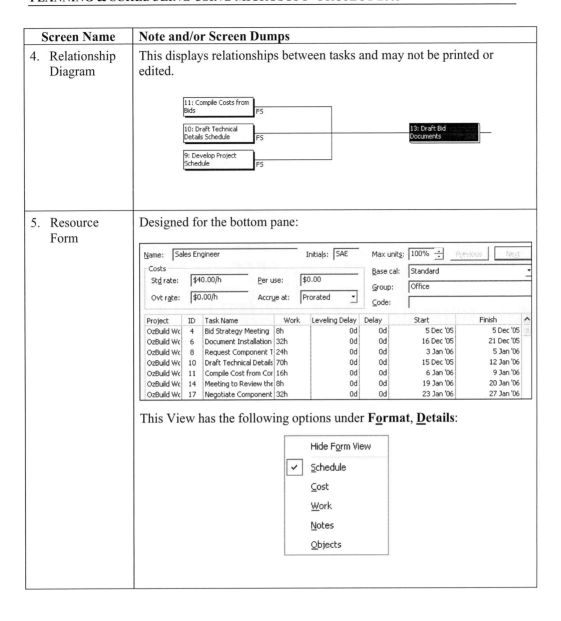

<table>
<tr><th>Screen Name</th><th>Note and/or Screen Dumps</th></tr>
<tr><td>6. Resource Graph</td><td>

This View is designed to be displayed in the bottom pane:

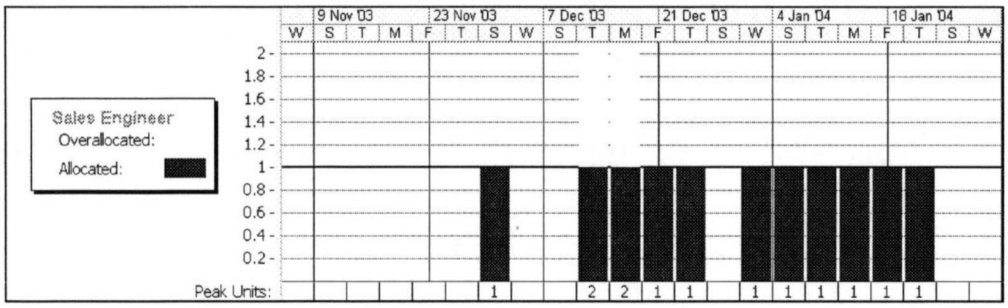

This View has the following options under **Format**, **Details**:

- ✓ Peak Units
- Work
- Cumulative Work
- Overallocation
- Percent Allocation
- Remaining Availability
- Cost
- Cumulative Cost
- Work Availability
- Unit Availability

</td></tr>
<tr><td>7. Resource Name Form</td><td>

This screen is designed for the lower pane with the **Resources Sheet** view in the upper pane:

Name: Sales Engineer   Previous   Next

<table>
<tr><th>Project</th><th>ID</th><th>Task Name</th><th>Work</th><th>Leveling Delay</th><th>Delay</th><th>Start</th><th>Finish</th></tr>
<tr><td>Network D</td><td>4</td><td>Bid Strategy Meeting</td><td>8h</td><td>0d</td><td>0d</td><td>1 Dec '03</td><td>1 Dec 'C</td></tr>
<tr><td>Network D</td><td>6</td><td>Document Installation</td><td>32h</td><td>0d</td><td>0d</td><td>12 Dec '03</td><td>17 Dec 'C</td></tr>
<tr><td>Network D</td><td>8</td><td>Request Component E</td><td>24h</td><td>0d</td><td>0d</td><td>2 Jan '04</td><td>6 Jan 'C</td></tr>
<tr><td>Network D</td><td>10</td><td>Draft Technical Detail</td><td>72h</td><td>0d</td><td>0d</td><td>12 Dec '03</td><td>24 Dec 'C</td></tr>
<tr><td>Network D</td><td>11</td><td>Compile Costs from C</td><td>16h</td><td>0d</td><td>0d</td><td>7 Jan '04</td><td>8 Jan 'C</td></tr>
<tr><td>Network D</td><td>14</td><td>Meeting to Review the</td><td>8h</td><td>0d</td><td>0d</td><td>14 Jan '04</td><td>14 Jan 'C</td></tr>
<tr><td>Network D</td><td>17</td><td>Negotiate Component</td><td>24h</td><td>0d</td><td>0d</td><td>16 Jan '04</td><td>20 Jan 'C</td></tr>
</table>

This View has the following options under **Format**, **Details**:

- Hide Form View
- ✓ Schedule
- Cost
- Work
- Notes
- Objects

</td></tr>
</table>

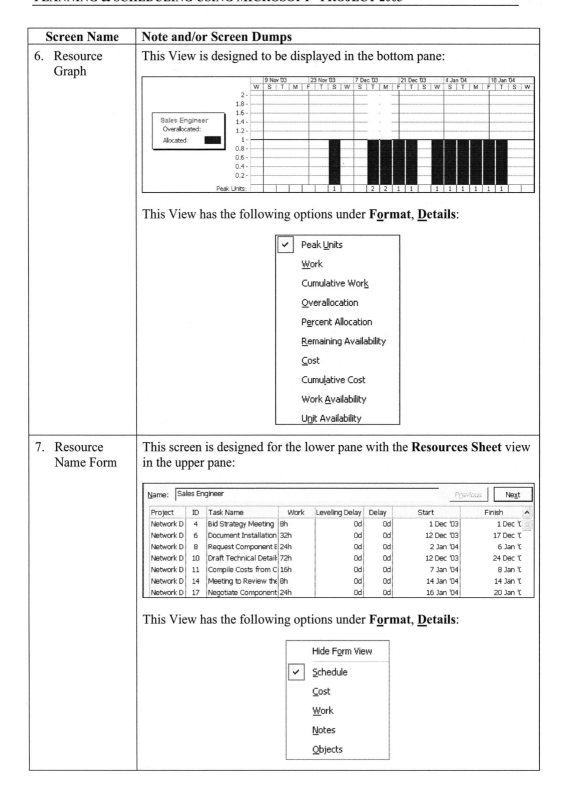

Screen Name	Note and/or Screen Dumps
8. Resource Sheet	This screen may be used in the top or bottom pane and the columns may be formatted with the **Table and Columns** functions:

	❶	Resource Name	Type	Materia Label	Initials	Group	Max. Units	Std. Rate	Ovt. Rate	Cost/Use	Accrue At	Base Calendar
1		Clerical Support	Work		CS	Office	2	$25.00/hr	$30.00/hr	$0.00	Prorated	Standard
2		Sales Engineer	Work		SALESENGI	Office	1	$40.00/hr	$0.00/hr	$0.00	Prorated	Standard
3		Scheduler	Work		SC	Office	1	$27.00/hr	$0.00/hr	$0.00	Prorated	Standard
4		System Engineer	Work		SYSENG	Office	2	$40.00/hr	$0.00/hr	$0.00	Prorated	Standard
5		Tender Manager	Work		TM	Office	1	$50.00/hr	$0.00/hr	$0.00	Prorated	Standard
6		Concrete	Material	m3	CONC	Material		$120.00		$50.00	Prorated	
7		Dozer	Work		DOZER	Equipment	6	$2,500.00/wk	$0.00/hr	$1,000.00	Prorated	Standard
8		Excavator	Work		EXCAV	Equipment	1	$400.00/day	$0.00/hr	$700.00	Prorated	Standard
9		Crushed Rock	Material	ton	CRCK	Material		$80.00		$0.00	Prorated	

Screen Name	Note and/or Screen Dumps
9. Resource Usage	This screen may be used in the top or bottom pane:

	Resource Name	Work	Details	'06 M	T	W	T	F	S	8 Jan '06 S	M	T
2	⊟ Sales Engineer	190 hrs	Work		8h	16h	16h	16h			16h	8h
	Bid Strategy Meeting	*8 hrs*	Work									
	Document Installation Requirements	*32 hrs*	Work									
	Request Component Tenders	*24 hrs*	Work		8h	8h	8h					
	Draft Technical Details Schedule	*70 hrs*	Work		0h	8h	8h	8h			8h	8h
	Compile Cost from Component Bids	*16 hrs*	Work					8h			8h	
	Meeting to Review the Draft Bid Document	*8 hrs*	Work									
	Negotiate Component Work Packages	*32 hrs*	Work									

The columns may be formatted with the **Table and Columns** functions and the **Details** area on the right-hand side It is formatted by selecting **Format**, **Detail Styles…** to open the **Detail Styles** form:

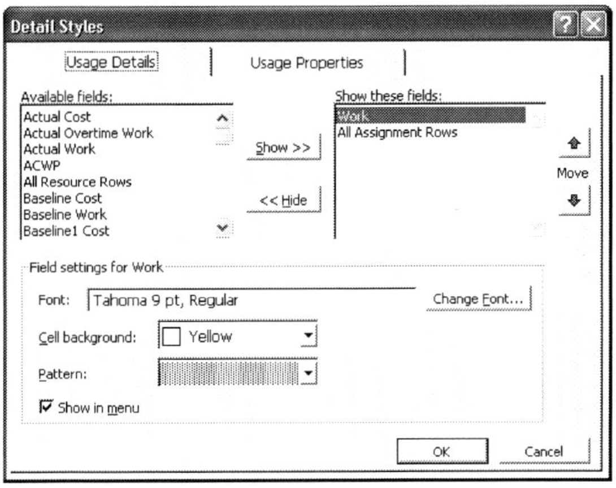

This View has the following options under **Format**, **Details**:

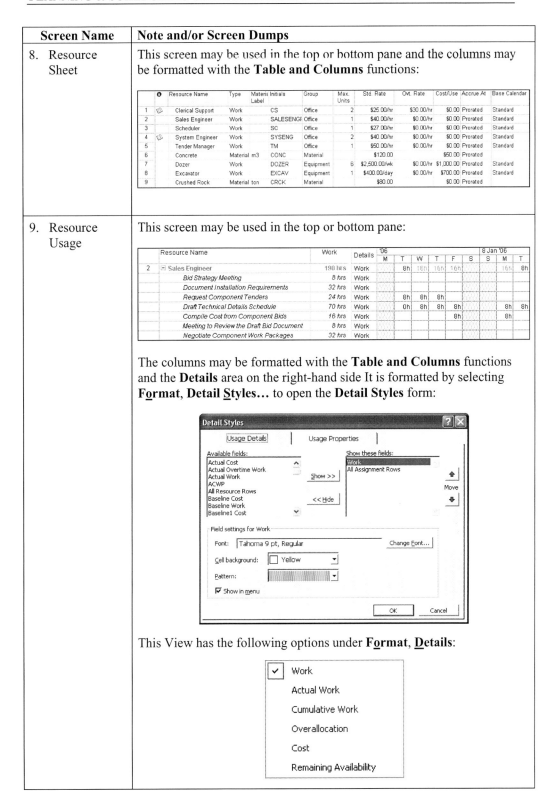

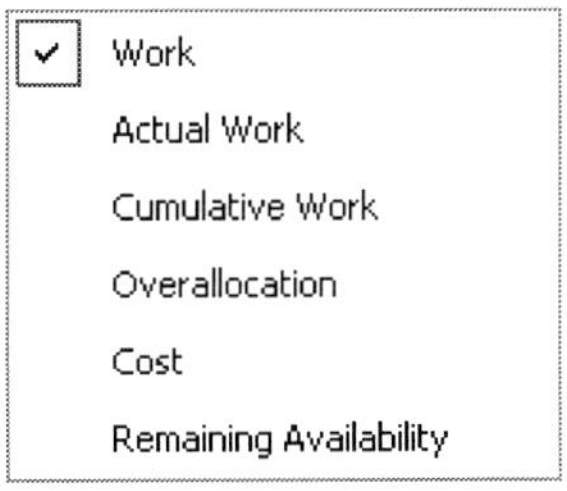

<table>
<tr><th>Screen Name</th><th>Note and/or Screen Dumps</th></tr>
<tr><td>10. Task Details Form</td><td>

This form is best displayed in the bottom pane and displays information about the task highlighted in the top pane:

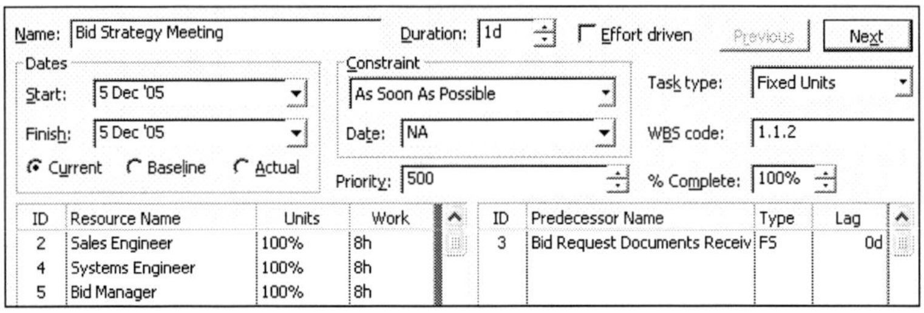

This View has the following options under **Format**, **Details**:

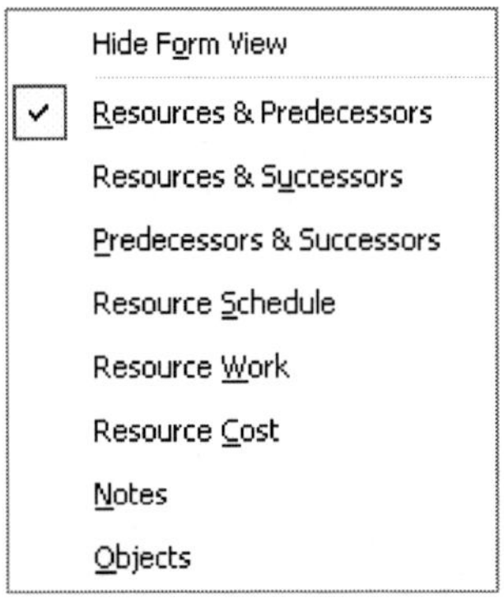

</td></tr>
<tr><td>11. Task Form</td><td>

This form is best displayed in the bottom pane.

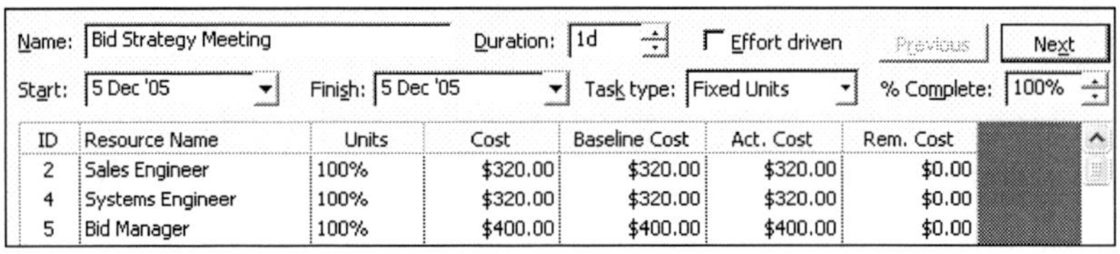

This View has the following options under **Format**, **Details**:

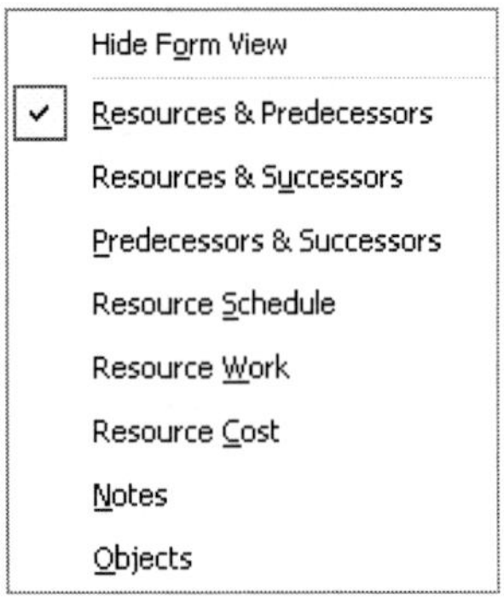

</td></tr>
</table>

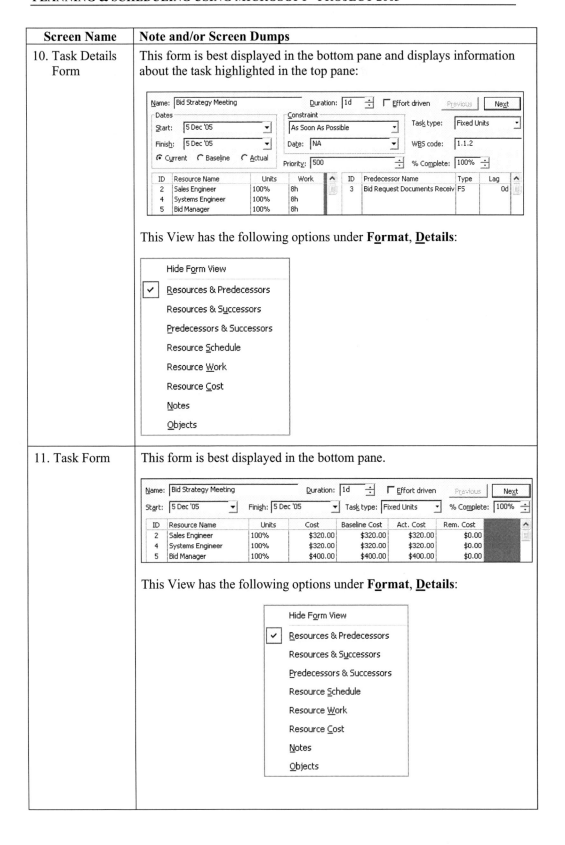

Screen Name	Note and/or Screen Dumps
12. Task Name Form	This form is best displayed in the bottom pane.

Name: Bid Strategy Meeting   Previous   Next

ID	Resource Name	Units	Work
2	Sales Engineer	1	8h
4	System Engineer	1	8h
5	Bid Manager	1	8h

ID	Predecessor Name	Type	Lag
3	Bid Request Documents Recei	FS	0c

This View has the following options under **Format**, **Details**:

- Hide Form View
- ✓ Resources & Predecessors
- Resources & Successors
- Predecessors & Successors
- Resource Schedule
- Resource Work
- Resource Cost
- Notes
- Objects

Screen Name	Note and/or Screen Dumps
13. Task Sheet	This View is similar to the Gantt but without displaying the bars. The columns may be formatted using the **Table** and **Columns** functions.

	Indicators	Task Name	Duration	Start	Finish	Predecessors	Resource Names
1		**OzBuild Bid**	**37.75 days**	**5 Dec '05**	**30 Jan '06**		
2	✓	**Research**	**13 days**	**5 Dec '05**	**21 Dec '05**		
3	✓	Bid Request Documents Received	0 days	5 Dec '05	5 Dec '05		
4	✓	Bid Strategy Meeting	1 day	5 Dec '05	5 Dec '05	3	Sales Engineer,Systems Engineer,Bid Manager
5	✓	Investigate Technical Feasibility	8 days	6 Dec '05	15 Dec '05	4	Systems Engineer[200%]
6	✓	Document Installation Requirements	4 days	16 Dec '05	21 Dec '05	5	Sales Engineer,Systems Engineer
7		**Estimation**	**19.75 days**	**15 Dec '05**	**16 Jan '06**		
8		Request Component Tenders	3 days	3 Jan '06	5 Jan '06	5	Sales Engineer
9		Develop Project Schedule	4 days	22 Dec '05	9 Jan '06	6	Scheduler
10		Draft Technical Details Schedule	11.25 days	15 Dec '05	16 Jan '06	5	Sales Engineer,Systems Engineer
11		Compile Cost from Component Bids	2 days	6 Jan '06	9 Jan '06	8	Sales Engineer

Screen Name	Note and/or Screen Dumps
14. Task Usage	This view displays the tasks in the columns and the resources and work on the right-hand side:

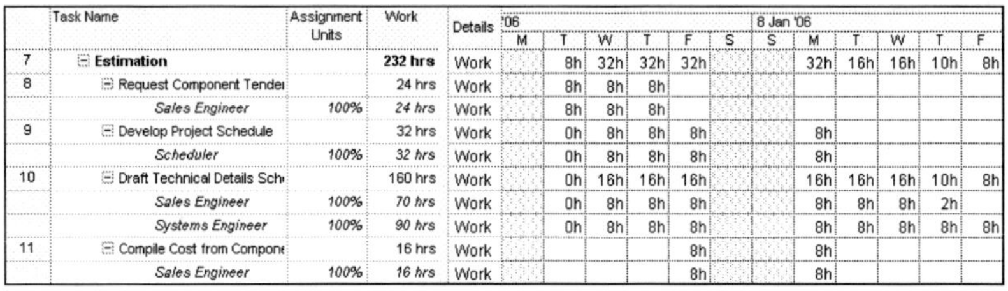

	Task Name	Assignment Units	Work	Details	'06						8 Jan '06					
					M	T	W	T	F	S	S	M	T	W	T	F
7	**Estimation**		**232 hrs**	Work		8h	32h	32h	32h			32h	16h	16h	10h	8h
8	Request Component Tender		24 hrs	Work		8h	8h	8h								
	Sales Engineer	*100%*	*24 hrs*	Work		8h	8h	8h								
9	Develop Project Schedule		32 hrs	Work		0h	8h	8h	8h			8h				
	Scheduler	*100%*	*32 hrs*	Work		0h	8h	8h	8h			8h				
10	Draft Technical Details Sch		160 hrs	Work		0h	16h	16h	16h			16h	16h	16h	10h	8h
	Sales Engineer	*100%*	*70 hrs*	Work		0h	8h	8h	8h			8h	8h	8h	2h	
	Systems Engineer	*100%*	*90 hrs*	Work		0h	8h	8h	8h			8h	8h	8h	8h	8h
11	Compile Cost from Compone		16 hrs	Work					8h			8h				
	Sales Engineer	*100%*	*16 hrs*	Work					8h			8h				

The columns may be formatted using the **Table** and **Columns** functions.

The information displayed on the right-hand side of the screen may be set by selecting **Format**, **Details**:

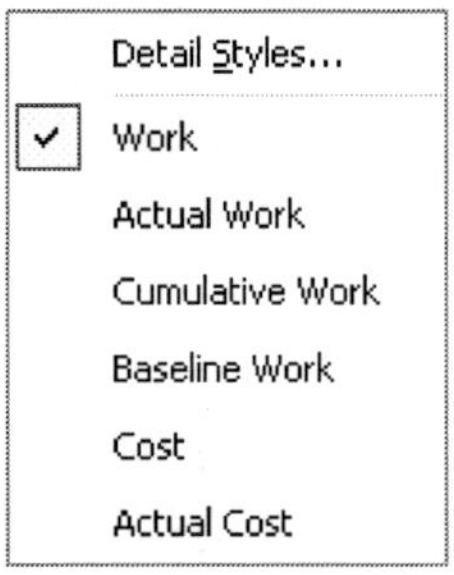

The formatting of the right-hand side of the screen may be changed by selecting **Format**, **Detail Styles…** to open the **Detail Styles** form:

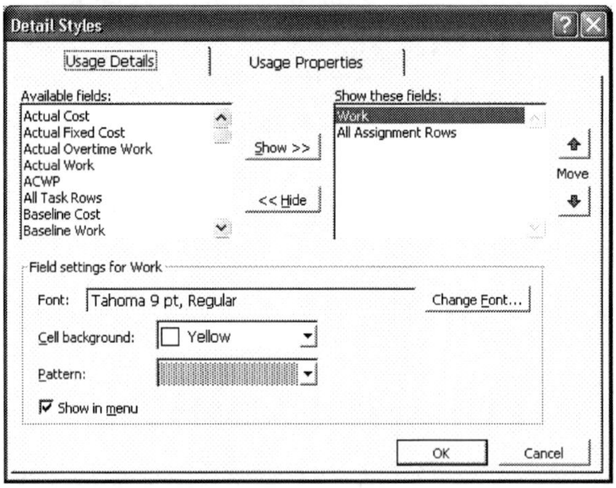

26 INDEX